Faults and Subsurface Fluid Flow in the Shallow Crust

Geophysical Monograph Series

Including

IUGG Volumes
Maurice Ewing Volumes
Mineral Physics Volumes

Geophysical Monograph 113

Faults and Subsurface Fluid Flow in the Shallow Crust

William C. Haneberg
Peter S. Mozley
J. Casey Moore
Laurel B. Goodwin

Editors

American Geophysical Union
Washington, DC

Published under the aegis of the AGU Books Board

Library of Congress Cataloging-in-Publication Data

Faults and subsurface fluid flow in the shallow crust / William C.
   Haneberg . . . [et al.].
         p.   cm. -- (Geophysical monographs ; 113)
      Includes bibliographical references.nd index.
      ISBN 0-87590-096-8
      1. Faults (Geology)   2. Fluids--Migration.   3. Earth--Crust.
      I. Haneberg, William C. II. Series
      QE606.F385   1999
      551.8'72--dc21   99-42450
                                                                              CIP

ISBN 0-87590-096-8
ISSN 0065-8448

Front cover: A view of the Sand Hill fault, a normal fault bounding the western margin of the Albuquerque Basin, Rio Grande rift, New Mexico, looking south. The sunlit wall is a portion of the hanging wall mixed zone that is preferentially cemented by calcite (see Heynekamp et al., this volume, for details). The sharp margin bounding the wall to the west is the fault core. The small tree in the center of the photo is roughly 5 m tall. Photograph by Laurel B. Goodwin

Back Cover (top): Map of the magnitude of simulated Darcy fluid flux for a fault zone model containing a fault core, damage zone, and protolith. Fracture apertures are set at 100 μm in the protolith and damage zone and 10 μm in the fault core. The direction of one-dimensional flow, normal to the fault zone and slip vector, is shown by arrows. The map represents the contoured data from a one meter thick slab projected onto a plane that is perpendicular to the fault zone and parallel to the slip direction. For additional details, see Caine and Forster, this volume.

Back Cover (bottom): Scanning electron micrograph image of a shear band developed in the post-failure stage of experimental deformation of Berea sandstone. Significant grain-size reduction through cataclasis is evident in the shear band; the sense of shear is shown by arrows and the field of view is 2.3 mm. Permeability reduction associated with such shear localization is described by Wong and Zhu, this volume. Reprinted from *Journal of Structural Geology*, Volume 18, Menéndez, B., Zhu, W., and Wong, T.-F., Micromechanics of brittle faulting and cataclastic flow in Berea sandstone, pp. 1-16. Copyright 1996, Elsevier Science, used with permission.

CONTENTS

CONTENTS

Geothermal Studies

PREFACE

This volume offers a sample of the diversity of research on faults and fluid flow in the late 1990s. It describes detailed surface and subsurface characterization of fault-zone structure and diagenesis with implications for hydrology and petroleum geology; the role of faults in geothermal systems; laboratory studies of rock mechanics, permeability, and geochemistry of faults and fault rocks; and mathematical modeling of fluid flow through faulted and fractured rocks.

The most striking and appealing feature of the volume, as well as the general research topic of faults and subsurface fluid flow, is its interdiscplinary nature. The authors are drawn from the fields of structural geology, engineering geology, geohydrology and hydrogeology, sedimentology, petroleum geology, geothermal geology, rock mechanics, and geochemistry. Likewise, the emphasis on faults rather than simple open fractures raises issues not addressed in much of the literature on flow through fractured rocks. Although faults are a type of fracture and semantics can confuse the issue, faults are generally more complicated than the simple fractures that are the focus of most work in fractured rock hydrology. Most notably, faults can have very large displacements (up to many kilometers) and develop complicated tectonic fabrics, gouge zones, and juxtaposition of rocks or sediments of different types.

Materials contained in this volume were developed in conjunction with the presentation of a 1997 Geological Society of America Penrose Conference convened in Taos, New Mexico, and titled "Faults and Subsurface Fluid Flow: Fundamentals and Applications to Hydrogeology and Petroleum Geology." The volume could not have been prepared without the logistical and financial support of the New Mexico Bureau of Mines & Mineral Resources, especially the secretarial efforts of Rita Case. We are also indebted to the many individuals who reviewed manuscripts for this volume: Mark Austin Person, Carol E. Renshaw, Richard H. Sibson, Larry W. Lake, Jeffrey A. Nunn, Leslie Smith, Peter J. Vrolijk, James Evans, Renata Dmowska, Alan E. Beck, Raymond A. Price, and William B. Durham.

William C. Haneberg
Haneberg Geoscience
South Colby, Washington

Peter S. Mozley
New Mexico Tech
Socorro, New Mexico

J. Casey Moore
University of California
Santa Cruz, California

Laurel B. Goodwin
New Mexico Tech
Socorro, New Mexico

Introduction

Laurel B. Goodwin and Peter S. Mozley
Department of Earth and Environmental Science, New Mexico Tech, Socorro, NM

J. Casey Moore
Earth Sciences Board, University of California, Santa Cruz, CA

William C. Haneberg
Haneberg Geoscience, Port Orchard, WA

The study of fluid-fault interactions is burgeoning across a variety of disciplines in the earth sciences. As the number of researchers studying these interactions grows, it becomes increasingly apparent that the questions raised can best be addressed by interdisciplinary teams. The intent of the Penrose conference that launched this volume was to assemble this diverse community of researchers from industry, the national labs, and academia, to synergize their interaction, and to develop visions for further investigations. Although this volume does not represent the full variety of topics discussed at the conference [see *Goodwin et al.*, 1998, for the conference report], it does address many aspects of the evolution of faults and the relationships among fault-zone architecture, permeability, and fluid flow. In general, these papers consider fluid-fault interactions in the brittle crust. In the following sections, we introduce many of the contributions made by the authors of this compilation. For simplicity, we cite these papers by author only, rather than repeatedly refer to this volume.

FAULT-ZONE ARCHITECTURE AND PERMEABILITY STRUCTURE

Fault zones, whether developed in sediments, sedimentary rock, or crystalline rock, are structurally and hydrologically heterogeneous. Such heterogeneities and anisotropies are part of the geologic framework created by faults and adjacent undeformed material, which influences a variety of fluid-fault interactions, such as how hydrocarbons and groundwater accumulate or migrate along faults

and the development of ore deposits in fault zones. Thus, studies of fault-zone architecture are essential to both academic and applied geology. A number of key research questions focus on fault-zone heterogeneity: How can we evaluate whether a fault will act as a barrier, conduit, or barrier-conduit system with respect to flow? Can we generalize structural and hydrologic data sufficiently to create adequate fault-zone conceptual models given the variability in fault rocks, protolith materials, and conditions of deformation? Can basin-scale flow models incorporate the spatial variability of faults to achieve more realistic simulations? How do we incorporate data collected at small scales into larger scale models? Work presented in this volume suggests that a fruitful avenue to pursue in answering these questions lies in a) understanding how deformation mechanisms are affected by such petrophysical properties as porosity, and b) determining the correlation between specific structures, such as fractures and deformation bands, and changes in permeability with respect to parent material.

Faults in Low Porosity Rocks

Geologists and hydrogeologists have identified two primary components, or architectural elements, of faults in low porosity sedimentary and crystalline rocks: a core zone, which has accommodated the majority of fault slip, and a damage zone, which brackets the fault core [*Chester and Logan*, 1986; *Smith et al.*, 1990; *Caine et al.*, 1996]. Each element can contain a variety of structures related to fault movement. The more strongly deformed core may include, for example, gouge or cataclasite; the damage zone contains subsidiary structures such as minor faults, fractures, and folds. In general, deformation processes operating in the core zone result in features, such as gouge, that reduce permeability in the core with respect to the protolith, whereas fracture networks in the damage zone result in an

Faults and Subsurface Fluid Flow in the Shallow Crust
Geophysical Monograph 113

increase in permeability. *Caine et al.* [1996] broadened this simple model to accommodate field observations made by a variety of workers, showing that different fault zones fit into a conceptual scheme in which both the core zone and the damage zone may be either poorly or well developed. This scheme does a good job of describing many fault zones, including those studied by contributors *Levy et al.*, *Nelson et al.*, and *Tamanyu*.

The structure that exerts the primary control on permeability in this low porosity rock model is the fracture. *Caine and Forster* demonstrate the construction of idealized fault end members that range from a single fracture to a fault with well developed core and damage zones. In their models, each fault zone is composed solely of a fracture or fractures, fracture spacing increases both with proximity to the center of the fault zone and with increasing displacement, and fault-zone fractures are kinematically compatible. These end members could represent individual faults or a sequence in the evolution of a single fault with increasing displacement. The permeability contrast between the core and damage zones is represented by a difference in fracture aperture. Three-dimensional numerical simulations of fluid flow through these end-member fault zones demonstrate the importance of permeability contrast between the protolith and fault-zone architectural elements, as well as the effects of fault-zone permeability anisotropy. Fracture sealing, which *Nelson et al.* and *Tamanyu* remind us is typically an episodic process related to the seismic cycle, could be represented in such a model by a progressive decrease in aperture.

Faults in High Porosity Rocks and Sediments

In the 1970s, studies of faults in porous sandstone resulted in the recognition of a very different type of fault than that illustrated by the conceptual model described above, in which the primary structural 'building block' is a deformation band rather than a fracture [*Aydin*, 1978; *Aydin and Johnson*, 1978]. More recent work has indicated that deformation bands and zones of deformation bands result in a reduction in permeability of up to three orders of magnitude with respect to the protolith [e.g., *Antonellini and Aydin*, 1994, 1995], rather than the increase in permeability that accompanies production of connected fractures. Similar reductions in permeability are evident in deformation bands in poorly lithified sand [*Antonellini and Aydin*; *Sigda et al.*]. Contributors *Wong and Zhu* further illustrate this connection between deformation mechanisms and permeability by contrasting experimental studies of low porosity crystalline rock with those of porous sandstone and unlithified granular material. All of these materials exhibit dilatancy during brittle failure, but this dilatancy is accompanied by an increase in permeability of low porosity rock

and a decrease in permeability of high porosity materials. The inverse correlation between dilatancy and permeability in the latter is related to cataclasis and shear-enhanced compaction within deformation bands.

Porous sediments do not, of course, consist solely of sands. *Heynekamp et al.* and *Sigda et al.* describe faults that cut poorly lithified sediments in which displacement is greater than bed thickness, so that the faults juxtapose sands and other sediments, such as clays, silts, and gravels. Both fractures and deformation bands are absent in the finer grained silts and clays, but the poorly lithified character of the sediments facilitates tectonic mixing from the outcrop to the grain scale. *Heynekamp et al.* provide a conceptual model for 'fracture-free' fault zones in these high porosity, poorly lithified sediments. The model divides fault zones into roughly fault-parallel, tabular architectural elements: a narrow core zone flanked by wider footwall and hanging wall mixed and damage zones. Material from adjacent sedimentary layers is incorporated into the mixed zones during slip. Field observations indicate that sediments transported away from contributing source beds are thinned and tectonically mixed within these zones. Transport distances are greatest for clay and least for sand. A variety of observations indicates that grain-scale mixing acts to destroy bedding and produce materials such as clay/sand mixtures within the mixed and core zones. This conceptual model is fundamentally different than that described by *Caine et al.* [1996] in that processes operating within a fault zone in poorly lithified sediments – grain-scale mixing (particulate flow), grain reorganization and pore collapse, and cataclasis within deformation bands – act to decrease, rather than increase, permeability with respect to the parent sediments.

Faults are dynamic systems (Figure 1). Because mechanical, geochemical, and hydrologic properties vary with such factors as lithology, temperature, pressure, and deformation rate, the effect of a given fault on fluid flow will vary in both space and time [e.g., *Sibson*, 1990; *Knipe*, 1993]. This variation reflects, in part, the spatial distribution of structures and the evolution of structures within the fault zone over time. With increasing displacement, the number of architectural elements within a fault zone can increase, and the structures within individual elements can be modified through deformation, mineralization, and alteration of fault-zone materials [compare, e.g., *Sigda et al.* with *Heynekamp et al.*]. These modifications will result in changes in fault-zone permeability, which can occur both progressively with increasing displacement and episodically through a seismic cycle [e.g., *Caine and Forster*; *Nelson et al.*; *Tamanyu*]. A number of our contributors remind us that the character of fault zones can vary significantly both laterally and vertically, and that these variations can control a fault's influence on fluid flow [*Antonellini and Aydin*;

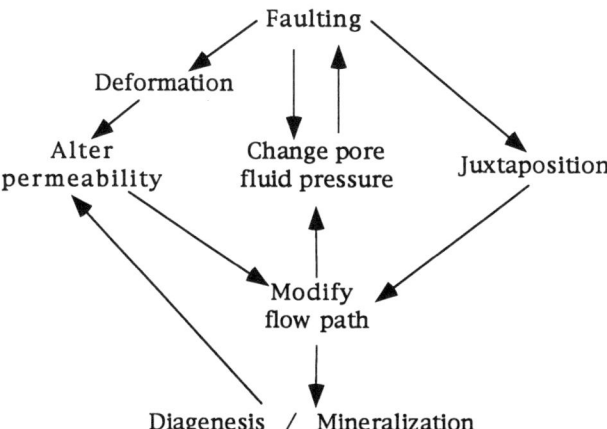

Figure 1. Illustration of the dynamics of a fault / fluid system. The earthquake cycle [cf. *Sibson*, 1990] is encompassed in this diagram: Faulting results in a fluid pressure drop. Subsequent diagenesis and mineralization gradually decrease the permeability of the fault system, resulting in a progressive increase in fluid pressure. A high pore fluid pressure decreases the effective normal stress across the fault, facilitating faulting.

Heynecamp et al.; *Reiter*; *Sigda et al.*; *Tamanyu].* *Heynekamp et al.* discuss the link between fault-zone structure and width, and parent sediment type, and *Caine and Forster* and *Reiter* emphasize the importance of the contrast between fault-zone and protolith permeability, and lateral and vertical changes in protolith permeability, respectively.

FLOW THROUGH SATURATED VERSUS UNSATURATED MEDIA

Although most hydrologic studies of faults have focused on saturated flow, it is also important, particularly in arid and semi-arid areas, to consider the impact of faults on flow in the unsaturated, or vadose, zone, above the water table. This volume includes papers that address vadose zone flow in both types of fault zone discussed above – those with fractures and those without. *Levy et al.* consider the former in their analysis of flow paths in faults and fractures at Yucca Mountain. They find that bomb-pulse ^{36}Cl records fast flow – as much as 300 m into the mountain during the last 50 years – within faults cutting poorly fractured, non-welded tuffs. Flow locally extends into subsidiary fractures and breccia zones. Examination of diagenetic minerals and textures, surprisingly, shows no difference between fractures and faults with a bomb pulse signature (the fast paths) and those without.

Sigda et al. explore variations in saturated permeability within small-displacement fault zones in poorly lithified sand, which consist of variably spaced deformation bands, and the reasons for these variations. They note that significant reduction in both grain and pore size with respect to

the parent sand should, because of an increase in capillarity, result in an increase in unsaturated permeability within deformation bands and zones of deformation bands, even though saturated permeability is decreased. Deformation band fault zones thus might also act as fast paths for flow and transport in the vadose zone.

TRACKING FLOW PATHS THROUGH PATTERNS OF DIAGENESIS AND GEOCHEMICAL ANALYSES

Geochemical and diagenetic studies offer a better understanding of the regions over which faults can channel flow and controls on fault-zone cementation [e.g. *Burley et al.*, 1989; *Knipe*, 1993]. This is well illustrated by a number of papers in this volume. *Kharaka et al.* conducted detailed chemical and isotopic analyses from springs and wells located on or near the San Andreas fault system. The waters analyzed are predominantly meteoric in origin, and record shallow to moderate circulation depths of up to 6 km. In contrast, compositions and isotope abundances of noble gases and carbon isotopic data indicate high fluxes of CO_2 from the deep crust and mantle. These fluxes are substantial – sufficient to effect lithostatic pressures at time scales of the same order of magnitude as seismic cycles. They could also impact fault-zone dissolution and precipitation, modifying fault-zone mechanics through diagenesis, reminding us once more of the interconnectedness of fault-zone processes (e.g., Fig. 1), and showing the value of geochemical studies in providing constraints on the role of fluids in the faulting process. As indicated in the previous section, isotopic analyses can also provide a better understanding of near-surface flow [*Levy et al.*].

Patterns of fault-zone diagenesis – the chemistry, mineralogy, texture, geographic distribution, location within a fault zone, and morphologic features of cements - offer a window into paleo-fault-zone plumbing and processes controlling mineralization. Once again, the papers in this volume illustrate this point for fault zones with and without fractures. *Tamanyu* looks at the origin of productive fractures that vary in orientation with depth in geothermal fields in Japan. These fractures are typically characterized by hydrothermal veins, and the author considers the history of fracture formation, reactivation, and mineralization in terms of strike-slip faulting, and both high- and low-angle reverse faulting. His interpretation of this complex history is founded in *Sibson's* [1990] model of fault-valve behavior.

Nelson et al. evaluate the relative timing of mineralization and deformation in fault-fracture networks, demonstrating a complex history of slip, flow, and cementation. They provide a case study of faults in which the diagenetic history of the core zone is distinct from that of the damage zone, and matrix cementation halos have formed around the fault zones. The fault cores contain hydrocarbons, and *Nel-*

son et al. include an analysis of how such faults might affect hydrocarbon reservoirs and production. Deformation band faults can also affect reservoirs, as demonstrated by *Antonellini and Aydin* in a comprehensive study of faults in a shallow production field in poorly lithified sandstone. Zones of cementation in this case extend up to 30 m from a given fault. *Antonellini and Aydin* point out that sealing potential must be evaluated for a given reservoir, considering pore fluid pressure, hydrocarbon columns, and the fact that capillary pressures in cataclastic deformation bands may be 1-2 orders of magnitude higher than in the protolith. *Nelson et al.* and *Antonellini and Aydin* emphasize the importance of fault-zone structure and diagenesis in affecting both hydrocarbon migration and production issues; these are useful case studies that allow a comparison of faults with and without fractures.

Heynekamp et al. consider patterns of cementation within a single fault, evidence of timing of cementation versus deformation, and regional distributions of cements within and outside fault zones to evaluate paleoflow. They present evidence that flow was reoriented from subhorizontal to subvertical within the fault zone. Mixing of fluids from different structural levels through a combination of cross-fault and fault-parallel flow is inferred to have resulted in preferential cementation of the hanging wall mixed zone where it is coarse-grained. This is, of course, just one possible cementation process. A second process was evaluated by *Whitworth et al.* Although it is known that clays can act as geologic membranes, which can facilitate cement precipitation through solute sieving [e.g., *Whitworth and Fritz*, 1994], the potential membrane characteristics of fine-grained, clay-free material have not previously been investigated. *Whitworth et al.* present the results of an experiment in which an undersaturated calcium carbonate solution was forced through a layer of clay-sized quartz particles, intended to represent clay-poor fault gouge. Calcite crystals precipitated on the synthetic fault gouge, which suggests that solute-sieving may result in selective cementation of a fault zone, even when the clay content of the gouge is negligible.

GEOPHYSICAL TECHNIQUES

Geophysical approaches to evaluating fluid-fault interactions are many and varied. The authors of this volume touch on a few of these approaches, with some very interesting results. *Reiter* uses subsurface temperatures made in boreholes to constrain ground-water flow paths across several faults, including a traverse near a fault intersection. These data are interpreted to record significant lateral and vertical variations in cross-fault flow, related to variations in both fault-zone and protolith hydraulic conductivity. Fault-zone transmissivity varies between three traverses, in

which: 1) the primary fault acts as a seal, 2) the faults cut by the traverse are transmissive, and flow across the fault occurs along subhorizontal paths, and 3) data are inconclusive, but suggest that one fault is acting as a flow channel.

Antonellini and Aydin and *Nelson et al.* utilize neutron porosity and seismic velocity measurements, respectively, in conjunction with characterization of petrophysical parameters of the rocks studied to evaluate the effects of deformation and diagenesis on rock properties. *Antonellini and Aydin* ask the important question of whether or not faults with less than 10-20 m displacement, which are not seismically detectable, can be identified through borehole geophysical techniques. They report that the signature of deformation bands is very difficult to detect with conventional logging methods. *Nelson et al.*, however, find seismic velocity a useful tool in detecting mineralization haloes around faults and veins. Further investigation of techniques that permit characterization of fault zones in areas of poor exposure has obvious benefits for evaluation of both petroleum and water resources.

SUMMARY

A variety of papers that address fluid-fault interactions in the upper crust are included in this volume. They contribute to a growing body of data on fault-zone deformation processes and resulting structures; fault-zone permeability; migration of fluids within fault zones; fault-zone diagenesis; fault mechanics; and approaches to constraining subsurface structure. Although written independently, these papers work together to produce a picture of the current level of understanding of fault-zone processes, and suggest productive directions for future research. Read and enjoy.

REFERENCES

Antonellini, M. and A. Aydin, Effect of faulting on fluid flow in porous sandstones: geometry and spatial distribution, *AAPG Bull.*, 79, 642-671, 1995.

Antonellini, M. and A. Aydin, Effect of faulting on fluid flow in porous sandstones: petrophysical properties, *AAPG Bull.*, 78, 355-377, 1994.

Aydin, A., Small faults formed as zones of deformation bands and as slip surfaces in sandstone, *Pure Applied Geophys.*, 16, 931-942, 1978.

Aydin, A. and A.M. Johnson, Development of faults as zones of deformation bands and as slip surfaces in sandstone, *Pure Applied Geophys.*, 116, 913-930, 1978.

Burley, S.D., J. Mullis, and A. Matter, Timing of diagenesis in the Tartan Reservoir (UK North Sea): Constraints from combined cathodoluminescence microscopy and fluid inclusion studies, *Marine Petroleum Geology*, 6, 98-120.

Caine, J.S., J.P. Evans, C.B. Forster, Fault-zone architecture and permeability structure, *Geology*, 24, 1023-1028, 1996.

Chester, F.M., and J.M. Logan, A classification scheme for permeability structures in fault zones, *EOS*, 24, 677, 1993.

Goodwin, L.B., W.C. Haneberg, J.C. Moore, and P.S. Mozley, Penrose Conference Report: Faults and subsurface fluid flow: fundamentals and applications to hydrogeology and petroleum geology, *GSA Today*, 8, 25-27, 1998.

Knipe, R.J., The influence of fault zone processes and diagenesis on fluid flow, *in Diagenesis and Basin Development*, edited by A.D. Horbury and A.G. Robinson, AAPG Studies in Geology, 36, 135-148, 1993.

Sibson, R.H., Conditions for fault-valve behavior, *in Deformation Mechanisms, Rheology and Tectonics*, edited by R.J. Knipe and E.H. Rutter, Geological Society Special Publication 54, 15-28, 1990.

Smith, L., C. Forster, and J. Evans, Interaction of fault zones, fluid flow, and heat transfer at the basin scale, *in Hydrogeology of Low Permeability Environments*, edited by S.P. Neuman and I. Neretnieks, Verlag Heinz Heisse, 41-67, 1990.

L.B. Goodwin, Department of Earth and Environmental Science, New Mexico Tech, Socorro, NM 87801-4796 (e-mail: lgoodwin@nmt.edu)

W.C. Haneberg, Haneberg Geoscience, 10411 SE Olympiad Drive, Port Orchard, WA 98366 (e-mail: bill@haneberg.com)

J.C. Moore, Earth Science Board, University of California, Santa Cruz, CA 95064 (e-mail: casey@emerald.ucsc.edu)

P.S. Mozley, Department of Earth and Environmental Science, New Mexico Tech, Socorro, NM 87801-4796 (e-mail: mozley@nmt.edu)

Outcrop-Aided Characterization of a Faulted Hydrocarbon Reservoir: Arroyo Grande Oil Field, California, USA

Marco Antonellini and Atilla Aydin

Department of Geological and Environmental Sciences, Stanford University, Stanford, California

Lynn Orr

Department of Petroleum Engineering, Stanford University, Stanford, California

The Arroyo Grande Oil Field in Central California has been in production since 1905 from the Miocene-Pliocene Edna member of the Pismo Formation. The Edna member is a massive, poorly consolidated sandstone with an average porosity of 0.2 and a permeability of 1000-5000 md. The producing levels are shallow, 100 to 500 m from the earth's surface. We mapped the major structures of the oil field in outcrops and determined the distribution and orientation of small faults (deformation bands) both in cores and outcrops. We established the relationship between deformation band density and major faults by detailed mapping, and we used image analysis to determine the petrophysical properties of the sandstone in outcrop and in cores. Outcrop, aerial photo, and core data analysis provided insight into the fault distribution and the petrophysical properties of the reservoir rocks, showing a good correspondence between subsurface and outcrop data. The permeability of faults is on average three orders of magnitude lower than that of the host rock and capillary pressure is 1-2 orders of magnitude larger in faults than in the host rock. Faults with tens of meters offsets are associated with an high density of deformation bands (10 to 250 m^{-1}) and with zones of cement precipitation that extend up to 30 m from the fault. This implies that cement precipitation together with cataclasis could form permeability barriers or baffles. We used synthetic logs based on core data and neutron porosity detector characteristics to simulate the log signature of deformation bands in sandstone and we show that the detection of these small faults is very difficult with the conventional logging methods.

1. INTRODUCTION

The detailed subsurface architecture of a petroleum reservoir is usually difficult to interpret because well data are sparse and the resolution of the seismic surveys is too coarse to provide information about small to medium scale (0.01-20 m) reservoir heterogeneity. A combination of subsurface and outcrop or analogue data improves the interpretation of the structure of a reservoir and, in particular, the distribution of faults or fractures which may affect fluid flow. In this paper, we study a reservoir that is partially exposed in outcrop to establish the effect that faults have on the sealing and migration of the hydrocarbons.

Faults and Subsurface Fluid Flow in the Shallow Crust
Geophysical Monograph 113

The control exerted by faulting on fluid flow in a rock mass is a problem of great interest in both applied [*Antonellini and Aydin*, 1994, 1995; *Byerlee*, 1993; *Haneberg*, 1995; *Hardmaan and Booth*, 1991; *Hippler*, 1993; *Hooper*, 1991] and regional studies [*Knipe et al.*, 1991; *Logan*, 1991; *McKaig*, 1989; *Reynolds and Lister*, 1987; *Roberts*, 1990]. Faulting may either inhibit (in a direction normal to their plane) or enhance (in a direction parallel or normal to their plane) fluid flow. A field example from the Scapa fault in Scotland [*Hippler*, 1993] has shown that hydrocarbons may migrate vertically along the fractured and cataclastic material of a fault zone in sandstone. Similar behavior has been predicted for faults in sandstone by *Seeburger* [1981] and *Seeburger et al.* [1991], for faults in carbonate rocks by *Kastning* [1977] and *Mollema and Antonellini* [1998], and in crystalline rocks by *Caine et al.* [1996]. The mechanism by which fluids penetrate a fault zone is a matter of debate [*Rice*, 1992; *Byerlee*, 1993]. *Sibson et al.* [1975] propose that fluids can be channeled along a fault zone as a result of dilatancy-induced pumping during seismic events in a cataclastic fault zone. Fluids may also move into a fault zone through the extensive fracture network localized near the fault [*Alastair et al.*, 1996; *Caine et al.*, 1996; *Kastning*, 1977; *McAllister and Knipe*, 1996; *Valenta et al.*, 1994], or because of the existence of an open discontinuity (e.g. a slip plane) between the rocks of the hanging wall and the rocks of the footwall [*Antonellini and Aydin*, 1994, 1995; *Caine et al.*, 1991]. Fault zones may have the opposite effect on fluid flow in a direction normal to their plane. It is, in fact, possible that the crushed and recrystallized material of the fault zone forms a permeability barrier or a seal to the movement of the fluids across the fault plane [*Antonellini and Aydin*, 1994; *Edwards et al.*, 1993; *Fowles and Burley*, 1994; *Gibson*, 1994; *Knipe*, 1992; *Nelson*, 1985; *Pittman*, 1981]. Diagenesis in the fault rock and surrounding areas in the form of cement precipitation, illitization, and mass transfer may also degrade the porosity and permeability around a fault zone [*Knipe*, 1993; *Mozley and Goodwin*, 1995; *Sibson*, 1987].

Recognition of the behavior of faults as seals or conduits is of strategic importance when evaluating hydrocarbon traps and in predicting fluid flow patterns during production [*Gibson*, 1994; *Hippler*, 1993]. The distribution and connection of the fault segments and the in-situ effective stress determine how effective the faults are in sealing or channeling the fluids [*Mollema and Antonellini*, 1998]. Fault sealing has been invoked to explain the existence of compartments with different fluid pressures in hydrocarbon reservoirs [*Downey*, 1984; *Harding and Tuminas*, 1988; *Knott*, 1993; *Smith*, 1980]. The processes generally believed to generate a fault seal are geometric juxtaposition of

permeable units against impermeable units [*Allen*, 1988; *Bouvier et al.*, 1989; *Yielding et al.*, 1997], clay smearing [*Downey*, 1984; *Gibson*, 1994; *Harding and Tuminas*, 1988; *Lindsay et al.*, 1993; *Nybakken*, 1991; *Weber and Mandl*, 1978], diagenetic healing [*Hippler*, 1993; *Nybakken*, 1991; *Smith*, 1980; *Watts*, 1987], and cataclasis [*Antonellini and Aydin*, 1994; *Downey*, 1984; *Knott*, 1993; *Nybakken*, 1991; *Seeburger*, 1991]. Also pressure seals have been recognized in large compartmentalized gas reservoirs with anomalous overpressures [*Tigert and Al-Shaieb*, 1990]; in these compartments, faults provide horizontal sealing, whereas vertical sealing is provided by subhorizontal lithologic permeability barriers (shale units).

The importance of the cataclastic fault zones as an exploration concept for trap formation has been discussed by *Nybakken* [1991] and *Gibson* [1994], who argue that this kind of faults (intra-reservoir faults) have less potential for sealing over geologic time with respect to clay-smear seals. *Knott* [1993] has made an analysis of faults sealing over geologic time in the North Sea, and has presented a series of empirical relationships that evaluate, on a statistical basis, the potential for sealing a fault. His analysis shows that most of the faults with offsets comparable to the reservoir thickness (extra-reservoir faults) are sealing because of the juxtaposition of permeable units against impermeable units. On the other hand, one-third of faults with an offset less than the reservoir thickness (intra-reservoir faults) are sealing as well, which can be interpreted as an indication of cataclastic fault zones to be sealing within a sandstone reservoir. The potential for sealing of these latter faults increases with displacement [*Knott*, 1993; *Yielding*, 1997]. Intra-reservoir sub-seismic resolution (< 10-20 meter offset at the current state of seismic technology) faults are important because they can seal off compartments of pay that are by-passed during production; this phenomenon has been well documented in recent years by time-lapse seismic [*Anderson*, 1996]. If the small faults are not sealing, they can still cause permeability barrier effects that degrade the performance of the wells during production and diminish the recovery efficiency.

In this paper, we present examples of fault seals and fault permeability reduction in sandstone from the Arroyo Grande oil field, California. Photo-lineaments, which are parallel to the major regional faults, represent major boundaries between tar-impregnated and tar-free sandstone. These photo-lineaments can be correlated to the presence of small-medium offset (1-100 m) high angle reverse faults that compartmentalize the reservoir. Small faults (below the limit of seismic detectability; < 10-20 m offset) observed in outcrops and in cores appear to be sealing with respect to a non-wetting phase (hydrocarbon). We account for the microscopic structure of these faults and we relate it to the effect faults exert on fluid flow. Based on this information, we construct a reservoir description for the oil field, which accounts for the effects of faults on hydrocarbon flow.

2. TECTONIC SETTING AND GENERAL GEOLOGY

The oil field and associated outcrops for this study are located in the Pismo Basin about 100 km west of the San Andreas Fault Zone in San Luis Obispo County (Central California), 5 km northeast of the city of Pismo Beach (Figure 1). The Pismo Basin is a narrow syncline with an axial trace oriented NW-SE that extends for 20-30 km. The Pismo syncline is bounded to the NE by the Edna fault zone that is a NE-dipping thrust accommodating about 300 m of reverse slip [*Hall*, 1973]. The Edna fault zone merges with the West Huasna fault zone about 12 km SE of the town of Edna. The West Huasna fault zone is a NE-dipping thrust trending sub-parallel to the San Andreas Fault Zone and is the major structural element of the study area.

The principal reservoir section in the field and exposed at the outcrop is a sandstone from the Edna member of the Pismo formation. Most of the surface exposures belong to the "Main Pismo Tar Sand" level. These sands were deposited during the Mio-Pliocene and subsequently folded to form the Pismo syncline [*Hall*, 1973]. Faults in the reservoir section formed during this phase of deformation.

The surface exposures of the reservoir section are located along Price Canyon road within the Edna-Arroyo Grande oil field. The field, currently operated by Calresources, started production around 1905. The tar-impregnated sandstone of the Pismo formation is more resistant to erosion than the tar-free sandstone; scarps and ridges in this area are often located at the contact between tar-impregnated and tar-free sandstone. Geomorphically, these ridges are rectilinear features aligned parallel to the trend of the Edna Fault Zone (WNW-ESE). A map of the area compiled by *Page et al.* [1944] documents how the bituminous outcrops are aligned along scarp-forming lineaments with an WNW-ESE trend.

Two cores used in this study were taken adjacent to the outcrops. Core 9J-1 has been taken from a well 200 meters south of outcrop 3 and core 17L has been taken 500 meters south of outcrop 1 (Figure 2).

During the peak of development, in the mid-fifties, the Edna-Arroyo Grande oil field had about 50 wells producing at the same time. Initial production from each well varied from 30 to 400 barrels per day. After that period a series of enhanced recovery programs was started: water injection,

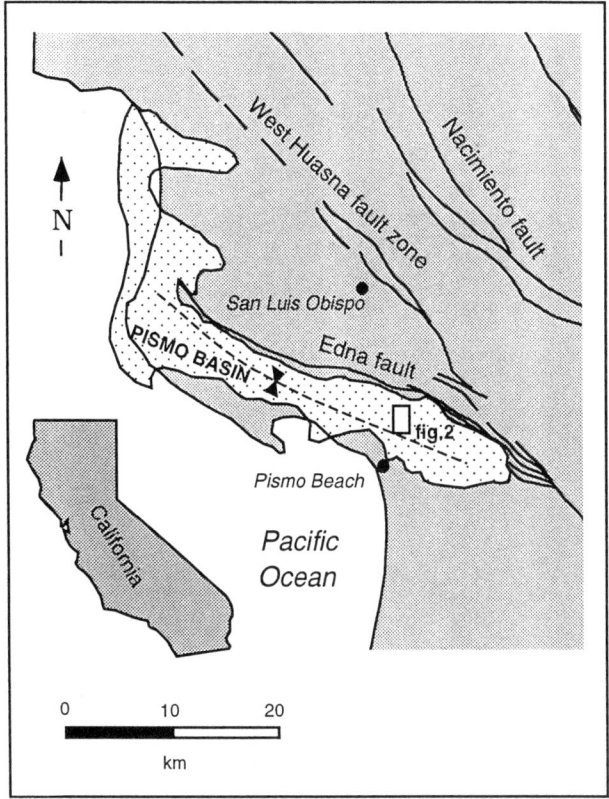

Figure 1. Location map for the Pismo Basin (dotted pattern). The Arroyo Grande Oil Field is located in the East end of the basin. The field is shown in more detail in Figure 2.

carbon dioxide injection, and steam injection. The oil characteristics range from 14 to 22 API gravity [*Lawrence*, 1958].

3. OUTCROP CHARACTERIZATION

Sandstones and conglomerates belonging to the Miocene Edna member of the Pismo Formation are cut by two sets of small-offset faults. The sets of faults are shown in Figures 3a and 3b (location 3 in Figure 2). Both sets strike WSW-ENE and dip at low angle (15-30°) to the NNW and to the SSE respectively (Figure 3a). A fault solution on the north dipping set indicates a 80% component of reverse movement and a 20% component of left lateral strike-slip movement. The fault solution was obtained by using bedding and deformation bands belonging to the south-dipping set as offset markers. Mutual crosscutting relationships (Figure 4) indicate that both systems were developing at the

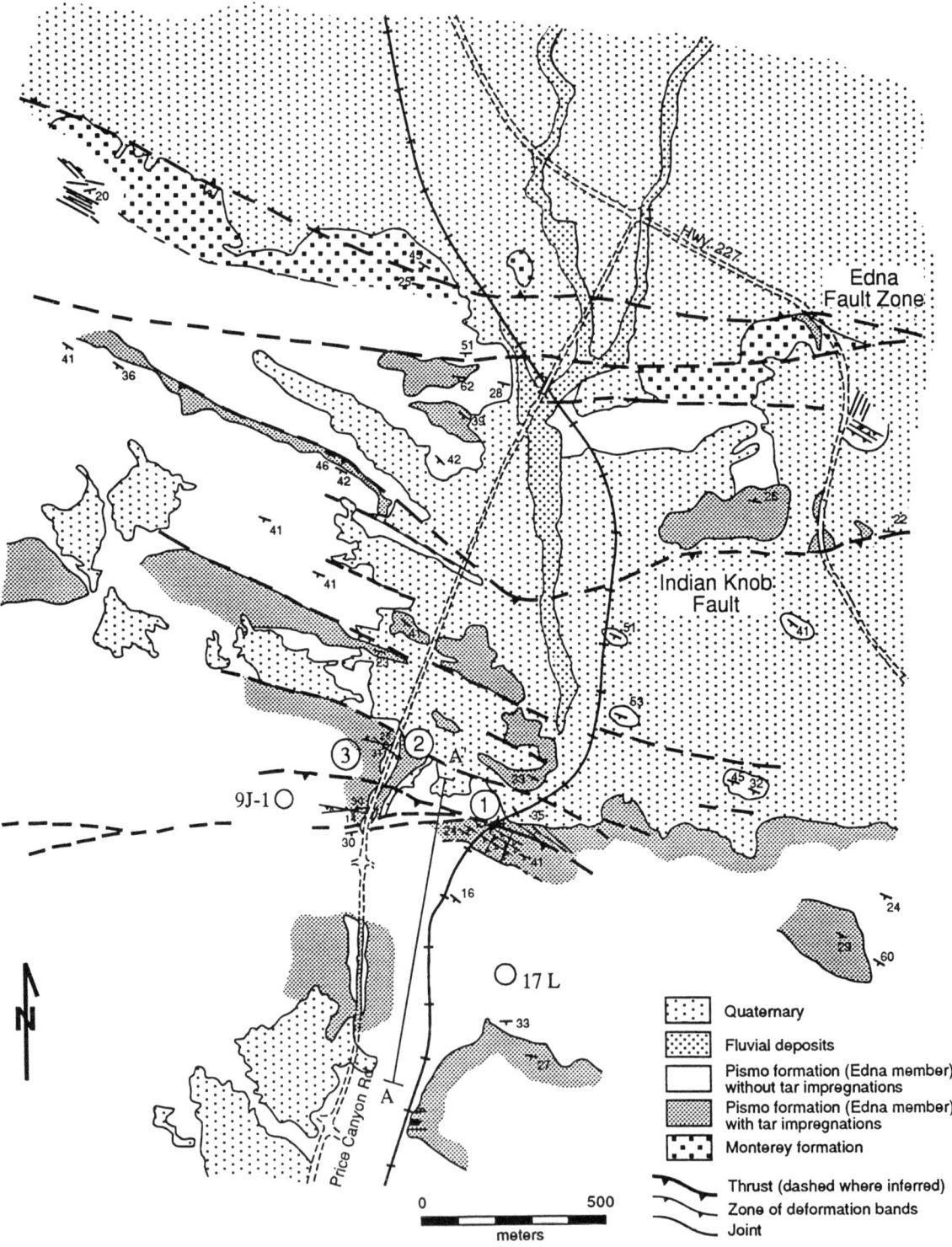

Figure 2. Geologic and tectonic map of the Arroyo Grande oil field area. A circle and well identification number on the side indicates the location of the well from which the cores have been studied. The circles with a number inside indicate the localities where the outcrop exposures have been mapped and studied in detail. The line AA' is the trace of the section shown in Figure 14.

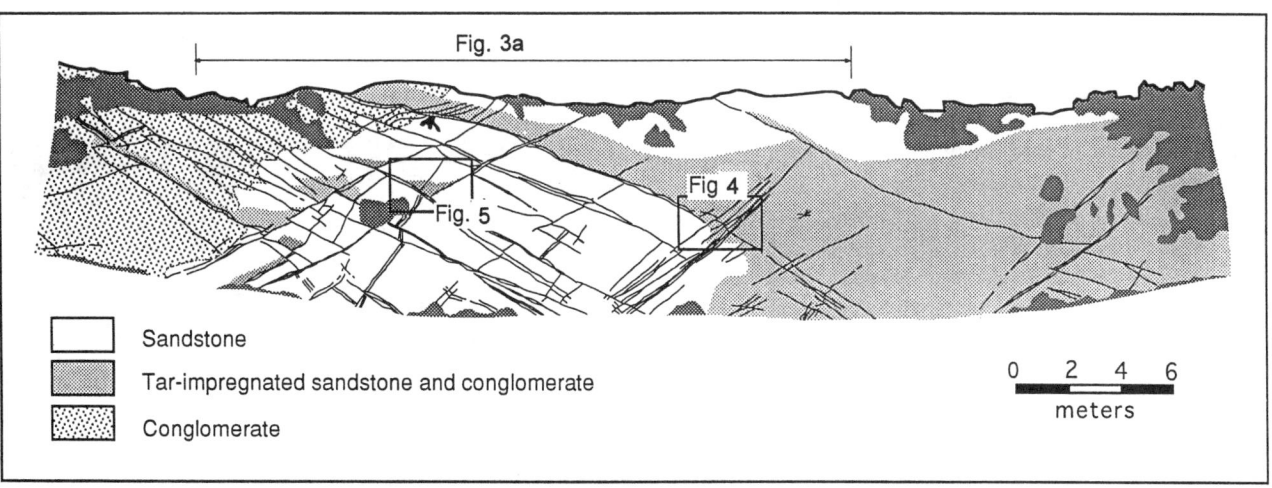

Figure 3. (a) Panoramic of the outcrop at location 3 in Figure 2. The stereonet inset gives the orientation of the two sets of deformation bands. (b) Map of the outcrop in (a). Two sets of deformation bands are developed in this outcrop. The effect of compartmentalization caused by the zones of deformation bands is well shown by the contrast between dark (tar-impregnated) and light-colored (tar-free) sandstone. The rectangles show the position of Figures 4 and 5.

Figure 4. Mutual cross-cutting relationships between deformation bands of two sets. The major zone of deformation bands is sealing also in the area where it has been faulted. The lens cap is 54 mm in diameter.

same time and probably were related to the same tectonic event.

The faults consist of zones of deformation bands [*Antonellini et al.*, 1994; *Aydin*, 1977, 1978; *Aydin and Johnson*, 1983; *Underhill and Woodcock*, 1987]. Each deformation band is a tabular structure identified by a 1-2 mm thick layer in which grains are crushed and porosity has been reduced. A deformation band typically accommodates an offset in the order of 1-2 cm. The bands in a zone may be very narrowly spaced (less than 1 mm) or more diffuse with 1-1.5 cm of intervening intact rock in between each deformation band. Each zone contains a variable number of deformation bands ranging from 2 to more than 20. The thickness of the zone of deformation bands, therefore, varies between 0.003 and 0.35 meters.

The sandstone impregnated with hydrocarbons appears dark gray or black in outcrop. The tar-free sandstone, on the other hand, has a light color. Most of the boundaries between tar-impregnated and tar-free sandstone are very sharp. The sharp boundaries consist of two sets of faults forming diamond-shaped compartments (Figure 5). The compartments formed by the faults have a width ranging between 1-20 meters and a dihedral angle in the range 30-60° (Figures 3 and 5). Thick zones of deformation bands (>10-15 deformation bands) are more effective in sealing hydrocarbons within compartments than thin zones of deformation bands (less than 10 deformation bands). The sealing properties of the faults do not change where the faults cross from sandstone to conglomerate (Figures 3a, 3b, 6a, and 6b).

Deformation bands within the sandstone saturated with hydrocarbons are white and stand out very clearly. Examinations of these deformation bands in thin section show that the fine pore space between the crushed grains does not contain hydrocarbons. At the fault boundary between hydrocarbon-saturated sandstone and hydrocarbon-free sandstone there are well-developed stains of iron oxides with an orange color (Figures 6a and 6b).

We investigated the relationships between major lineaments corresponding to the structural grain in the area and deformation band distribution by combining the study of aerial photographs and detailed outcrop mapping. Along the railroad near location 1 (Figure 2), we observe a section through one of the photo-lineaments. A scarp between tar-impregnated and tar-free sandstone created the lineament. Well 9J-1 is close to the strike of this structure, and was used to check the subsurface expression of the lineament. In the railroad outcrop, deformation bands are localized in 2-8 cm thick zones next to the scarp. Two sets of deformation bands are recognized. One is sub-parallel to bedding (WNW-SSE direction, 24° dip to the SSW) and to the contact between tar-impregnated and tar-free sandstone (location 1 in Figure 2). A second set is oriented at 70° to the first set (NNE-SSW direction, 40° dip to the SE).

South from the scarp (location 1 in Figure 2), we measured a 140 meters long transect (Figure 7) to characterize the distribution of the deformation bands away from the scarp. The transect shows a clustering of deformation bands within 90 meters from the scarp. Further along the railroad, 80 meters to the south of location 1 (Figure 2), there is a change from tar-impregnated to tar-free sandstone (Figures

Figure 5. Compartment bounded by two zones of deformation bands. The lens cap is 54 mm in diameter.

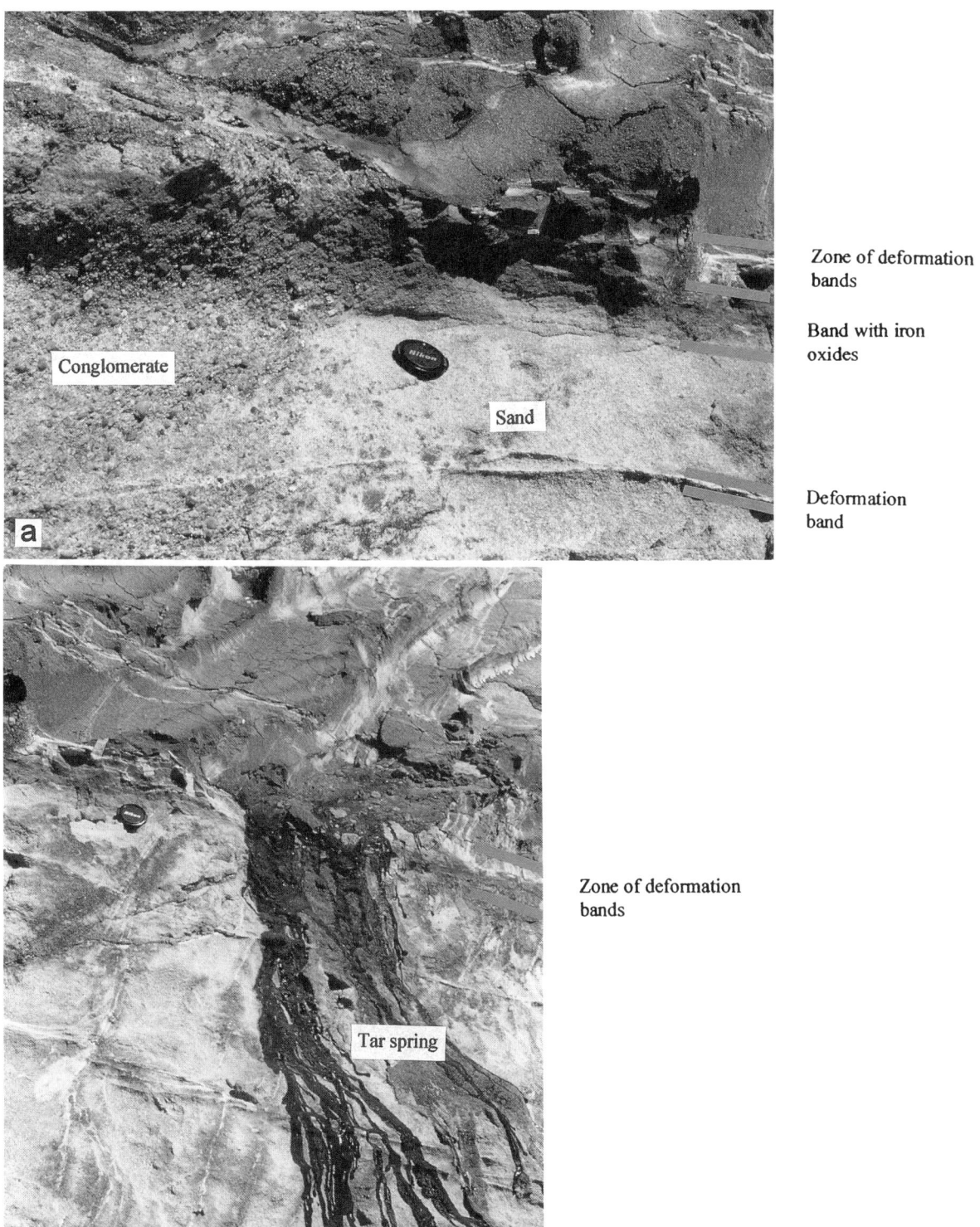

Zone of deformation
bands

Band with iron
oxides

Deformation
band

Zone of deformation
bands

Figure 6. (a) Band of iron oxides along a zone of deformation bands separating tar-impregnated from tar-free sandstone. Note the continuity of the sealing effect also in the coarser interval. (b) Same as above. Note the tar "spring" where the fault is sealing the hydrocarbon. This fault has only 20 cm offset.

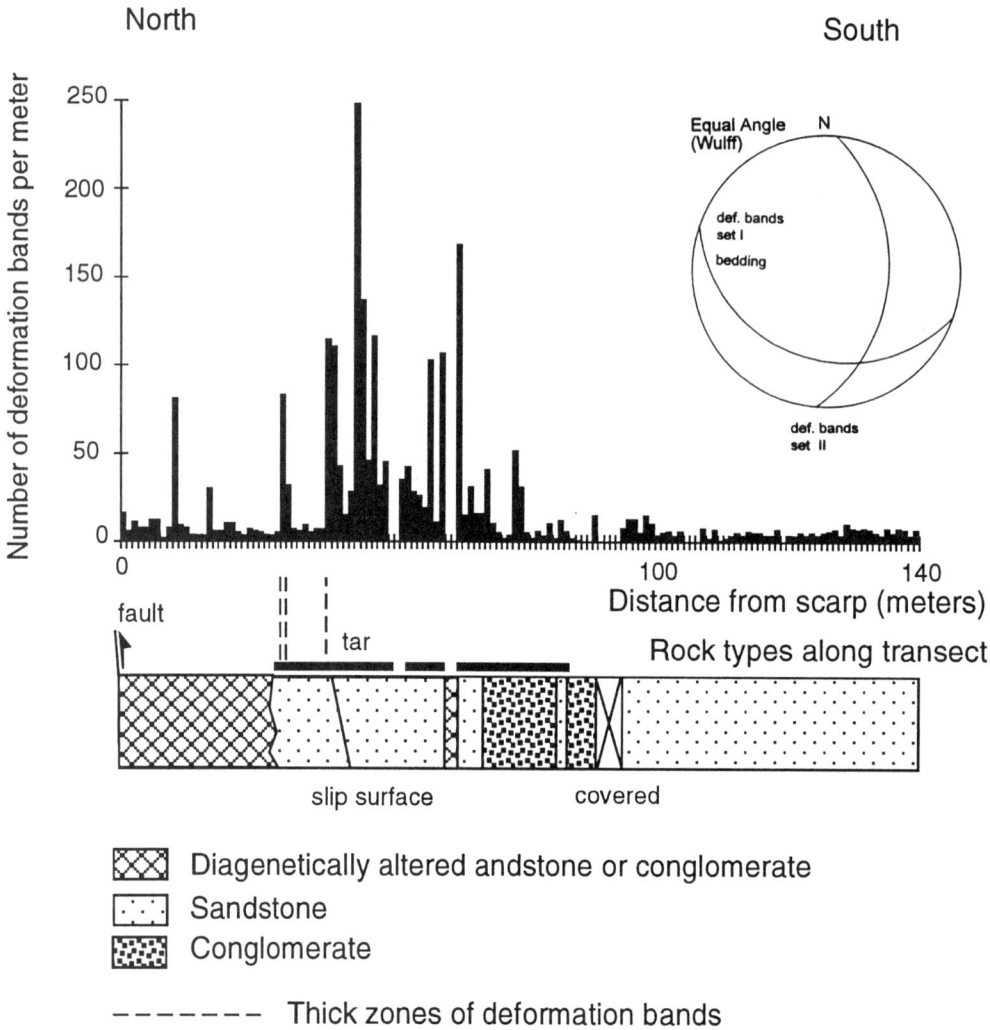

Figure 7. Histogram showing the density of deformation bands per meter vs. distance from a major fault along the railroad exposure. The deformation bands clusters near the fault scarp. Note the low density of deformation bands (but higher than background) in the area with carbonate cement in the pore space of the sandstone. Note also the change between tar-impregnated and tar-free sandstone. The stereonet inset gives the orientation of the two sets of deformation bands and of the bedding. The contact between tar-free and tar-impregnated sandstone is sub-parallel to bedding.

2 and 7). The contact dips 20-30° to the south and strikes E-W; the orientation of the contact is sub-parallel to the bedding and to one of the sets of deformation bands. The tar-free sandstone is very poorly cemented and has the consistency of loose wet sand. Small faults and deformation bands crosscut the tar-free sandstone. Most of the deformation bands have an apparent reverse slip component inferred from offset beds.

The deformation band density along the transect has a bell-shaped form (e.g. proportional to the form of a Gaussian density) that has its maximum value (250 m⁻¹) 40-45 m south of the scarp (Figure 7). Spikes in deformation band

density are in most cases related to the presence of thick zones of deformation bands. Within 30 m of the scarp the sandstone is not impregnated with tar (Figure 7), but the pore space is filled with carbonate cement. In this region, the deformation band density is 10-18 deformation bands per m, which is slightly higher than the background value of 5-6.

Near location 2 (Figure 2), which is in a road cut along Price Canyon Road, a set of deformation bands has a strike parallel to a small ridge that is at almost 90° from the trace of the road. The small ridge corresponds to a south-dipping fault, which is exposed at the northern termination of the

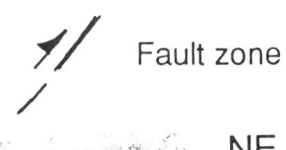

Fault zone

SW

NE

Tar-impregnated
sandstone

Tar-free
sandstone

Figure 8. Fault bounding tar-free from tar-impregnated sandstone on Price canyon road. The road cut is oriented NE-SW and the photograph has been taken looking NW. The fault is dipping to the west.

outcrop. The fault has an apparent thrust-slip component of 5-10 meters and juxtaposes hydrocarbon-impregnated sandstone in the hanging wall with hydrocarbon-free sandstones in the footwall (Figure 8). The fault zone consists of a 20-cm thick zone of deformation bands and an associated slip plane. This fault, that forms a clear lineament on the aerial photo, is sub-parallel to the lineament observed at location 1 in Figure 2. The outcrop expression of both photo-lineaments is a boundary between tar-impregnated and tar-free sandstone and a concentration of deformation bands in this boundary. A slip plane, however, has not been observed in outcrop 1.

4. CORE DESCRIPTIONS

The fault distribution in two non-oriented 4 1/2 inches (11.5 cm) cores from two different wells (9J-1 and 17L) of

the Arroyo Grande oil field was examined to establish if the outcrop data would have provided a good analog for sub-surface reservoir conditions. The two cores were the only available and they were recovered for more than 90%. The locations of the two wells, 9J-1 and 17L, is given in Figure 2. The density of the deformation bands was measured every three feet (0.9144 m) from photos of the cores, and the results were checked on selected segments of cores for consistency. The orientation of the deformation bands in the cores was measured to recognize the presence of different sets of deformation bands in each core segment examined. The lithology was also described. The deformation band density, lithologic description, neutron log porosity, and gamma ray log curves are presented in Figures 9 and 10 for the two cores. Synthetic neutron logs constructed from the core data are given in the same figures. These curves indicate the change in porosity due to the presence of deformation bands.

Core 9J-1

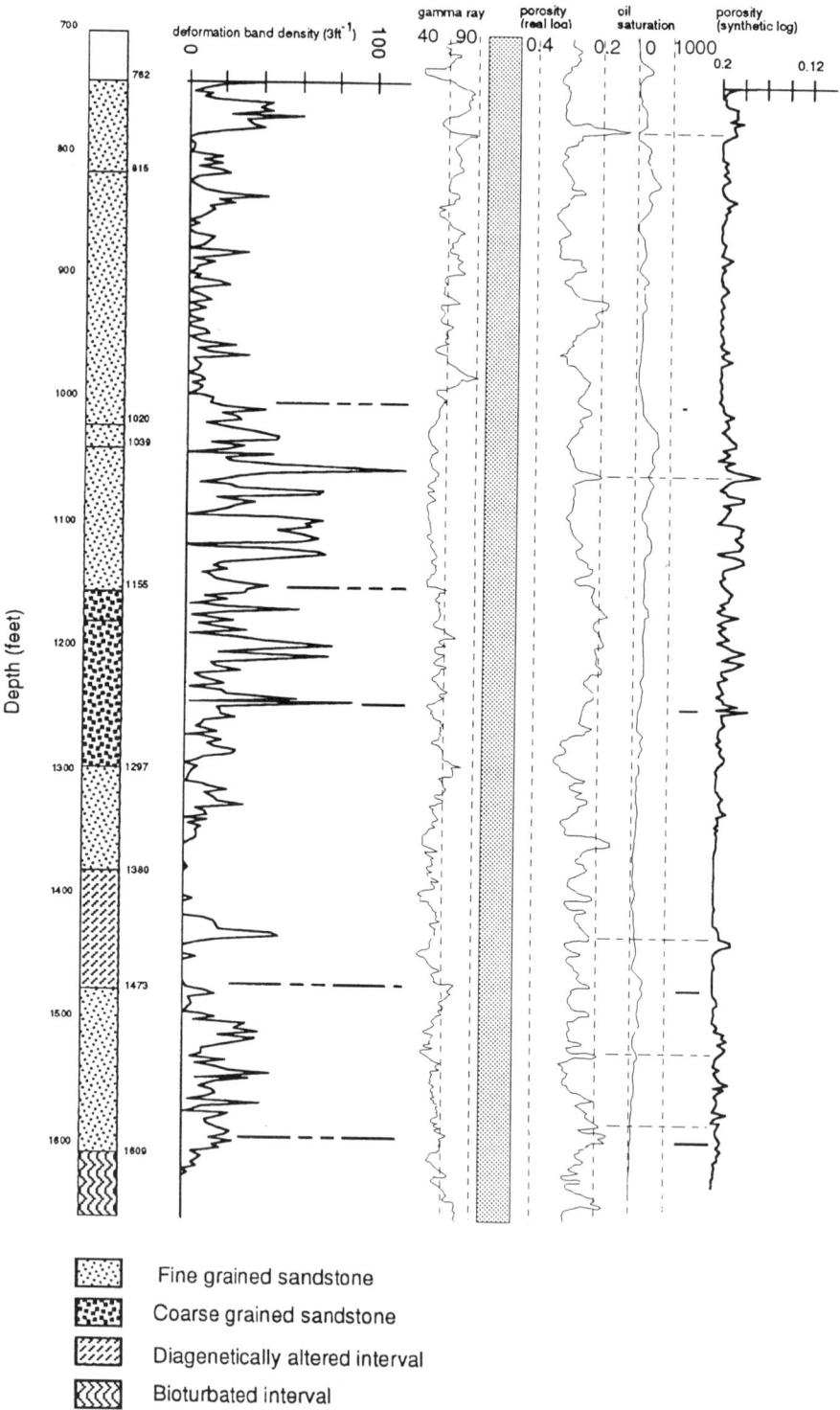

Figure 9. Logs from core 9J-1. From the left: lithology vs. depth, deformation bands density, gamma ray, CNL neutron porosity, oil saturation, synthetic CNL neutron porosity. Fine grained sandstone: 0.1 mm average grain size. Coarse grained sandstone: 0.25 mm average grain size. Horizontal dash/dotted lines represent correlation's between peaks in deformation bands density in the core and lows in the neutron log porosity.

Core 17L

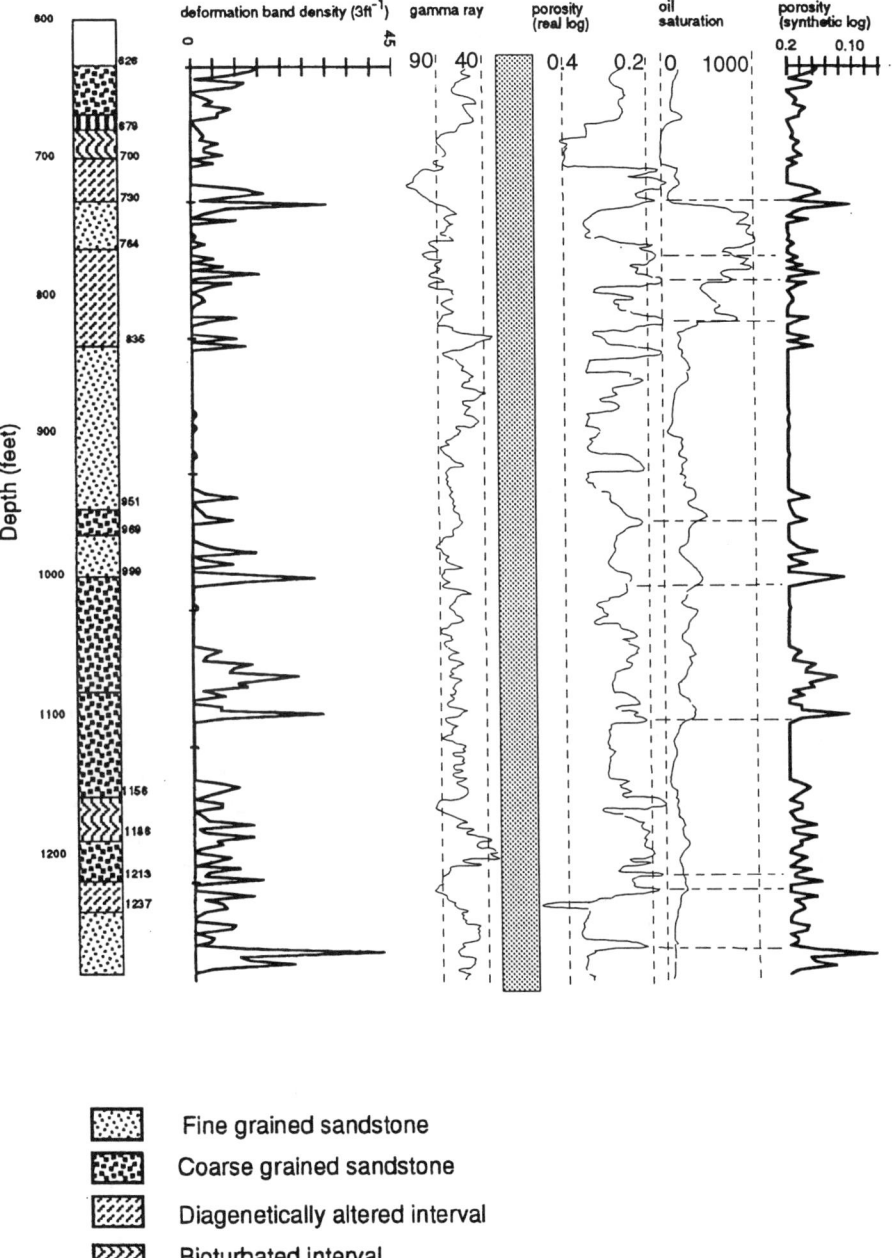

Figure 10. Logs from core 17L. From the left: lithology vs. depth, deformation bands density, gamma ray, CNL neutron porosity, oil saturation, synthetic CNL neutron porosity. Fine grained sandstone: 0.1 mm average grain size. Coarse grained sandstone: 0.25 mm average grain size. Horizontal dash/dotted lines represent correlation's between peaks in deformation bands density in the core and lows in the neutron log porosity.

The synthetic logs have been generated by considering the volume investigated by the neutron log detector, and the thickness, density, and porosity of deformation bands. The volume investigated by the neutron log detector is dependent on instrument characteristics [*Sclumberger Well Services*, 1972]. The thickness of the deformation bands and their density was measured directly from the cores; the porosity of the host rock and of the deformation bands was measured with image analysis of the thin sections (see sections on microstructure and petrophysical properties). The synthetic neutron log porosity was calculated by averaging (arithmetic mean) the bulk porosity in the detector's volume [*Antonellini and Aydin*, 1994].

4.1. Core Description From Well 9J-1 - Interval 762-1650 Feet (232.25-502.92 m)

The core is composed completely by sandstone. The sandstone is poorly cemented; a few cemented levels are present between 1400 and 1500 feet (426.72-457.2 m). A coarse grained interval (0.2-0.3 mm) is present between 1155 and 1297 feet (352-395.32 m). The well logs along the cored interval (Figure 9) show that the fine-grained intervals (~0.1 mm) have a larger porosity (about 5%) than the coarse grained intervals (porosity = 0.10-0.15). This happens because the coarse grained sandstone is poorly sorted with respect to the fine-grained ones. The natural radioactivity measured by the gamma rays is higher in the coarse grained intervals than in the fine-grained ones because of a larger amount of feldspars and volcanoclastic rock fragments in the former.

The largest number of deformation bands localized between 1000 and 1300 feet (304.8-396.24 m); in this interval the deformation band density can be as high as 120 every 3 feet (0.9144 m). The logarithm of the vertical density of deformation bands in well 9J-1 has a bell-shaped form centered on a depth of 1100 feet (335 m). The largest deformation band density (100-150 m^{-1}) is encountered in the intervals where the sandstone is fine grained (~ 0.1 mm) and it has a high porosity (~0.2). The deformation bands density in cemented layers is lower (0-50 m^{-1}) than in unconsolidated ones (0-150 m^{-1}).

Neutron log porosity and synthetic neutron log porosity (Figure 9) show porosity reduction of only a few percent where the deformation band density is large (~120 m^{-1}).

A zone of shallow dipping (10-15°) deformation bands (Figure 11a) at a depth of 1557 feet (474.57 m) is a seal between oil-impregnated and oil-free sandstone. Throughout the core there are zones of deformation bands that are boundaries between different degrees of hydrocarbon saturation in sandstone belonging to the same level.

Measurement of the dip of deformation bands from the cores shows two major sets of zones of deformation bands.

Both sets are dipping at low angle from the horizontal (~ 25-35° dip). A third higher angle set (45-70° dip) is also present. This set is only observed occasionally both in outcrop exposures and in cores. Sometimes the high angle set is localized between two major sub-parallel (and sub-horizontal) zones of deformation bands; *Antonellini and Aydin* [1995] have also described this structure at Arches National Park, Utah.

The deformation bands appear to be thick (1.5-3 mm) in the coarse grained intervals (0.2-0.3 mm average grain size), whereas they are thin (0.5-1 mm) in the fine grained sandstone (0.1 mm average grain size), and in the levels that have been extensively cemented.

4.2. Core Description From Well 17L - Interval 626-1250 Feet (190.8-381 m)

This core is completely composed by fine to coarse-grained sandstone. The pay zone is between 730 and 835 feet (222-255 m). There is local cement precipitation in the fine-grained intervals (0.1 mm average grain size). Cement precipitation is observed between 700-730 feet (213-222 m), 764-835 feet (232-254 m), and 1213-1237 feet (388-377 m); the porosity of these intervals is less than 0.1. Thick coarse grained (0.2-0.3 mm average grain size) low-porosity (~0.15) intervals are present between 626-679 feet (191-207 m), 951-969 feet (290-295 m), and 999-1213 feet (304-370 m). The rest of the core is made up by well sorted, fine-grained (~0.1 mm average grain size), high-porosity (0.2), unconsolidated sandstone. The high porosity sandstone makes up only 35% of this core, whereas in core 9J-1 it makes up about 70% of the core.

The maximum deformation bands density is 40 m^{-1} in the fine grained unconsolidated sandstone. High deformation band density is almost always localized in the fine-grained unconsolidated sandstone.

Joints filled with hydrocarbons are observed in the low porosity (< 0.1) cemented sandstone. The joint density can be up to 2 per centimeter (Figure 11b). The joints belong to two sets: one dips at a high angle (70°) and one at a low angle (30°). The pay zone corresponds to a zone of high joint density.

5. MICROSTRUCTURE

Fifteen samples were collected from the outcrop in the sealing fault zones and within the tar-impregnated, and tar-free sandstone. Thin sections were cut for optical microscopy analysis. Six more thin sections were cut from core samples (cores 9J-1 and 17L).

Grain sizes within the host rock range from 0.02 mm to 1.4 mm. The average grain size is 0.2-0.3 mm in coarse grain size levels, and 0.1 mm in fine-grain-size levels. The grain size distribution within the deformation band varies from 0.005 to 1.375 mm. The average grain size is 0.01 mm. The sorting within the fault rock is very poor (Figure 12). Porosity determined via image analysis [*Antonellini et al.*, 1994; *Ehrlich et al.*, 1985] and point counting ranges between 0.15 and 0.2 in the host rock and between 0.01 and 0.06 in the fault rock. Sand grains include monocrystalline quartz, polycrystalline quartz, chert, feldspars and volcaniclastic rock fragments. Feldspars and rock fragments can be in excess of 15%; these grains are the most commonly crushed grains in the deformation band and its proximity. Grains with an oblate shape tend to have their longer axis lying on the plane of the band. All these characteristics are common to the deformation bands in other sandstone types described by *Antonellini et al.* [1994].

6. PETROPHYSICAL PROPERTIES

Gas permeability [*Antonellini and Aydin*, 1994] and plug permeability [*Nelson*, 1985; *Pittman*, 1981] measurements from deformation bands in different sandstone types indicate that permeability of deformation bands may be 1 to 4 orders of magnitude less than in the surrounding host rock: 0.1-10 md in the deformation band, 100-10000 md in the host rock. This variability in permeability is due to different microstructures (i.e. different degrees of cataclasis development and volumetric deformation) present in deformation bands. Where a distinct slip plane localizes in the sandstone, the permeability in its wall rock can be more than 7 orders of magnitude smaller than in the surrounding host rock [*Antonellini and Aydin*, 1994]. The microstructure in the wall rock of a slip plane is characterized by compaction and recrystallization of quartz grains that reduce the porosity to levels below 0.01.

The cataclastic microstructure of the faults in the Pismo outcrop is very similar to that of deformation bands whose measured gas permeability is about 3-4 orders of magnitude less than in the surrounding host rock [*Antonellini and Aydin*, 1994]. We observed only two slip planes in the surface exposures of the Pismo formation; in none of them we were able to observe the microstructural characteristics.

We can calculate permeability with the Carman-Kozeny formula as modified by *Dullien* [1992] by using porosity (ϕ), the hydraulic grain diameter (D_h), the tortuosity of the flow path (τ), and the shape of the grains (s) measured from thin section observations:

$$K = [(\phi \, D_h^2) / (16 \, s \, \tau^2)] \qquad (1)$$

Assigning a tortuosity of 1, a porosity of 0.2 for the host rock, a porosity of 0.01 for the deformation bands, an average grain diameter of 0.2 mm for the host rock, and an average grain diameter of 0.01 mm for the deformation bands,

we can estimate the ratio between host rock (K_{hr}) and deformation band permeability (K_b): $K_{hr}/K_b \sim 6000$, which has the same order of magnitude of similar deformation bands from which the permeability was directly measured [*Antonellini and Aydin*, 1994].

The simple observation that hydrocarbon-saturated sandstone is in contact with hydrocarbon-free sandstone at a fault boundary implies that the relative permeability of the hydrocarbon through the fault zone is zero or nearly so. The mobility of a non-wetting hydrocarbon phase in the pore space is controlled by the capillary pressure [*Downey*, 1984; *Pittman*, 1981; *Purcell*, 1949; *Schowalter*, 1979]. The non-wetting phase can move through the pore space only if its driving pressure is larger than the capillary pressure at the water-oil interface, the magnitude of which is inversely proportional to the radius of the interconnected pore space [*Schowalter*, 1979]. The wetting-phase (water) prefers to occupy smaller spaces than the non-wetting phase. Where a rock with a large grain size, containing a wetting and non-wetting phase, is in contact with a rock with a smaller grain size, the wetting fluid tends to imbibe into the tighter rock.

The Young-Laplace equation modified by *Hubbert* [1956] relates the capillary pressure, Pc, in terms of grain size:

$$Pc = [(C \, \gamma \cos \theta) / d] \qquad (2)$$

where C is a proportionality constant, θ is the wettability, γ the interfacial tension and d the average grain diameter. The average grain diameter in the crushed zone of a fault from the Arroyo Grande outcrops is more than 1 order of magnitude smaller than in the surrounding host rock, which implies that the capillary pressure in a deformation band is more than one order of magnitude larger than in the host rock.

The large capillary pressure in a deformation band prevents the hydrocarbons from moving through the faults and causes the faults to be sealing with respect to the hydrocarbon.

7. STEAM INJECTION DATA

We analyzed steam breakthrough time data provided by Calresources to investigate the possibility that the orientation of the faults control the conductivity of the steam in the reservoir. A steam drive project was initiated in 1984 on 320 acres of land located 150 m south of location 1 in Figure 2. This project was expected to recover 69.4 million barrels of 14° API gravity oil from the Dollie sands by using a new modified 5-acre inverted 9-spot pattern [*Cumming et al.*, 1989]. The critical change in the pattern design is the addition of a "side" injector near each corner producer

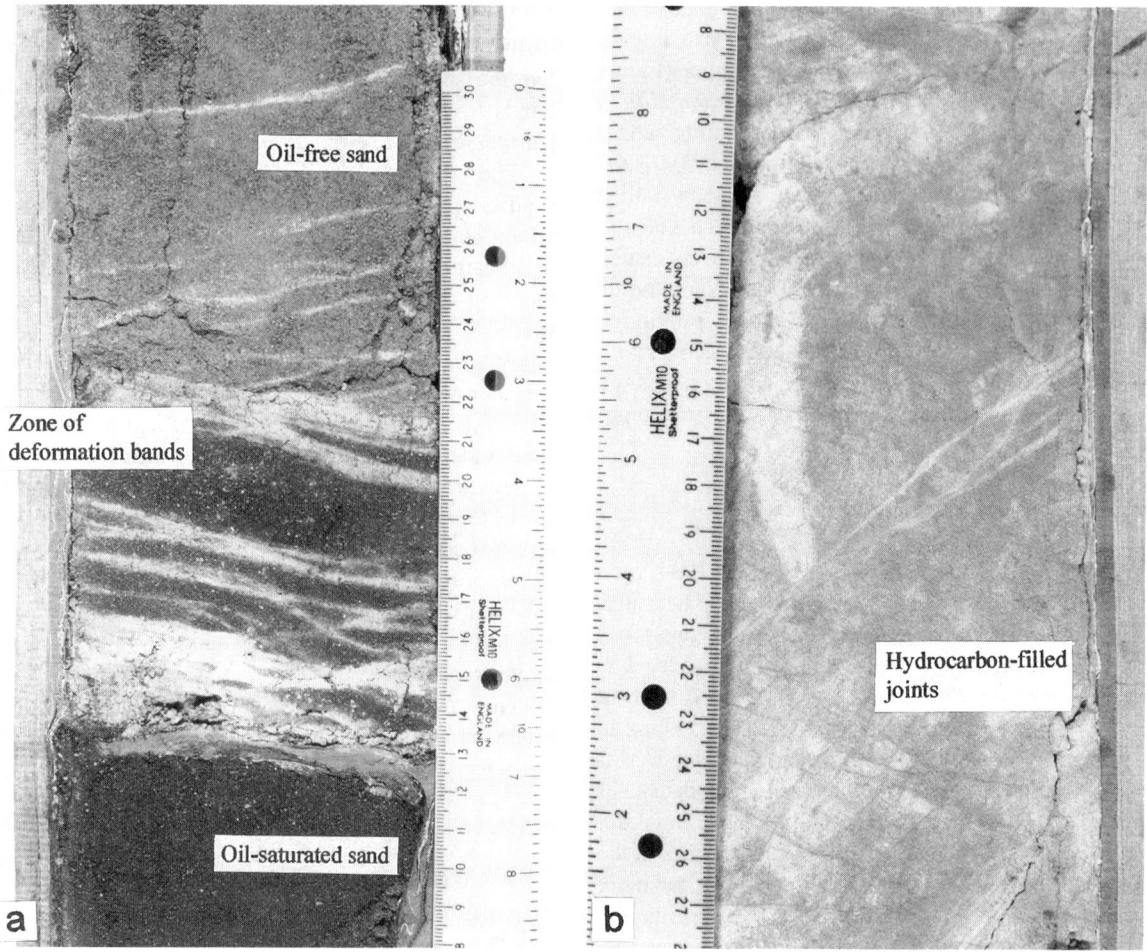

Figure 11. (a) Boundary between tar-impregnated and tar-free sandstone in correspondence of a zone of deformation bands in core 9J-1 at a depth of 1557 feet (474.5 m). (b) Open, tar-impregnated fractures in core 17L.

(Figure 13a). We analyzed data relative to 12 injector wells and 33 producer wells. By knowing the distance between injectors and producers, as well as their pattern design, we calculated the conductivity of the steam along different azimuths between injector and producer. The relationship between steam conductivity and azimuth of the line connecting injector with producer is shown in Figure 13b. Figure 13b shows that the minimum steam conductivity is in a direction 10-25° and that the maximum steam conductivity is in a direction 110-120°. The steam conductivity ellipsis is anisotropic; the maximum conductivity is equal to 9.1 times the minimum conductivity (Figure 13b). Note that the direction of minimum conductivity is parallel to the average dip azimuth of the faults in the area (Figures 2 and 13b).

8. DISCUSSION

8.1. Comparison Between Surface and Subsurface Data

The study of the aerial photographs and of the outcrop exposures shows a geometric relationship between photo-lineaments, tar impregnation, and deformation band density distribution. The traces of the Edna fault zone and Indian Knob fault were mapped from lithologic contacts and photo-lineaments (Figure 2). Other lineaments, detected from aerial photographs, typically correspond to changes in the dip of the strata or to scarps at the contact between tar-impregnated and tar-free sandstone. The photo-lineaments are shown as dashed lines in Figure 2. These lineaments are sub-parallel to the structural grain represented by the traces of the major faults (e.g. Edna fault). Examination of one of these lineaments in outcrop at location 2 in Figure 2 shows that the lineament is associated with a ridge and a small offset (5-10 m) reverse fault in the road cut.

Another photo-lineament at location 1 in Figure 2 is associated with a steep scarp (70-80°). The deformation band density is high within 90 meters from the scarp (Figure 7). *Antonellini and Aydin* [1995] have documented a relationship between increase in deformation band density and vicinity to a fault in porous sandstones. On the basis of the high density of deformation bands near the scarp as well as of its steep dip, we propose that the scarp is the surface expression of an high angle (70-80°) reverse fault with an offset not larger than 20-30 m (this value is based on field correlation of lithologies in the hanging wall and in the footwall). The location of the scarp on the map at location 1 in Figure 2 also corresponds, in the subsurface, to a permeability boundary for two hydrocarbon producing levels in the Elberta sand of the Pismo formation [*Lawrence*, 1958]. No producing wells have been drilled north of this boundary. The boundary, that is rectilinear on the structure map, could be interpreted as a fault cutting at depth through the levels of the Elberta sand. All these observations suggest the possibility that the lineament observed corresponds to a fault with sealing characteristics.

The vertical distribution of deformation bands in well 9J-1 (Figure 9) is very similar to the distribution of deformation bands observed in outcrop near the lineament interpreted as a fault (location 1 in Figure 2). Projection of this lineament in the subsurface would correlate the observed peak of deformation bands in the well to the same structural position where the peak in deformation band density is observed in outcrop exposure.

The zone of low-angle deformation bands (Figure 11a) at a depth of 1557 feet (474.57 m) in core 9J-1 is a seal between oil-saturated and oil-free sandstone. This seal is similar to the seals formed by the zones of deformation bands observed in outcrop (Figures 3a and 3b). Hence, the sealing phenomena observed at the surface apparently extend to reservoir depth.

8.2. Trap Mechanism

Synclines are generally not believed to form traps during migration of the hydrocarbons. The Arroyo Grande oil field, however, is located on the limb of a syncline (Figure 1). We used data from 6 wells, for which we had logs, along section AA' (Figures 2 and 14) to understand the factors that control hydrocarbon distribution and trap formation within the Pismo formation. The top of the reservoir

Figure 12. Thin section (plain polarized light) of a deformation band from the outcrop at location 3 in Figure 2. Note the cataclastic material in the deformation band and the absence of tar in the small fault. Note also abundant microcracking at grain contacts in proximity to the deformation band.

indicated by the well logs dips to the north (Figure 14), switching to the south just in proximity of the lineament that we interpret as a fault. The direction of dip of the top of the reservoir surface is inconsistent with the attitude of bedding observed on the section along the same profile AA' at the surface (Figure 14). The structural setting of the bedding does not seem to control the hydrocarbon distribution. A possible explanation of the conflicting data from the wells and from the outcrops, could be the presence of a change in sedimentary facies with low permeability deposits in the north, which prevents hydrocarbon flow out of the syncline structure. Given the lack of field observations for very low permeability deposits in the north as possible lateral sealing units, the presence of sealing faults with the same trend as the lineaments observed at the surface could offer an alternative explanation for the trapping mechanism in the syncline. The faults may have acted as vertical conduits for fluids (e.g. water) that caused the degradation of the hydrocarbon to a tar. The tar, in combination with the cataclastic material of the fault forms a seal preventing the lateral migration of the hydrocarbons (Figure 14).

The faults, however, may represent a trapping mechanism only for the hydrocarbons but not for the connate water. The hypothesis that the wetting-phase (water) is continuous across the faults is speculative but it is supported by field observations. The most interesting observation to this regard concerns well-developed stains of iron oxides at the fault boundary between hydrocarbon-saturated and hydrocarbon-free sandstone (Figures 6a and 6b). A possible

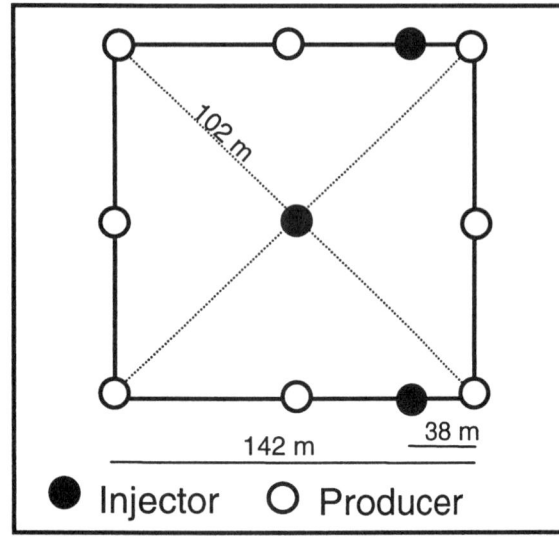

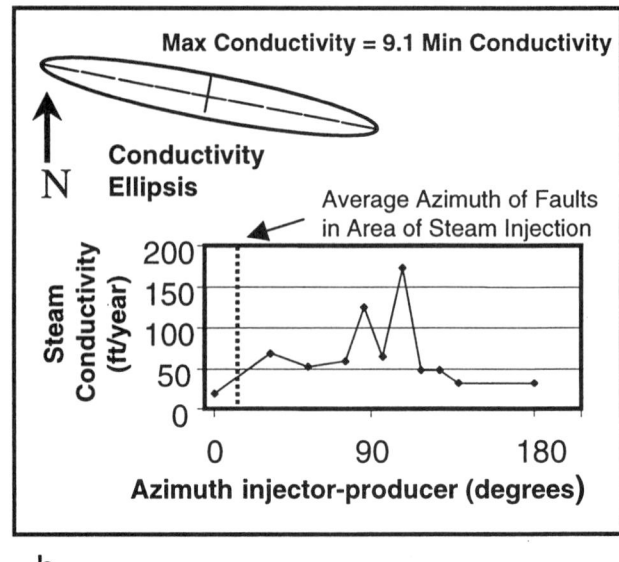

a   b

Figure 13. Steam drive project at Arroyo Grande. (a) Pattern design for injectors and producers. (b) Relationship between steam conductivity and the azimuth between injector and producer. Note that the minimum conductivity is also parallel to the average dip azimuth of the faults in the area. The steam conductivity ellipsis shows that the maximum conductivity is equal to 9.1 times the minimum conductivity.

mechanism for the formation of these iron oxide precipitates is explained in the following. The redox potential (*Eh*) within the hydrocarbon-saturated sandstone is negative (redox environment) and iron is probably present in the reduced form $Fe(OH)_2$. This form of iron is very soluble [*Blatt et al.*, 1980], so that the ions Fe^{+2} are easily transported in solution from within the hydrocarbon-saturated sandstone to the fault interface where they come in contact with the oxidizing environment of the hydrocarbon-free sandstone. At the interface the *Eh* becomes positive and the Fe^{+2} is oxidized and precipitated in the pore space adjacent to the fault as the stable authigenic mineral $Fe(OH)_3$. The presence of this oxidation reaction suggests that the hydrocarbon-free sandstone was saturated with water at the time of stain formation and that there was connection of the water phase through the sandstone. This connection may have involved advective flow or diffusion of ions (Fe^{+2}) in the wetting-phase.

8.3. Fault Detection With Well Logs

We tried to detect the presence of small cataclastic faults from the porosity logs, because high-resolution dipmeter or image logs for the Arroyo Grande Oil Field were not available. Porosity loss due to deformation bands localization is not large compared to the bulk porosity of the rock [*Antonellini and Aydin*, 1994]. Examination of the

neutron log response in core 9J-1 shows that some lows in porosity correspond to zones of high density of deformation bands. The porosity lows due to the deformation bands, however, are hard to distinguish from porosity changes due to small sedimentary structures. It is apparent, from Figures 9 and 10, that the porosity reduction due to deformation band localization is only a few percent in zones where a deformation band density in the order of 120 m^{-1} was observed in the core. The signatures of the deformation bands are shown both on the neutron log and on the synthetic neutron log. The synthetic neutron log response in the massive sandstone shows small (1-5%) variations in porosity caused by the deformation bands. These small variations in porosity can be correlated to the presence of a thick zone of deformation bands only if the cores are available, because small scale sedimentary structures can cause the same variability in porosity where no deformation bands are present (Figures 9 and 10).

8.4. Influence of Lithology on Fault Characteristics

The deformation bands observed in the Arroyo Grande oil field have similar microstructural characteristics (development of cataclasis, one order magnitude porosity reduction from host rock to fault rock, etc.) to those of the deformation bands observed at Arches National Park by *Antonellini et al.* [1994]. Their geometry, however, has some

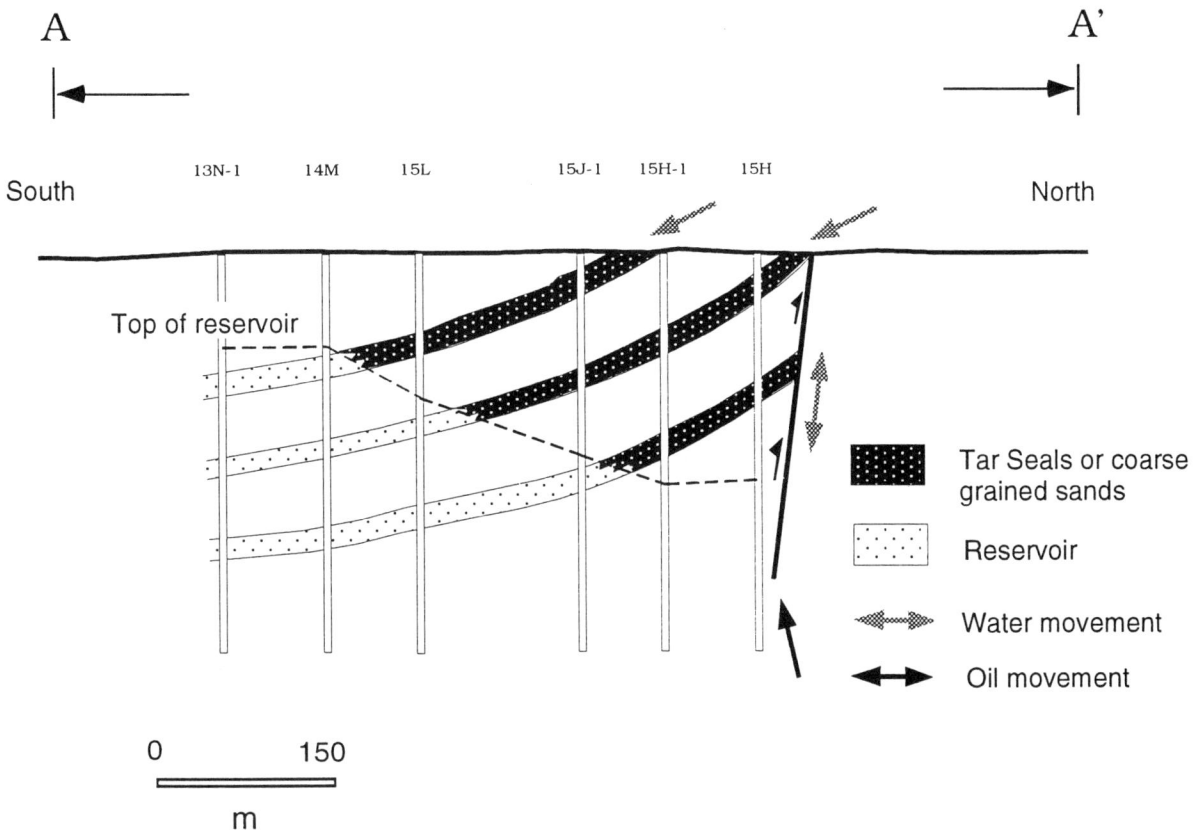

A   A'

South   North

13N-1   14M   15L   15J-1 15H-1   15H

Top of reservoir

■■■ Tar Seals or coarse
      grained sands

⋯⋯ Reservoir

⟷ Water movement

⟷ Oil movement

0   150

m

Figure 14. Cross section from wells and outcrop data. Outcrop data show that the beds dip to the west and they are on the north limb of the Pismo syncline. Well data show that the top of the reservoir dips to the north. An interpretation based on outcrop mapping, well data, and formation of tar seals as a result of water flow from the surface and along the fault plane, is proposed to explain the discrepancy. Grey arrows indicate the possible direction of water flow; black arrows indicate the possible direction of hydrocarbon flow.

notable differences. Deformation bands in the Pismo formation tend to be shorter and less interconnected than those present at Arches National Park. Another difference is that the faults studied in the Arroyo Grande area also display only two slip planes similar to those observed in the Entrada Sandstone of the San Rafael desert [*Aydin and Johnson*, 1978] and Arches National Park [*Antonellini et al.*, 1994].

Apparent differences between the outcrops in Utah and those described here include the different tectonic setting (shortening in the Pismo syncline - extension at Arches National Park), and the degree of cementation of the sandstone. The sandstones of the Pismo formation are almost completely loose if not impregnated by tar or cemented near the faults, whereas the sandstones cropping out in southeastern Utah are cemented because of pressure solution at grain contacts or by intergranular calcite cement.

The control of lithology on the localization of the deformation bands is also observed in cores 9J-1 and 17L. If we compare the density of deformation bands in core 9J-1 with that of core 17L, it is apparent that in the latter the average density of deformation bands is about a factor of 3-4 less than in core 9J-1. The lithology in core17L is similar to that of core 9J-1, however the pore space is filled by carbonate cement. The presence of cement in the pore space seems to have prevented to a certain extent the formation of deformation bands in core 17L. This, however, is not the unique explanation, different deformation band densities in cores 9J-1 and 17L could also be due to a different structural position; core 9J-1, in fact, cuts across the subsurface projection of a fault observed in outcrop, whereas core 17L does not. Given that the cataclastic material in the deformation bands is not cemented, it seems that the deformation bands formed after cement precipitation.

The well-cemented levels of core 17L contain joints. Joints are not present anywhere in core 9J-1. The hydrocarbon distribution in core 17L is controlled by the deformation bands and by the joints. These joints are filled with calcite cement and hydrocarbons (Figure 11b) and they are observed only in very low porosity sandstone (< 5%).

8.5. Influence of Faults on Reservoir Management

The distribution of the photo-lineaments, their spacing, and orientation (Figure 2) provide a framework to describe structural heterogeneity within the oil reservoir. The photo-lineaments correspond to traces of reverse faults with offsets of 5-50 m. These faults appear to be sealing structures or permeability baffles both in outcrop exposure and in the subsurface. The outcrop study shows how the lineaments are associated to clusters of deformation bands that contribute to locally degrade the permeability and porosity characteristics of the reservoir. The deformation bands may contribute to fault sealing and compartmentalization at a scale ranging from 0.1 to 20 m throw. By knowing the spacing of the major photo-lineaments (100-300 m), their orientation, the drainage area of the wells, the production vs. time performance curves, and the volumes of oil recovered one can attempt to compute how much of the reservoir has been effectively drained during production, estimate the size of the fault-bounded compartments, and identify the areas where some pay compartments may have been by-passed.

Currently, an enhanced oil recovery (EOR) project with steam vapor injection is under way in the Arroyo Grande oil field. The analysis of the relationship between steam conductivity and orientation of the azimuth between injector and producer wells has shown (Figure 13b) that the minimum conductivity direction is parallel to the average dip azimuth of the faults and that the ratio between maximum steam conductivity and minimum steam conductivity is 9.1 in the field. This means that faults are a baffle to the migration of the steam in the reservoir sands. Knowledge of the orientation and distribution of faults and deformation bands could be used a-priori to optimize drilling direction, plan the location of in-fill wells (by knowing the orientation and magnitude of the conductivity ellipsis), and to better plan the sites of the injecting and producing wells in a way to avoid low permeability baffles and areas with large reservoir heterogeneity. Optimization of the steam injection scheme allows to produce oil faster and cheaper.

9. CONCLUSIONS

(1) Reverse faults expressed in outcrop as surface lineaments are aligned with the major structural trends of the area and control the distribution of the tar in the sandstone above the reservoir of the Arroyo Grande Oil Field. The reverse faults are high angle and they are interpreted to have prevented in various degree the lateral migration of

the hydrocarbons. Diagenetic phenomena (carbonate cement precipitation in the pore space of the sandstone and tar seals) are associated with these faults, suggesting that the faults may have also acted as conduits for the migration of fluids (water or hydrocarbons) in the direction parallel (up and down dip) to their plane.

(2) Small, seismically undetectable (<10-20 m offset) faults consisting of zones of deformation bands are effective in sealing hydrocarbons at outcrop scale. This sealing effect can also be observed in cores and we suspect that sealing by small cataclastic faults is important at reservoir conditions, because of the low hydrostatic pressures (shallow reservoir) and small hydrocarbon columns (from a few meters to a few tens of meters). Even, if the hydrocarbon phase is continuous through these small faults, they still can reduce hydrocarbon flow during production because of their low permeability.

(3) Deformation band density in outcrop and in cores increases in proximity to a thrust fault with more than 20 m offset. The maximum density is in the hanging wall. The deformation bands density is larger (of a factor 3-4) in high-porosity (0.1-0.2), fine-grained (0.1 mm average grain size), well-sorted sandstone than in low-porosity (< 0.1), coarse-grained (0.2-0.3 mm average grain size), poorly-sorted sandstone.

(4) The sealing of the hydrocarbons is caused by the large capillary pressures in the fine cataclastic material of the fault zones (1-2 orders of magnitude larger in the fault zone) respect to the surrounding host rock. Only the non-wetting phase (hydrocarbon) is sealed; water may be continuous through the fault zone. Also cement precipitation in the pore space of the sandstone and formation of tar seals may contribute to fault sealing.

(5) Comparison of fault zones in outcrops and in core shows that they have the same characteristics. The deformation band distribution measured along the scanline in outcrop is very similar to that measured in core 9J-1.

(6) The neutron log signature of deformation bands is weak even where their density is high, because of the small effect they have on bulk porosity. The porosity variation associated with very thick (> 10 cm) and high density zones of deformation bands can be hardly distinguished from the porosity variation associated to small scale sedimentary heterogeneity in a relatively homogeneous reservoir.

(7) The relationship between steam conductivity and orientation of the azimuth between injector and producer wells shows that the minimum conductivity direction is parallel to the dip azimuth of the faults. This demonstrates that faults control the permeability of fluids in the subsurface. The knowledge of the distribution, orientation and petrophysical properties of the structural heterogeneity permits better plans of enhanced oil recovery techniques in the late phases of development of this field.

Acknowledgments. Thanks to Shell Oil Co. USA and to Larry Gibson for providing the two cores and the logs used in this study. We are also grateful to B. H. Huff at Calresources for making available the steam injection data. This work was funded by the Rock Fracture Project an industry affiliate program at Stanford University. Discussions with Rene' White and Patricia Phillips were useful to get us started on this project. Comments from Ken Cruikshank, Allan Gutjahr, William Haneberg, Pauline Mollema, and an anonymous reviewer have been very helpful in improving the quality of this manuscript.

REFERENCES

Alastair, I. W., A. Beach, P. J. Brockbank, L. J. Brown, S. D. Knott, and J. E. McCallum, Damage zone geometry: outcrop examples from Western Sinai and the United Kingdom, *Fault Sealing Meeting Proceed. Abstr.*, Leeds, 21, 1996.

Allen, U. S., Model for hydrocarbon migration and entrapment within faulted structures, *AAPG Bull.*, 73, 803-811, 1988.

Anderson, R., 4D seismic interpretation technologies, *AAPG Abstr. Programs*, 81, 456, 1996.

Antonellini, M. A., and A. Aydin, Effect of faulting on fluid flow in porous sandstones: petrophysical properties, *AAPG Bull.*, 78, 355-377, 1994.

Antonellini, M. A., and A. Aydin, Effect of faulting on fluid flow in porous sandstones: geometry and spatial distribution, *AAPG Bull.*, 79, 642-671, 1995.

Antonellini, M. A., A. Aydin, and D. D. Pollard, Microstructure of deformation bands in porous sandstones at Arches National Park, *J. Struct. Geol.*, 16, 941-959, 1994.

Aydin, A., *Faulting in sandstone*, Ph. D. thesis, Stanford Univ., Stanford, CA, 1977.

Aydin, A., Small faults formed as deformation bands in sandstone, *Pure Appl. Geophys.*, 116, 913-930, 1978.

Aydin, A., and A. M. Johnson, Development of faults as zones of deformation bands and as slip surfaces in sandstone, *Pure Appl. Geophys.*, 116, 931-942, 1978.

Aydin, A., and A. M. Johnson, 1983, Analysis of faulting in porous sandstones, *J. Struct. Geol.*, 5, 19-31, 1983.

Blatt, H., G. Middleton, and R. Murray, *Origin of sedimentary rocks*, Prentice and Hall, N. J., 1980.

Bouvier, J. D., K. Sijpesteifn, D. F. Kluesner, C. C. Onyejekwe, and R. C. Van der Pal, Three-dimensional seismic interpretation and fault sealing investigations, Nun River field, Nigeria, *AAPG Bull.*, 73, 1397-1414, 1989.

Byerlee, J., Model for episodic flow of high-pressure water in fault zones before earthquakes, *Geology*, 21, 303-306, 1993.

Caine, J. S., D. R. Coates, N. P. Timoffeef, and W. D. Davis, *Hydrogeology of the Northern Shawangunk Mountains*, New York State Geol. Surv. Open-File Report 1-G-806, 1991.

Caine, J. S., J. P. Evans, and C. B. Forster, Fault zone architecture and permeability structure, *Geology*, 24, 1025-1028, 1996.

Cumming, E. W., A. E. Mueller, and H. A. Orndorff, Arroyo Grande steam drive project; San Luis Obispo County, California, *AAPG Geol. Petrophys. Conference Meeting Volume*, 1107-1116, 1989.

Downey, M. W., Evaluating seals for hydrocarbon accumulations, *AAPG Bull.*, 68, 1752-1763, 1984.

Dullien, F. A. L., *Porous media; fluid transport and pore structure*, Academic Press, San Diego, CA, 1992.

Edwards, H.E., A. D. Becker, and J. A. Howell, Compartmentalization of an aeolian sandstone by structural heterogeneities: Permo-Triassic Hopeman Sandstone, Moray Firth, Scotland, *in Characterization of fluvial and aeolian reservoirs*, edited by C. P. North, and D. J. Prosser, *Geol. Soc. Spec. Publ.*, London, 73, 339-365, 1993.

Ehrlich, R., S. J. Crabtree, S. K. Kennedy, and R. L. Cannon, Direct estimation of flow properties from petrographic thin section image analysis, in *Applied mineralogy*, edited by W. C. Park, D. M. Hausen, and R. D. Hagni, pp. 205-221, Freeman, Los Angeles, CA, 1985.

Fowles, J., and S. Burley, Textural and permeability characteristics of faulted, high porosity sandstones, *Marine Petrol. Geol.*, 11, 608-623, 1994.

Gibson, R. H., Fault-zone seals in siliciclastic strata of the Columbus Basin, offshore Trinidad, *AAPG Bull.*, 78, 1372-1385, 1994.

Hall, C. A., Geology of the Arroyo Grande quadrangle, *Calif. Divis. Mines Geol.*, Map no. 24, scale 1:56000, 1 sheet, 1973.

Haneberg, W. C., Steady state groundwater flow across idealized faults, *Water Resour. Res.*, 31, 1815-1820, 1995.

Harding, T. P., and A. C. Tuminas, Interpretation of footwall (low side) fault traps sealed by reverse faults and convergent wrench faults, *AAPG Bull.*, 72, 738-757, 1988.

Hardmann, R. F. P., and J. E. Booth, The significance of normal faults in the exploration and production of North Sea hydrocarbons, in *The geometry of normal faults*, edited by A. M. Roberts, G. Yielding, and B. Freeman, *Geol. Soc. Spec. Publ.*, London, 56, 1-13, 1991.

Hippler, S. J., Deformation microstructures and diagenesis in sandstone adjacent to an extensional fault: Implications for the flow and entrapment of hydrocarbons, *AAPG Bull.*, 77, 625-637, 1993.

Hooper, E. C. D., Fluid migration along growth faults in compacting sediments, *J. Petrol. Geol.*, 14, 161-180, 1991.

Hubbert, M. K., Entrapment of petroleum under hydrodynamic conditions, *AAPG Bull.*, 37, 1954-2026, 1953.

Kastning, E. H., Faults as positive and negative influences on ground-water flow and conduit enlargement, in *Hydrologic problems in karst regions*, edited by R. R. Dilamarter, and S. C. Csallany, pp. 193-201, West Kentucky University, Bowling Green, KY, 1977.

Knipe, R. J., S. M. Agar, and D. J. Prior, The microstructural evolution of fluid flow paths in semi-lithified sediments from subduction complexes, *R. Soc. Lond. Phil. Trans.*, A335, 261-273, 1991.

Knipe, R. J., The influence of fault zone processes and diagenesis on fluid flow, in *Diagenesis and Basin Development*, edited by A. D. Horbury, and A. G. Robinson, *AAPG Studies in Geology*, 36, 135-148, 1993.

Knott, S. D., Fault seal analysis in the North Sea, *AAPG Bull.*, 77, 778-792, 1993.

Lawrence, E. D., Arroyo Grande, (Edna) oil field, *Calif. Oil Fields*, 44, 41-45, 1958.

Lindsay, N. G., F. C. Murphy, J. J. Walsh, and J. Watterson, Outcrop studies of shale smears on fault surfaces, in *The geological modeling of hydrocarbon reservoirs and outcrop analogues*, edited by S. S. Flint, and I. D. Bryant, *IAS Spec. Publ.*, 15, 113-123, 1993.

Logan, J. M., The influence of fault zones on crustal-scale fluid transport, *AAPG Bull.*, 75, 623, 1991.

McAllister, E., and R. Knipe, A detailed analysis of sub-seismic faults: implications for up-scaling, *Fault Sealing Meeting Proceed. Abstr.*, Leeds, 24, 1996.

McCaig, A. M., Fluid flow through fault zones, *Nature*, 340, 600, 1989.

Mollema, P. N., and M. A. Antonellini, Influence of fault architecture on fluid flow: a comparison between faults in dolomite and aeolian sandstone, *AAPG Abstr. Programs*, 2, A468, 1998.

Mozley, P. S., and L. B. Goodwin, Patterns of cementation along a Cenozoic normal fault: A record of paleoflow orientations, *Geology*, 23, 539-542, 1995.

Nelson, R., 1985, *Geologic analysis of naturally fractured reservoirs*, Gulf Publishing Company, Houston, TX.

Nybakken, S., Sealing fault traps; an exploration concept in a mature petroleum province; Tampen Spur, northern North Sea, *First Break*, 9, 209-222, 1991.

Page, B. M., M. D. Williams, E. L. Henrickson, C. N. Holmes, and W. J. Mapel, Geology of the bituminous sandstone near Edna, San Luis Obispo County, California, *U.S. Geol. Surv. Oil Gas Invest.*, Preliminary Map 16, scale 1:14000, 1 sheet, 1944.

Pittman, E. D., Effect of fault-related granulation on porosity and permeability of quartz sandstones, Simpson Group (Ordovician), Oklahoma, *AAPG Bull.*, 65, 2381-2387, 1981.

Purcell, W. R., Capillary pressure - their measurements using mercury and the calculation of permeability therefrom, *AIME Petrol. Trans.*, 186, 39-48, 1949.

Reynolds, S. J., and G. S. Lister, Structural aspects of fluid-rock interactions in detachment zones, *Geology*, 15, 362-366, 1987.

Rice, J. R., Fault stress states, pore pressure distributions, and the weakness of the San Andreas Fault, in *Earthquake mechanics and transport properties of rocks*, edited by B. Evans, and T. F. Wong, pp. 475-503, Academic Press, London, 1992.

Roberts, G., Structural controls on fluid migration through the Rencurel thrust zone, Vercors, French Sub-Alpine Chains, in *Petroleum migration*, edited by W. A. England, and A. J. Fleet, *Geol. Soc. Spec. Publ.*, London, 59, 245-262, 1990.

Schlumberger Well Services, *The essential of log interpretation practice*, Schlumberger Publications, Ridgefield, CT, 1972

Schowalter, T. T., Mechanics of secondary hydrocarbon migration and entrapment, *AAPG Bull.*, 63, 723-760, 1979.

Seeburger, D. A., *Studies of natural fractures, fault zones permeability and a pore space permeability model*, Ph. D. thesis, Stanford Univ., Stanford, CA, 1981.

Seeburger, D. A., A. Aydin, and J. L. Warner, 1991, Structure of fault zones in sandstone and its effect on permeability, *AAPG Bull.*, 75, 669.

Sibson, R. H., J. M. Moore, and A. H. Rankin, Seismic pumping - a hydrothermal fluid transport mechanism, *J. Geol. Soc. Lond.*, 131, 653-659, 1975.

Sibson, R. H., Earthquake rupturing as a hydrothermal mineralizing agent, *Geology*, 15, 701-704, 1987.

Smith, D. A., Sealing and non-sealing faults in Louisiana Gulf Coast Salt Basin, *AAPG Bull.*, 64, 145-172, 1980.

Tigert, V., and Z. Al-Shaieb, 1990, Pressure seals: their diagenetic banding patterns, *Earth Sci. Rev.*, 29, 227-240, 1990.

Underhill, J. R., and N. H. Woodcock, Faulting mechanisms in high-porosity sandstone; New Red Sandstone, Arran, Scotland, in *Deformation of sediments and sedimentary rocks*, edited by M. E. Jones, and R. M. F. Preston, *Geol. Soc. Spec. Publ.*, London, 29, 91-105, 1987.

Valenta, R., I. Cartwright, and N. H. S. Oliver, Structurally controlled fluid flow associated with breccia vein formation, *J. Metam. Geol.*, 12, 197-206, 1994.

Watts, N. L., Theoretical aspects of cap-rock and fault seals for single- and two-phase hydrocarbon columns, *Marine Petrol. Geol.*, 4, 274-307, 1987.

Weber, K. J., and G. Mandl, The role of faults in hydrocarbon migration and trapping in Nigerian growth fault structures, *OTC Abstr. Programs*, 4, 2643-2653, 1978.

Yielding, G., B. Freeman, and D. T. Needham, Quantitative Fault Seal Prediction, *AAPG Bull.*, 81, 897-917, 1997.

M. Antonellini, Terra Porosa Geosciences, 11839 Cedar Pass Dr., Houston, TX, 77077, USA.

A. Aydin, Department of Geological and Environmental Sciences, Stanford University, Stanford, CA, 94305-2115, USA.

L. Orr, Department of Petroleum Engineering, Stanford University, Stanford, CA, 94305-2115, USA.

Controls on Fault-Zone Architecture in Poorly Lithified Sediments, Rio Grande Rift, New Mexico: Implications for Fault-Zone Permeability and Fluid Flow

Michiel R. Heynekamp[1], Laurel B. Goodwin, Peter S. Mozley

Department of Earth and Environmental Science, New Mexico Tech, Socorro, New Mexico

William C. Haneberg[2]

New Mexico Bureau of Mines and Mineral Resources, New Mexico Tech, Albuquerque, New Mexico

We have studied the geometry and continuity of structures and diagenetic features of a normal growth fault in poorly lithified sediments. Fault-zone width and complexity vary spatially with the grain-size distribution of faulted beds. The fault zone is narrow and structurally simple where it cuts either thick beds with >20% clay and silt, or thin beds that alternate between >20% and ≤20% clay and silt. Where the majority of beds juxtaposed by the fault are ≥80% sand and gravel, and clay beds are thin and rare, the fault zone is wide and structurally complex. In all cases, the fault zone can be divided into three architectural elements. The core includes the primary slip surface(s) and a nearly continuous clay smear 0.3 - 32 cm wide. It is flanked by structurally and lithologically heterogeneous mixed zones, which include material derived from adjacent sediments during fault movement. Mixed zone sediments vary from little deformed to well foliated, tectonically mixed material within which bedding has been destroyed. The mixed zones are bound by damage zones, within which deformation was confined to minor faults and folds. Grain-size and structural variations among these elements lead us to conclude that they have hydrologic significance. In addition, the fault zone is preferentially cemented with respect to adjacent sediments. We use degree of cementation as a proxy for fluid flux, and patterns of cementation as a record of paleo-flow pathways. Extensive sparry calcite cement is typically confined to coarse-grained sediments in the hanging wall (basinward) mixed zone. Steeply plunging, elongate patterns of cement are interpreted to record subvertical groundwater flow at the time of precipitation. As regional flow is inferred to have occurred roughly from the margins to the center of the basin at the time of cementation, these relationships indicate a combination of cross-fault and subvertical, fault-parallel flow.

[1] Now at New Mexico Bureau of Mines and Mineral Resources, Socorro, New Mexico
[2] Now at Haneberg Geoscience, Port Orchard, Washington

Faults and Subsurface Fluid Flow in the Shallow Crust
Geophysical Monograph 113

INTRODUCTION

Faults can function as high permeability pathways that enhance fluid flow, as low permeability barriers that impede cross-fault flow, or as complex barrier-conduit systems [cf. *Caine et al.,* 1996]. In addition, a high permeability fault that served initially as a conduit for fluid flow can evolve over time into a low permeability barrier due to cementation or pore collapse, or may even alternate between periods of high and low permeability as a consequence of deformation during seismic cycles [*Sibson*, 1990; *Knipe*, 1993]. A given fault's influence on fluid flow will depend on the type and spatial distribution of adjacent rock types, fault-zone structures, and diagenetic features [e.g., *Davis and DeWeist*, 1966; *Knipe*, 1993; *Gibson*, 1994; *Caine et al.*, 1996] as well as the externally imposed flow regime (boundary conditions).

Previous workers have divided fault zones into a central gouge or core zone and bounding damage zones [e.g., *Sibson*, 1977; *Aydin*, 1978; *Chester and Logan*, 1986; *Smith et al.*, 1990; *Byerlee*, 1993; *Caine et al.*, 1996]. The structurally, lithologically, and morphologically distinct core zone accommodates most of the displacement. It may contain slip surfaces, clay-rich gouge, brecciated and geochemically altered rock, and cataclasite [*Sibson*, 1977; *Anderson et al.*, 1983; *Chester and Logan*, 1986; *Caine et al.*, 1996]. The damage zones consist of subsidiary structures that bound the fault core and may include minor faults, veins, fractures, cleavage, and folds. *Caine et al.* [1996] incorporated numerous field, laboratory, and numerical models of flow within and near faults into a general model for fault zone architecture and permeability of faults in rock. They divided fault zones into three basic architectural elements: a low permeability fault core, a less deformed damage zone, within which permeability is generally enhanced by a relatively high fracture density; and the protolith or host rock, characterized by intermediate permeability values. In this model, the thicknesses of core and damage zones determine a fault's influence on fluid flow.

Antonellini and Aydin [1995] offered a conceptual model of fault-zone permeability based on detailed field study of faults in porous sandstone, material more similar to the sediments we are studying than any other rock studied to date. Their model is founded on the observation that faults in porous sandstone consist of a hierarchy of structures, from isolated deformation bands to zones of deformation bands and adjacent slip surfaces [*Aydin*, 1978; *Aydin and Johnson*, 1978]. This conceptual model is distinctly asymmetric, with zones of deformation bands in the footwall and less deformation in the hanging wall. Each of the structural elements from which the model is constructed exhibits a discrete range in permeability that has been incorporated into an idealized 3-D permeability distribution model [*Antonellini and Aydin*, 1994, 1995]. Slip surfaces, foliations, and deformation bands all modify the original porosity and permeability of sedimentary rocks [*Pittman*, 1981; *Jameson and Stearns*, 1982; *Underhill and Woodcock*, 1987; *Antonellini and Aydin*, 1994; *Fowles and Burley*, 1994; *Antonellini and Aydin*, 1995].

All of the studies of fault-zone structure, cementation, and permeability noted above have focused on rock, not sediment. Yet poorly lithified sediments constitute many of the most important aquifers and reservoirs in the United States and elsewhere [e.g., *Anderson et al.*, 1988; *Mifflin*, 1988]. Within the Basin and Range Province, a variety of clastic sediments have been and are currently being deposited within subsiding, fault-bounded basins. Growth faults are common features of these basins, within which both the major population centers of the southwestern United States and their groundwater supplies are situated. Small-displacement faults in poorly lithified sand reduce permeability up to three orders of magnitude through cataclasis, and result in a fault-parallel permeability anisotropy [*Sigda et al.*, this volume]. Our study of the large-displacement Sand Hill fault zone of New Mexico suggests that the permeability characteristics of fault zones in poorly lithified sediment vary with the grain size and clay content of the host sediments. We also find that the distribution of lithologically different sediments within and adjacent to the fault zone influences both the location and degree of fault-zone cementation.

Our work has been directed toward the identification of field-scale units that exhibit distinct structural and lithologic characteristics that may subsequently be related to permeability. Both meso- and micro-structural relationships allow us to consider the possible effects of deformation and cementation on permeability. A long-term objective is to identify relationships that will allow us to predict fault-zone architecture and permeability from knowledge of the stratigraphy on either side of a fault. We have therefore focused on relationships between grain size and clay content of faulted sediments and fault-zone structure. One of the most interesting results of this work is that the structures found in faulted sediments are in some ways similar to those found in fault zones in rock, but also exhibit some profound differences. From the standpoint of investigating fault-zone permeability, the most important of these differences is that poorly lithified sediments can be mixed within the fault zone at the grain scale.

GEOLOGIC SETTING

The Sand Hill fault is located on the western margin of the Albuquerque Basin. The study area is situated approximately 30 km northwest of Albuquerque (Figure 1a) along an escarpment formed by incision of the Rio Puerco. The fault is one of the major normal faults bounding the Rio Grande Rift, and separates synrift sediments of the middle Miocene to Oligocene lower Santa Fe Group from those of the Pliocene-Pleistocene upper Santa Fe Group [*Wright*, 1946; *Hawley and Haase*, 1992; *Hawley et al.*, 1995; Figure 1]. The Santa Fe Group varies in thickness from ~1200 m along the basin margins to >4500 m in the central area of the basin. Field relationships indicate that the Sand Hill fault, like other faults associated with the Rio Grande Rift, is a growth fault. Consequently, sediments of different ages that are cut by the fault have experienced different amounts of deformation. Interpretation of core sam-

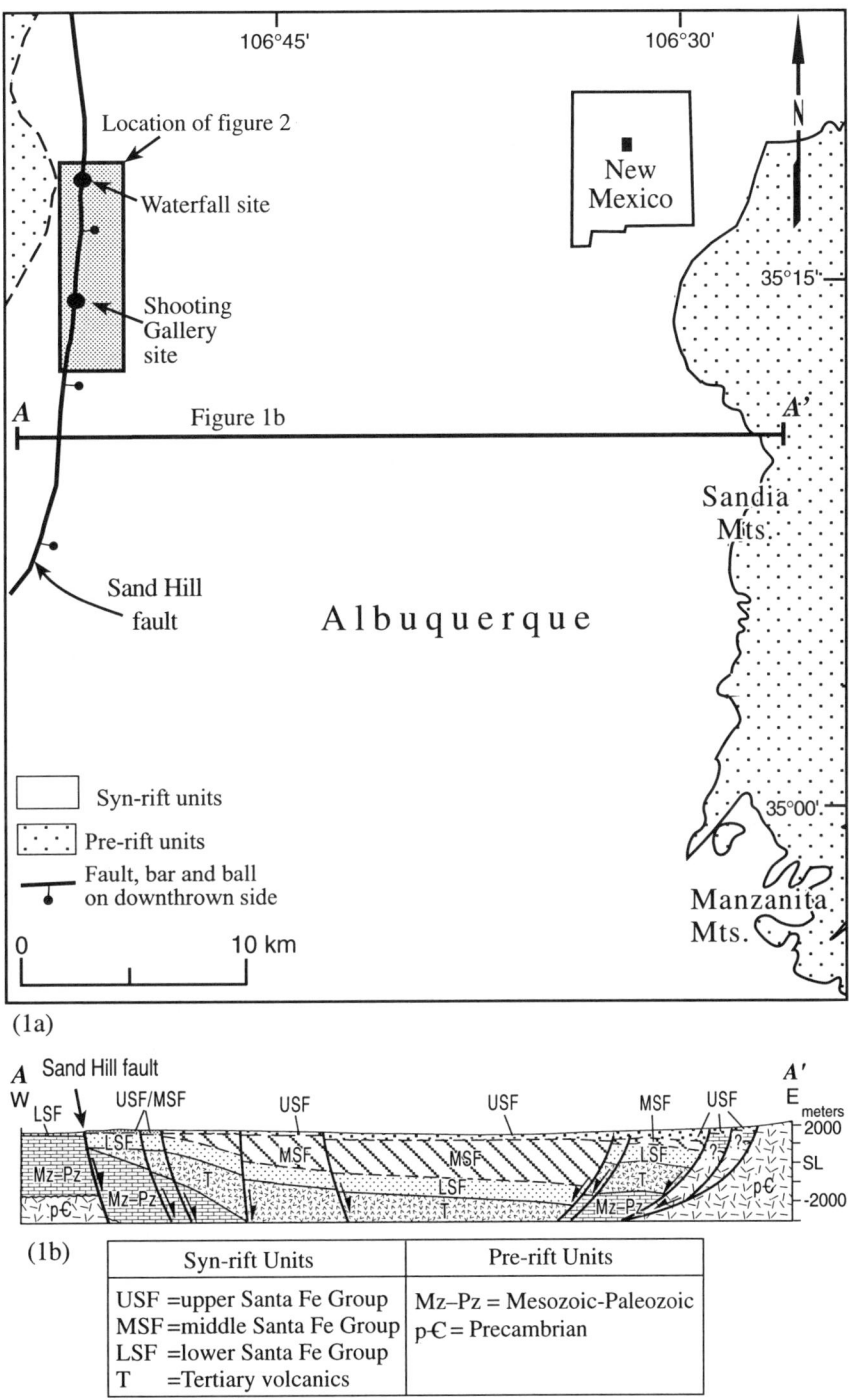

Figure 1. (a) Map of Albuquerque Basin, Rio Grande Rift, central New Mexico, showing Sand Hill fault zone, detailed study sites, and position of cross section shown in (b). Shaded box shows location of Fig. 2. (b) Cross section illustrating major faults and lithologic units. Modified from Hawley et al. (1995).

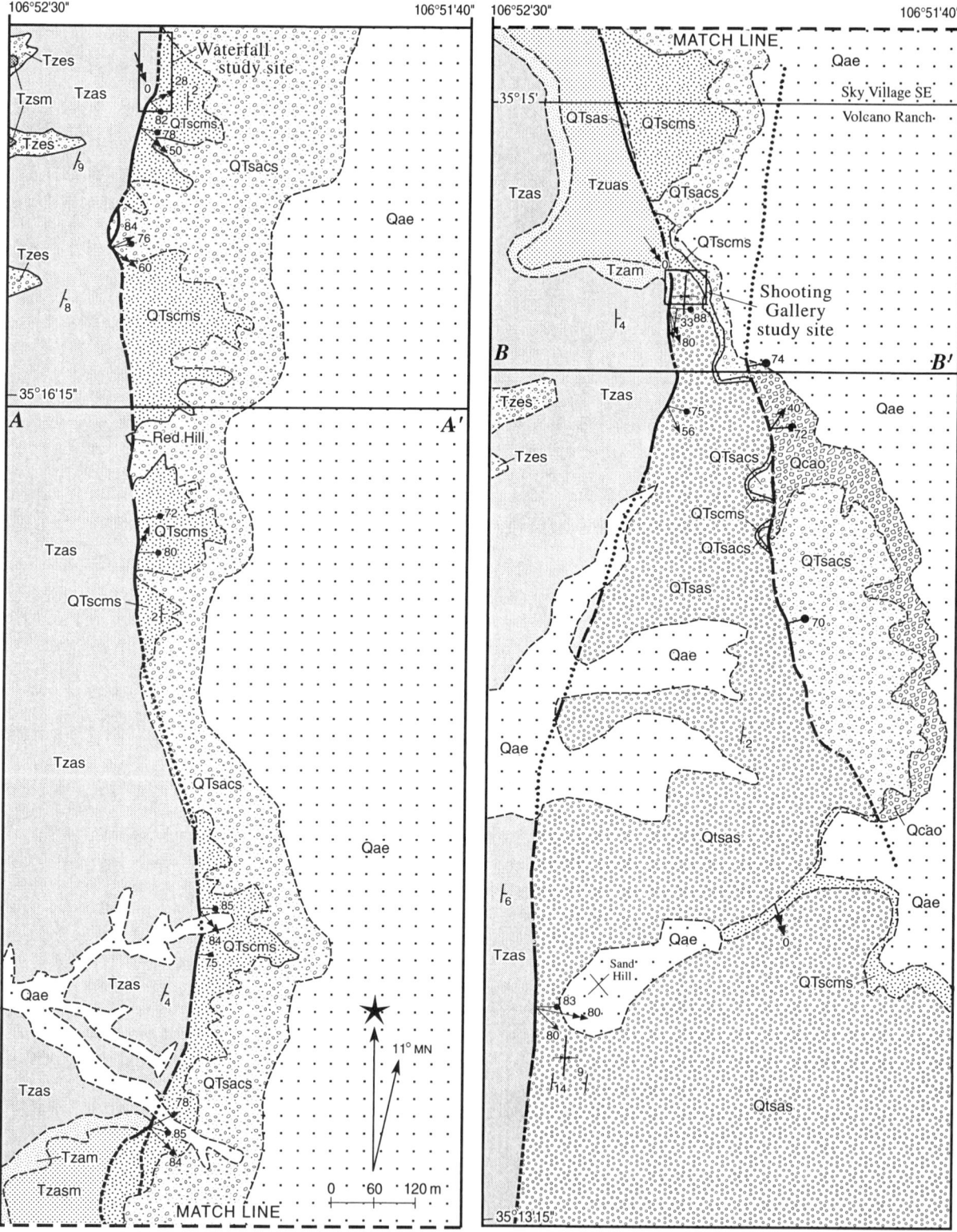

Figure 2. Map (1:6,000) and cross-sections of the Sand Hill fault zone, showing along-strike variations in geometry and degree of cementation. Strip maps join at the Match Line . Note locations of detailed study sites, from which cross sections shown in Figs. 7, 9, and 10 were produced.

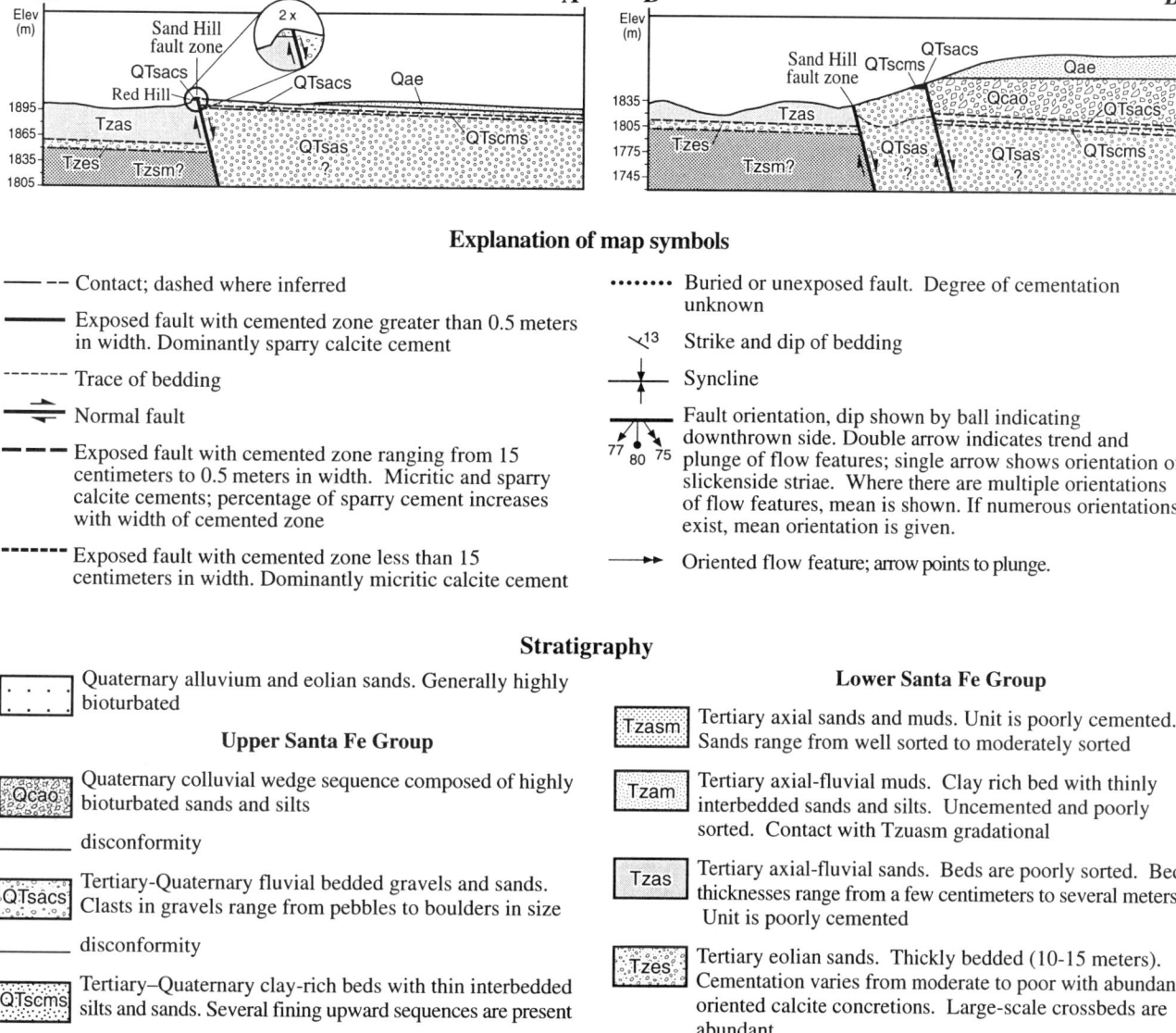

Explanation of map symbols

— — — Contact; dashed where inferred

——— Exposed fault with cemented zone greater than 0.5 meters in width. Dominantly sparry calcite cement

------- Trace of bedding

⇄ Normal fault

— — — Exposed fault with cemented zone ranging from 15 centimeters to 0.5 meters in width. Micritic and sparry calcite cements; percentage of sparry cement increases with width of cemented zone

········· Exposed fault with cemented zone less than 15 centimeters in width. Dominantly micritic calcite cement

········· Buried or unexposed fault. Degree of cementation unknown

⌐13 Strike and dip of bedding

⊥ Syncline

Fault orientation, dip shown by ball indicating downthrown side. Double arrow indicates trend and plunge of flow features; single arrow shows orientation of slickenside striae. Where there are multiple orientations of flow features, mean is shown. If numerous orientations exist, mean orientation is given.
77 80 75

⟶ Oriented flow feature; arrow points to plunge.

Stratigraphy

▢ Quaternary alluvium and eolian sands. Generally highly bioturbated

Upper Santa Fe Group

[Qcao] Quaternary colluvial wedge sequence composed of highly bioturbated sands and silts

——— disconformity

[QTsacs] Tertiary-Quaternary fluvial bedded gravels and sands. Clasts in gravels range from pebbles to boulders in size

——— disconformity

[QTscms] Tertiary–Quaternary clay-rich beds with thin interbedded silts and sands. Several fining upward sequences are present

⋁⋁⋁ angular unconformity

[QTsas] Tertiary-Quaternary alluvial sands. Mainly sand sized fluvial sediments. Some pebble layers present. Unit is poorly cemented to uncemented. Largest clasts are cobble-sized but rare. Fluvial channel structures showing strong pebble imbrications are common

Lower Santa Fe Group

[Tzasm] Tertiary axial sands and muds. Unit is poorly cemented. Sands range from well sorted to moderately sorted

[Tzam] Tertiary axial-fluvial muds. Clay rich bed with thinly interbedded sands and silts. Uncemented and poorly sorted. Contact with Tzuasm gradational

[Tzas] Tertiary axial-fluvial sands. Beds are poorly sorted. Bed thicknesses range from a few centimeters to several meters. Unit is poorly cemented

[Tzes] Tertiary eolian sands. Thickly bedded (10-15 meters). Cementation varies from moderate to poor with abundant oriented calcite concretions. Large-scale crossbeds are abundant

[Tzsm] Tertiary silty muds. Thick fluvial clay beds with interbedded sands and silts. Unit is poorly cemented to uncemented

Figure 2. Continued.

ples from wells south of the study area suggests that the base of the lower Santa Fe Group, the oldest rift-related sediment cut by the fault, is offset ~600 m [Figure 1b; *Hawley and Haase*, 1992; *Hawley et al.*, 1995]. In contrast, our work indicates that the uppermost unit of the upper Santa Fe Group is vertically displaced only ~10 m [Figure 2, cross section A-A']. Throw of units exposed at the surface does not vary substantially along strike within

the study area (compare cross sections A-A' and B-B', Figure 2).

The Santa Fe Group synrift deposits include fluvial, eolian, and lacustrine sediments. The group is divided into two unconformity-bounded formations within the study area: the Zia Formation of the lower Santa Fe Group and the Sierra Ladrones Formation of the upper Santa Fe Group [*Tedford*, 1982; *Cather et al.*, 1997]. The Zia Formation, exposed in the footwall, is dominated by fine-grained sand, with medium-grained sand, silty sand, and mud present in lesser quantities. Gravel interbeds are rare. The Zia Formation is estimated to be up to 200 m thick, although it is not fully exposed within the study area. Most of the unit is poorly lithified, but some of the sands are strongly cemented by calcite [*Beckner and Mozley*, 1998].

The Sierra Ladrones Formation is exposed in the hanging wall of the Sand Hill fault. *Wright* [1946] inferred a history of episodic movement on the fault based on observations of angular unconformities and intermittent sediment accumulation in hanging wall sediments. We have identified three map units separated by unconformities within the exposed portion of the Sierra Ladrones Formation. These units are 1) dominantly sand, overlain by 2) a sequence including sand, silt, and mud, which is locally capped by 3) gravel.

Patterns of cementation in the Santa Fe Group can be used to infer local details of past flow conditions. Calcite cementation in the Santa Fe Group appears to have been largely controlled by fluid flow, with initially high permeability units preferentially cemented by calcite due to a greater flux of Ca^{2+} and/or HCO^{3-} [*Mozley et al.*, 1995]. Such beds are currently low permeability units as a result of the cementation. Furthermore, elongate calcite concretions that record the orientation of groundwater flow at the time of calcite precipitation are present throughout the Santa Fe Group, including the study area (e.g., Figure 2). The close relationship between the orientation of elongate concretions and the direction of maximum permeability suggests that permeability anisotropy controls flow orientation in fluvial portions of the Santa Fe Group [*Mozley and Davis*, 1996].

Beckner [1996] and *Beckner and Mozley* [1998] examined calcite cementation and concretion orientations in the Zia Formation (footwall sediments) in the study area. They suggest that the dominant flow direction in fluvial portions of the Zia Formation was from NW to SE, toward the roughly N-striking Sand Hill fault. Elongate concretions are also present in the Sierra Ladrones Formation at the study site. Though a systematic analysis of these oriented concretions has not yet been conducted, the concretions that have been mapped also show consistent NW-SE orientations. SE-directed flow is consistent with both field observations and numerical models of flow in continental rift basins [e.g., *Titus*, 1963; *Person and Garven*, 1994; *Person et al.*, 1996], in which flow generally proceeds from the margin (such as the location of the Sand Hill fault; Figure 1) toward the center of a given basin. Elongate patterns of cement similar to those found in the Santa Fe Group are locally evident on the margins of strongly cemented portions of the Sand Hill fault zone [flow features, *Mozley and Goodwin*, 1995]. These flow features are found both in naturally exhumed exposures and in areas of the fault zone we have excavated with a shovel. They are largely subvertical, indicating that fluid flow was reoriented within the fault zone.

The Sand Hill fault has been mapped for over 50 km along strike [*Kelley*, 1977]. For at least 11 km of its extent [*Wright*, 1946], the fault is locally marked by an erosionally resistant sparry calcite cemented zone, originally believed to be a clastic dike, that has been exhumed along arroyos cutting into the Ceja de Rio Puerco west of Albuquerque (Figure 2). Similar laterally continuous zones of fault-related calcite cement are common in the Albuquerque Basin, and may serve as barriers to current groundwater movement [*Thorn et al.*, 1993; *Haneberg*, 1995].

THE SAND HILL FAULT ZONE

Reconnaissance mapping at 1:24,000 [*Cather et al.*, 1997] and more detailed mapping at 1:6,000 (Figure 2) were done to evaluate variations in displacement along strike, to investigate the distribution of fault-zone cements, and to evaluate the relationship, if any, between sediment age and fault zone characteristics. Large-scale map units were designated by age, formation, depositional system, and textural lithofacies [*Cather*, 1997; Table 1]. Two key areas were subsequently mapped at a scale of 1:100 using a theodolite with an EDM attachment. These sites were chosen to evaluate the effects of variations in sand content of beds, thickness of bedding, and curvature of the fault on fault-zone architecture.

Slickenside surfaces and striae locally cut cemented portions of the fault zone (Figure 3). The NNE-striking fault dips steeply east, and striae and stratigraphic separation record dominantly normal motion. Striae plunge from 44° to 88°, indicating local components of sinistral or dextral strike-slip movement. Elongate flow features found within strongly cemented portions of the fault zone are typically slightly inclined with respect to slickenside striae at any given site, though they are roughly similar in orientation and are locally parallel (Figs. 2 and 3).

Architectural Elements

The fault zone varies from 1 to 10 m in width and can be divided into mappable architectural elements, described in the following sections, based on the distribution and character of mesoscopic structures (Figure 4).

Damage zones. The hanging wall and footwall damage zones extend from the first fault-related structure visible as the fault is approached to the boundary of the mixed zones (Figure 4). Bedding is largely intact; structures include minor faults and gentle folds. Deformation bands [cf. *Aydin*, 1978; also known as granulation seams *Pittman*, 1981; or cataclastic slip bands, *Fowles and Burley*, 1994], zones of deformation bands, and minor slip surfaces (across which displacement is less than bedding thickness) are common features. Zones of deformation bands may be situated on either the hanging wall or footwall side of the slip surfaces. Many of the deformation bands are resistant to weathering,

Table 1. Key to map units for Fig. 2. Classification system from Cather (1997).

Unit Name	Age of Unit	Unit Formation	Depositional System	Textural Lithofacies
QTsacs	Quaternary-Tertiary	Sierra Ladrones	Axial-fluvial	conglomerate-sandstone
QTsasm	Quaternary-Tertiary	Sierra Ladrones	Axial-fluvial	sandstone-mudstone
QTsas	Quaternary-Tertiary	Sierra Ladrones	Axial-fluvial	sandstone
Tzasm	Tertiary	Zia	Axial-fluvial	sandstone-mudstone
Tzam	Tertiary	Zia	Axial-fluvial	mudstone
Tzas	Tertiary	Zia	Axial-fluvial	sandstone
Tzes	Tertiary	Zia	Eolian	sandstone
Tzsm	Tertiary	Zia	Axial-fluvial	silty mudstone

indicating that they are either preferentially cemented or better indurated as a result of deformation [cf. *Sigda et al.*, this volume]. Beds in the damage zones may be 'dragged' or extended up to several meters vertically along the boundaries with the mixed zones. The mesoscopic fractures and veins that are common features of damage zones in rock [e.g., *Bruhn et al.*, 1994; *Caine et al.*, 1996] are absent in these poorly lithified sediments.

Mixed zones. The hanging wall and footwall mixed zones typically contain sediments that have undergone a structural change: bedding is either modified and overprinted by a deformational fabric or completely destroyed through tectonic mixing. The outer margins of the mixed zones are defined by displacement greater than bed thickness across slip surfaces subsidiary to the main fault (Figure 4). This is a mappable distinction; no evidence for tectonic mixing is observed except where displacement is greater than the mean bed thickness. The damage zone/mixed zone boundaries therefore reflect a visible change in degree of deformation. This definition of mixed

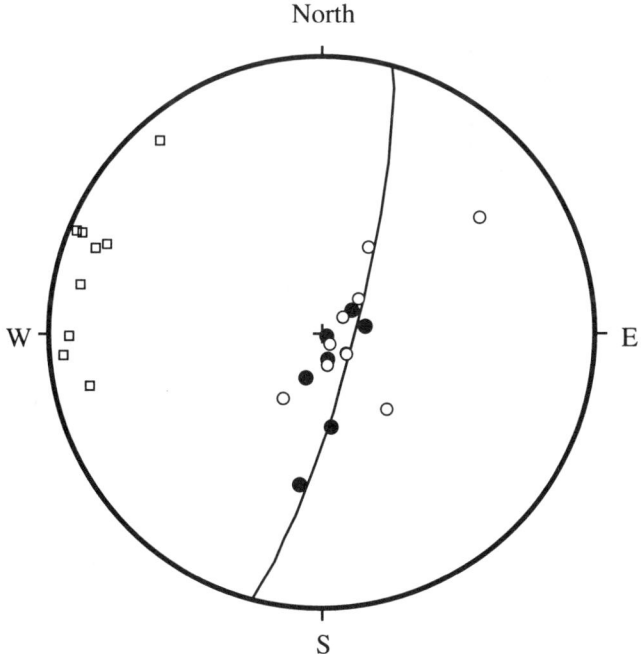

North

W — E

S

○ Flow feature long axis orientatation.

● Slickenside striae

□ Poles to fault plane.

Figure 3. Lower hemisphere, equal area net plot of fault zone structures and flow features.

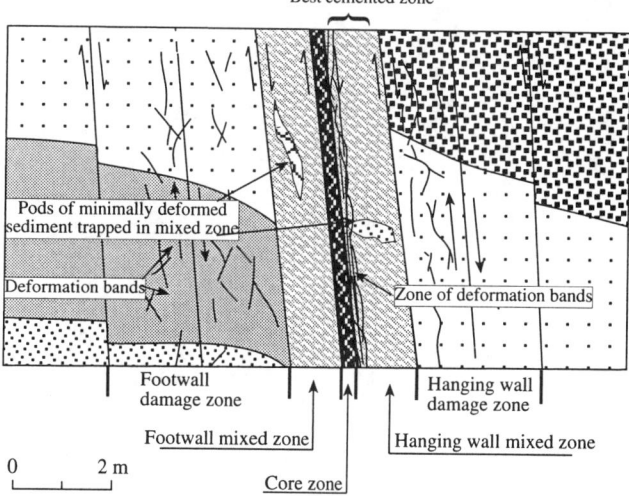

Best cemented zone

Pods of minimally deformed sediment trapped in mixed zone

Deformation bands

Zone of deformation bands

Footwall damage zone

Hanging wall damage zone

Footwall mixed zone

Hanging wall mixed zone

0   2 m

Core zone

Figure 4. Schematic diagram based on photos of the fault zone at the Shooting Gallery site (see Figs. 1 and 2), showing main structural features and fault-zone architectural elements.

Figure 5. (a) Cemented hanging wall mixed zone. Main slip surface (S) and zone of deformation bands (between slip surface and z) bound mixed zone to the left (west). Gravel bed in sand is cut by zone of deformation bands in upper left portion of photo, extended in pinch-and-swell boudins, and cut by post-cementation shear fractures (sf). Zone of deformation bands is 12 cm wide. **(b)** Photo of footwall mixed zone taken looking east with Chris Dimeo for scale. A zone of deformation bands and associated slip surface form the boundary between the mixed and damage zones (m). The striated main slip surface (S) is subparallel to the photo. Note slip surfaces (ss) and elongate compositionally distinct bands indicated by color changes (e.g., c) which parallel the pervasive foliation in the mixed zone. Foliation parallels the fault at the mixed zone boundaries, and is inclined in the center of the zone. **(c)** Photo looking north at contact (dashed line) between hanging wall damage zone (right) and mixed zone (left). Solid line shows location of main slip surface. Clay in mixed zone contains a string of lens-shaped sand bodies (s), which may be either boudins or asperities removed from the hanging wall during faulting. **(d)** Fault core, viewed to north at Shooting Gallery site. Boundaries of the ~10 cm wide core zone fall between the arrows showing shear sense. Sand ribbons in core (s) are partially cemented with micritic calcite. Both sand and clay are foliated, but both fault parallel and inclined foliations are more visible in clay ribbons. **(e)** Close-up of hand sample of clay-rich core. Sand grains and blobs (s) are clearly visible in the clay. Fault-parallel (subhorizontal in this photo) and inclined foliations are marked with arrows.

zones also emphasizes the fact that displacements greater than bed thickness can juxtapose beds with different permeabilities, which may significantly impact fluid flow.

The mixed zones are composed of material derived from hanging wall and footwall beds during slip, and their internal structure reflects fault-zone deformation processes. They contain rootless pods of intact bedding, rotated and extended beds, multiple slip surfaces, and areas of tectonically mixed sediment (Figure 5a-d). Where the incorporated beds are variable in composition and competence, such as a gravel bed surrounded by sand, boudinage may be evident (Figure 5a). Bedding and original sedimentary textures such as cross-bedding, imbrication, and grading were in places completely destroyed through tectonic mixing (Figure 5b). Tectonic mixing was accomplished by attenuation of beds, shear along minor faults and foliation, and mixing from meter-scale blocks to the grain scale – resulting, for example, in sand and pebbles in a clay-rich matrix. The change from relatively intact to tectonically mixed sediment in which no original features are apparent is transitional, with the degree of mixing increasing with increasing proximity to the core zone / mixed zone boundary. As this boundary is approached, thinned beds are typically reoriented from subhorizontal to sub-parallel to the fault. A fault-zone foliation is defined by attenuated beds, which form 0.5 to 10 cm wide ribbons, by the elongate axes of sand grains and pebbles, and by aligned clay minerals (Figs. 5b and c). Slickenside striae are pervasive on foliation planes in clay. Foliation within the Sand Hill fault zone, in general, varies from fault-parallel to inclined (Figure 6b). Where inclined, the foliation typically (we have found just one exception) dips to the east more shallowly than the fault – a relationship similar to that described in some cataclastic foliations in faults in rock [*Cowan and Brandon*, 1994]. Because the foliation generally exhibits a consistent inclination with respect to the known shear sense, it can be used as a kinematic indicator.

Core zone. The core zone, which accommodated the majority of slip within the fault zone, varies in thickness from 0.5 to 40 cm and is bound by slip surfaces. It locally includes multiple slip surfaces separated by foliated clays and sands (Figure 5d). The strongly foliated clays form one or more veneers that are both vertically and laterally extensive; where a veneer is absent, its gap is in most places less than 30 cm. Sand and clay may be mixed within the core, producing a clay-rich material with sand grains and irregular blobs of sand surrounded by a clay matrix (Figure 5e).

Detailed Mapping

Two key areas along the fault have been mapped in detail: the Shooting Gallery and Waterfall sites (Figure 2). These sites allow consideration of the effects of variations in sand content of beds, thickness of bedding, and curvature of the fault zone on fault-zone characteristics.

Grain-size analysis. Samples for grain-size analysis were taken from outside the fault zone and within the mixed and core zones. Several methods were employed to ensure representative analyses. Within beds several meters thick, samples were spaced approximately 0.5 m apart in a traverse perpendicular to bedding. Generally the grain-size distribution of the bulk sample from each bed was determined; in some cases, variations from top to bottom of individual beds were examined. For beds less than ~ 1.5 m thick, the entire bed was exposed through excavation and several samples taken and combined.

Areas of extensive tectonic mixing were sampled in horizontal traverses across the mixed zones. Because the mixed zones are so heterogeneous, distinct compositional bands and structural features such as deformation bands were sampled proportionately and the separate samples combined. The fault core is typically relatively narrow, so a representative section spanning the zone could generally be removed by hand. Where it is wider, a proportional sample of each structural / compositional element was taken, and the samples from a given locality combined. Grain-size distributions were determined using wet-sieving and settling time techniques.

Figure 6. Photos of Sand Hill fault. **(a)** Well cemented section, where sediments outside the fault zone are dominantly sands. First author for scale, fault viewed to south. Hanging wall, or basinward side of fault, to left in photo. Main slip surface (S) separates well cemented hanging wall mixed zone, left, from footwall mixed zone, right. The latter contains little cement, with the exception of preferential cementation of narrow zones of deformation bands, z. Note pod of little deformed sandy sediment in footwall (u). **(b)** Poorly cemented section, where sediments outside the fault zone are dominantly clay, silt, and fine-grained sand. Fault viewed to north. Hanging wall, or basinward side of fault, to right in photo. Small splays bound lens-shaped bodies adjacent to the main slip surface. White, micritic calcite cement is developed in patches. **(c)** Foliation (arrow) in footwall varies in orientation from inclined to roughly fault-parallel. Inclination records right-side-down shear sense.

The Shooting Gallery. Footwall sediments of the Zia Formation at the Shooting Gallery are dominantly fluvial, generally beds of sand alternating with mud or sandy mud (Figure 7). These beds are locally well cemented, but generally poorly lithified. Detailed map units are 1 to 6 m thick (Figure 7). In contrast, hanging wall map units are greater than 7 m thick. An angular unconformity in the hanging wall separates sandy units (QTsas; Figure 2) from overlying clay-rich units (QTscms) of the Sierra Ladrones Formation. There is little cementation of the hanging wall sediments outside the fault zone; they are generally un-

lithified to poorly lithified. Thus it is possible to compare fault-zone characteristics where there is a thick section of sand or mud with areas in which sand and mud layers are relatively thin and interbedded. For the purpose of this comparison, footwall or hanging wall fault-zone thickness is defined as the distance from the center of the core zone to the outer margin of the footwall or hanging wall damage zone, respectively.

The total fault-zone width ranges from 2.5 to 10 m at this site. Hanging wall or footwall fault-zone width generally appears to increase where units cut by the fault have

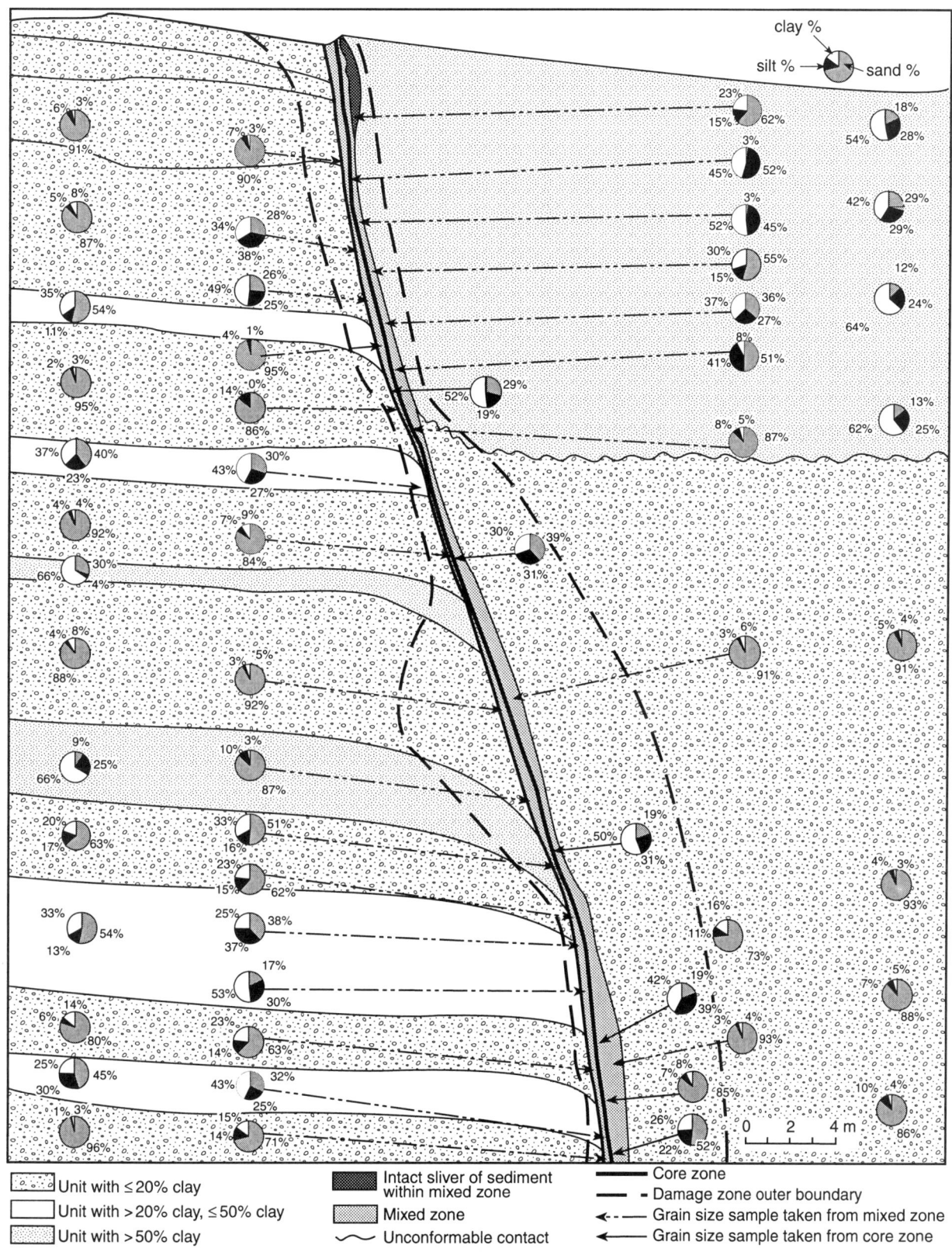

Figure 7. Vertical cross-section of Shooting Gallery site drawn perpendicular to fault, viewed to NNE, constructed from 1:100 scale theodolite map of fault zone. Pie charts summarize grain size analyses of parent sediments and material within mixed and core zones. Shading of beds outside the fault zone is keyed to sand, silt, and clay content, as indicated.

clay and silt grain size fractions less than 20% (Figs. 7 and 8a). The magnitude of this increase is highly variable, and some clay-rich layers also appear to correlate with small increases in fault-zone thickness. The latter are mostly thin clay-rich beds sandwiched by sand, suggesting that the composition of adjacent layers influences fault-zone thickness where bedding is thin. The sole exception is the lower part of the bed just above the unconformity in the hanging wall (Figure 7), in which fault-zone thickness appears to be partly influenced by the underlying sand.

The variation in fault-zone thickness is largely accomplished by changes in thickness of the damage zone; variations in thickness of the mixed zones are relatively minor. Variations in damage zone thickness are easiest to see in the hanging wall (Figure 7). The mixed zone is generally, but not everywhere, thinner where adjacent sediments are finer grained. Where the mixed zone is thickest, it typically includes relatively intact slivers of little deformed sand and gravel beds (e.g., Figure 5c).

The grain-size distribution of the hanging wall mixed zone reflects contributions of material from both adjacent and structurally lower beds, which moved past the mixed zone during faulting. Just below the unconformity, the mixed zone is dominated by sand; the clay content increases upward as the contribution of material from clay-rich beds increases (Figure 7). This variation in grain size with proximity to source beds is nonlinear, which reflects the heterogeneity of the previously discussed mixing processes. Observations made at a scale too detailed to map indicate that clay-rich units generally were extended further in the fault zone than coarse-grained units.

The grain-size distribution of the footwall mixed zone correlates most strongly with that of the adjacent beds (Figure 7). Departures from this correlation appear to reflect the incorporation of material from structurally higher beds. The thicknesses of the footwall damage and mixed zones appear overall to be less than the thicknesses of the hanging wall damage and mixed zones.

The core zone of the fault is mainly composed of a virtually continuous 0.3 to 32 cm wide strongly foliated clay veneer (Figure 6b). Two gaps in the veneer were noted. In some areas, the core zone includes multiple, thin clay veneers and foliated bands of sand (Figure 5d).

The Waterfall. Sediments juxtaposed by the fault at the Waterfall site are distinctly different from those at the Shooting Gallery site (Figs. 9 and 10). The majority of the Zia Formation sediments exposed in the footwall are fluvial, trough cross-bedded, medium- to coarse-grained, moderately to well sorted, thinly bedded sands. A thickly bedded, well sorted, eolian cliff-forming unit is present in the lowest exposure. This footwall section includes several laterally continuous, preferentially well-cemented beds. Just two beds contain more than 20% clay and silt, and only one of these has more than 50% clay and silt (Figs. 9 and 10). Hanging wall sediments include a lowermost unit (QTscms) that correlates with the beds above the

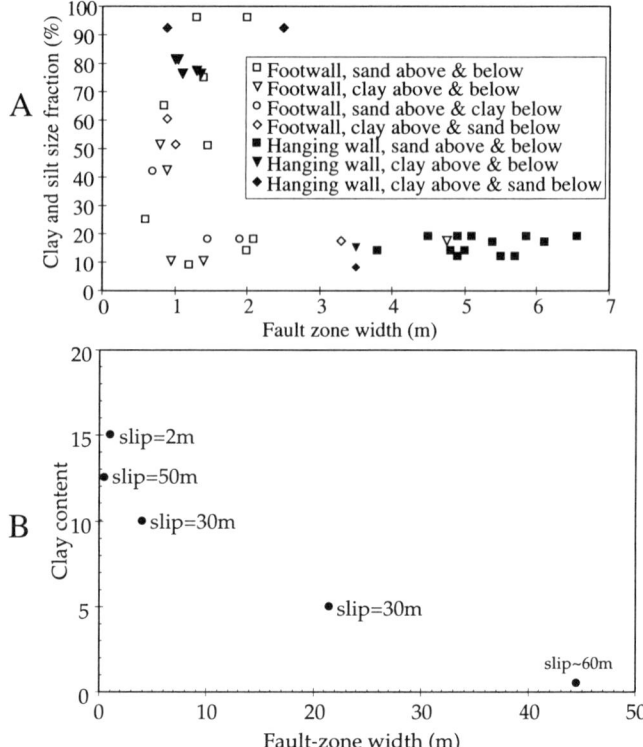

Figure 8. **(a)** Relationship between clay and silt size fraction and fault-zone width at the Shooting Gallery site. The clay and silt size fraction of the sediment must be greater than 20% for deformation to be localized within a narrow fault zone. **(b)** Relationship between clay content and fault-zone width in faults in sandstone (data from Antonellini and Aydin, 1995). Magnitude of slip of each fault is indicated. A clay content of 10% is sufficient to localize deformation (data on silt content not available). Note the difference in horizontal scale between the top and bottom plots. The greatest fault-zone width observed at the Shooting Gallery is narrow relative to fault-zone widths in sandstone.

unconformity at the Shooting Gallery. Sediments within this unit coarsen north of the Shooting Gallery, and at the Waterfall site they include well sorted, fine- to medium-grained sands and thin gravel interbeds. Bioturbation has completely eliminated bedding in the area of the cross sections shown in Figs. 9 and 10. Coarse sands and gravels (QTsacs) unconformably overlie this unit. The Waterfall site therefore represents areas in which sediments cut by the fault are substantially coarser grained than at the Shooting Gallery.

The entire fault zone at the Waterfall is 1.5 to ~10 times wider than at the Shooting Gallery site (compare Figs. 7, 9, and 10). The damage zones are relatively constant in width throughout the Waterfall map area. The footwall mixed zone, however, increases in thickness along strike within the vicinity of an S-shaped bend in the fault; thus fault geometry also influences fault-zone thickness. This increase

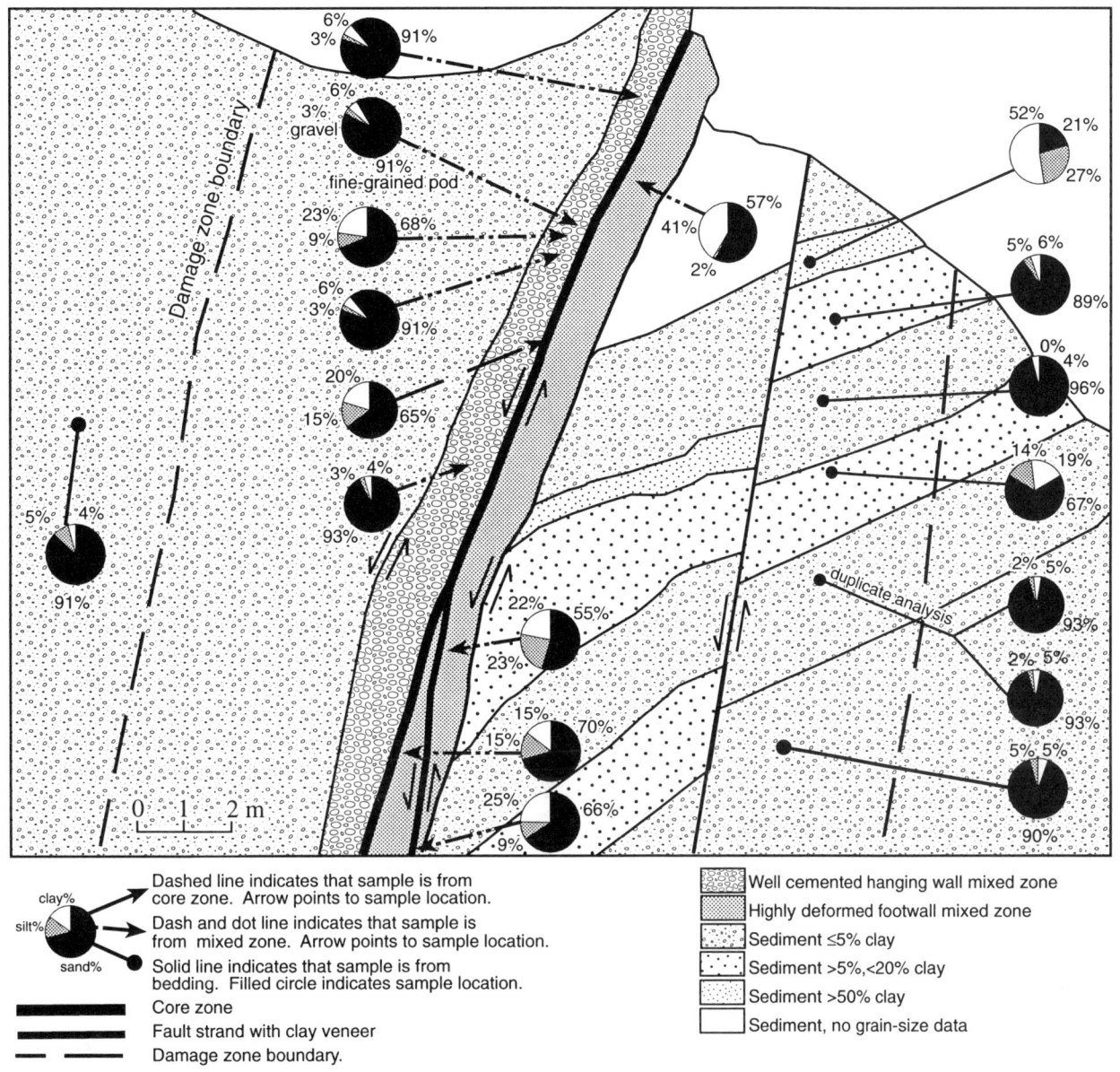

Figure 9. Vertical cross section of Sand Hill fault at the Waterfall site, viewed to south. Pie charts summarize grain size analyses of parent sediments and material within mixed and core zones. Shading of beds outside the fault zone is keyed to sand, silt, and clay content, as indicated. Compare degree of mixing and structural complexity of footwall mixed zone with that illustrated in Figure 10.

in thickness is due to both incorporation of lens-shaped slivers of relatively intact coarse-grained sediment into the mixed zone and development of small splays in the main slip surface. The slivers are typically bound by slip surfaces that are lined with thin clay veneers. Splays of the main slip surface locally anastomose around highly mixed sediments, resulting in significant variations in mixed zone structures and thicknesses over distances of a few meters (compare Figs. 9 and 10). The hanging wall mixed zone

also increases in thickness locally, but the increase is not as large as in the footwall.

The outer margin of the footwall mixed zone consists of a zone of deformation bands and an associated slip surface (Figure 5b). The mixed zone itself varies from highly mixed with a well-developed foliation (Figure 9) to relatively intact bedding with a narrower section of highly mixed sediments near the core (Figure 10). The mixed zone includes many well-foliated, tectonically mixed re-

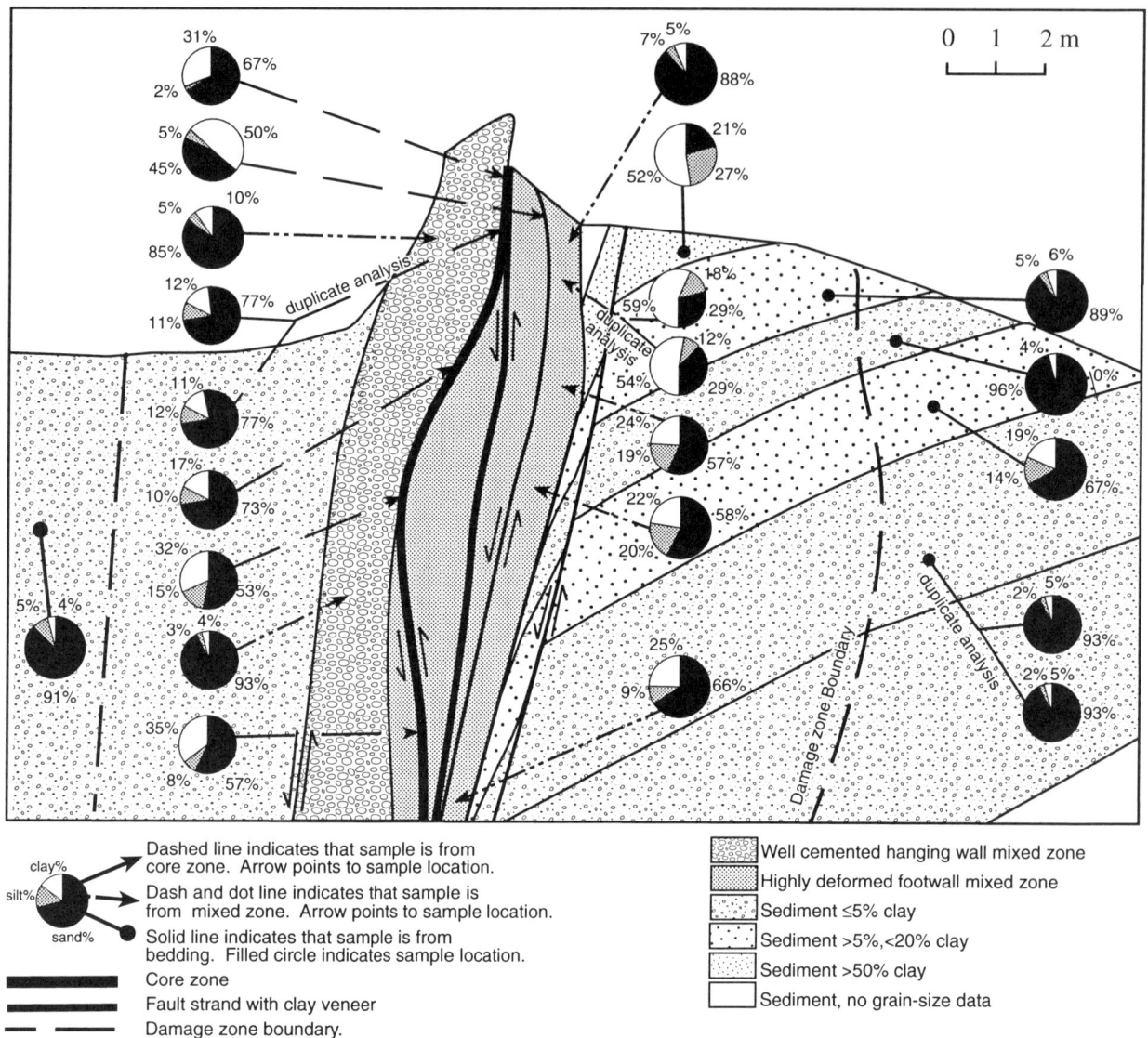

Figure 10. Vertical cross section of Sand Hill fault at the Waterfall site, viewed to north ~5meters north of Figure 9 cross section. See caption for Fig. 9 and text for further explanation.

gions that are compositionally different and contain more clay than beds immediately adjacent to the fault exposure. Typically these regions are bound by slip surfaces. In general, the clay content of the footwall mixed zone is higher than that of the adjacent sediments (Figs. 9 and 10). In contrast, the hanging wall mixed zone exhibits less tectonic mixing than the footwall. It is generally sand-rich, consistent with the hanging wall sediments, but also contains sediments that have more gravel than adjacent beds. The gravel incorporated into the hanging wall mixed zone is compositionally similar to gravel beds of the lower unit of the Sierra Ladrones Formation.

Variations in fault-zone width with grain size of individual beds are not evident at the Waterfall site; however, the variation in grain size between beds is not substantial. As mentioned above, only two beds in the footwall contain sufficient clay and silt to potentially effect a reduction in fault-zone width, using the criterion established in Figure 8, and these are relatively thin beds surrounded by much thicker sequences of sand. The vertical exposure at the Waterfall site is not as great as at the Shooting Gallery site, so that variations in mixed zone composition with changes in adjacent bed composition are difficult to investigate. A more complete, ~30 m high exposure of footwall sediments is, however, located nearby. Within the latter, beds are typically sandy, though three clay-rich layers ≤25 cm thick are located structurally above the footwall exposures illustrated in Figs. 9 and 10. It appears that these few clay-rich beds had a significant impact on the clay content of the footwall mixed zone.

The core zone of the fault at the Waterfall site is sandier than at the Shooting Gallery site (compare grain size plots in Figs. 9 and 10 with those in Figure 7). Gaps in the clay veneers of the core zone and fault splays are, collectively, more numerous than at the Shooting Gallery site. There are, however, never gaps in all veneers in a given transect across the fault.

Cementation

Fault-zone cements are almost exclusively confined to the hanging wall mixed zone. The hanging wall is on the basinward side of the fault (Figure 6); two faults outside the map area – the Jemez fault to the east and the Pilares fault to the west [cf. *Cather et al.*, 1997] – are also cemented on their basinward sides. Cemented regions within the Sand Hill fault zone are locally cut by shear fractures and slickenside surfaces (Figure 5a), indicating that cementation was broadly synchronous with faulting.

The width of the cemented area of the fault zone ranges from a few millimeters to 6 m (Figs. 2, 6). Sparry calcite cement, which results in a resistant outcrop wall typically greater than 25 cm wide, is found only in sand and gravel-rich sediments (Figure 6a). In areas where hanging wall sediments are relatively fine-grained and clay-rich, only micritic calcite cement is found within the fault zone (Figure 6b) and the cemented zone is less than 25 cm wide. Cements are locally absent where footwall sediments are very fine grained. Cemented areas are therefore widest and best developed where both footwall and hanging wall sediments are coarse grained, such as at the Waterfall site, where the cemented zone is generally more than 2m wide.

Core-zone cementation is limited and, where present, is restricted to foliated sand ribbons where the clay veneer is thin (e.g., Figure 5d). The cements are sparry calcite where well developed and micritic calcite where poorly developed. Hanging wall mixed-zone cements begin abruptly at the contact between the core and mixed zones. Cementation in the adjacent hanging wall damage zone, if present, is best developed at the contact with the mixed zone, and diminishes with increasing distance from the fault, never extending more than 4 m into the damage zone. The patterns of elongate cementation present along the margins of many cemented areas of the fault are typically subvertical, but vary where the fault bends from vertical to nearly horizontal (Figs. 2 and 6).

Cemented beds locally extend from adjacent sediments into the footwall damage zone; fault-zone cements, however, are generally absent from footwall fault-zone architectural elements. Where present, cement in the footwall mixed zone is generally diffuse, micritic calcite; it occurs only where the mixed zone is sand and/or gravel. Sparry calcite cement is rare.

Microscopic Deformation Features

The petrographic work presented below has been focused on sediments within the footwall mixed zone because: 1) with the inferred NW to SE paleo-flow pattern, the footwall mixed zone would have been the first strongly deformed fault-zone material encountered by flowing groundwater 2) the mixed zones exhibit a variety of structures, including features mesoscopically similar to those found in both the damage and core zones, and 3) the footwall mixed zone is generally better exposed than the hanging wall mixed zone. This is not a comprehensive analysis of fault-zone microstructures; instead, we describe features that we believe are most likely to affect fault-zone permeability. These features result from both deformation of grains and grain-scale mixing of different sediments. As indicated previously, the degree of deformation and mixing of sediments varies within the mixed zone. We begin by considering the least deformed sands, which should have the highest permeabilities of the material we have sampled.

Bedding laminae are still preserved in little deformed sands within the mixed zones, though they have typically been rotated during faulting and cut by deformation bands. Microstructures record minor deformation associated with this rotation of bedding, including minor grain fracturing, grain rotation, and grain reorganization (Figure 11a). Elongate grains are typically oriented subparallel to bedding in undeformed sands, but are reoriented where mesoscopically visible bedding laminae are folded adjacent to deformation bands and other structural contacts. Within these folded areas, volcanic rock fragments appear to be preferentially fractured, though fracturing is also locally evident in quartz and feldspar grains.

Deformation bands and zones of deformation bands in sands are characterized by the preferred orientation of elongate grains roughly subparallel to deformation band boundaries (Figs. 11b and c). The boundaries are not sharp at the microscopic scale. Within the deformation bands and zones of deformation bands, the grain size is smaller, grains are more angular, and lithic clasts are less common than within parent sands (Figure 11b and c). Both porosity and pore size are much less within than outside the deformation bands. Pore space is largely filled by brown clay-sized material. Elongate grains adjacent to deformation bands and zones of deformation bands are typically oriented subparallel to deformation band boundaries and rotated bedding laminae are evident. Some fracturing of grains, particularly lithic clasts, is evident. The latter results in formation of fine-grained brown material similar to that filling pores in deformation bands (Figs. 11a and b). Areas of the mixed zone that lack bedding and are dominantly sand are characterized by a foliation defined by roughly fault-parallel grain alignment.

Sands in the footwall mixed zone exhibit little cementation, although authigenic zeolites are evident in the pore spaces of some samples (Figure 11a). Unlike zeolite ce-

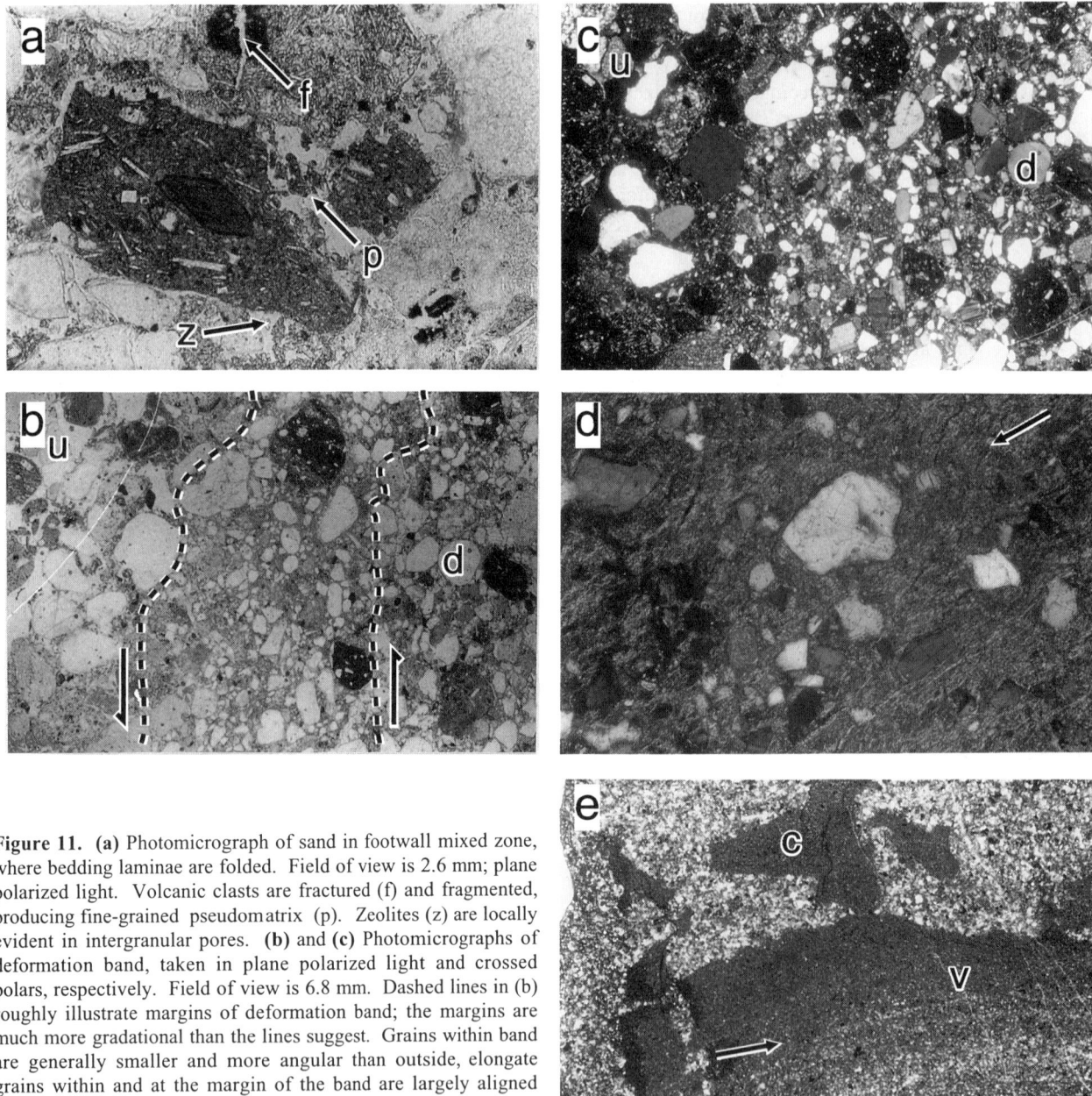

Figure 11. **(a)** Photomicrograph of sand in footwall mixed zone, where bedding laminae are folded. Field of view is 2.6 mm; plane polarized light. Volcanic clasts are fractured (f) and fragmented, producing fine-grained pseudomatrix (p). Zeolites (z) are locally evident in intergranular pores. **(b)** and **(c)** Photomicrographs of deformation band, taken in plane polarized light and crossed polars, respectively. Field of view is 6.8 mm. Dashed lines in (b) roughly illustrate margins of deformation band; the margins are much more gradational than the lines suggest. Grains within band are generally smaller and more angular than outside, elongate grains within and at the margin of the band are largely aligned parallel to slightly inclined to the boundary, and pore space in the zone is filled with pseudomatrix. Evidence of compaction, pseudomatrix, and some fine-grained, angular material are evident in deformed areas adjacent to the deformation band (in this case, largely in the footwall; e.g., d). More volcanic clasts are evident in sand adjacent to the band than within the band. In virtually undeformed sand (u), porosity is higher than within band. Sense of shear is shown by arrows. **(d)** Mixture of sand and clay. Field of view is 6.8 mm; crossed polars. Note variation in sand content. Clays are generally aligned within a fault-parallel foliation that extends from lower left to upper right of image. Clays locally anastomose around sand grains, or are contorted around clusters of grains (arrow). Elongate sand grains are generally aligned in the foliation.; isolated grains are generally less likely to be aligned.

(e) Compositional variations form a millimeter-scale banding that defines a fault-parallel (steeply dipping in outcrop) foliation in the mixed zone (arrow). Where grain-size contrast along the boundaries between these bands is high, the margins are locally irregular and blobs of material have been separated from their source and intermixed. This photo, taken in plane polarized light with a field of view of 6.8 mm, shows irregular clasts of clay and silt-rich sediment (c) enclosed in fine-grained sand and silt. Note fine calcite veins (v), which are subhorizontal.

ments we have observed in undisturbed sediments elsewhere, these generally do not radiate from grains, but instead are irregular in orientation. Illitic clay coatings are common in some compositional layers and bands, and calcite cements are rare and poorly developed. In some areas of the mixed zone, however, sands are mixed with finer-grained, silt and clay-rich sediments. These highly mixed areas are typically well foliated; the foliation is defined by the alignment of clay particles and, where the percentage of clay is relatively small, by elongate clasts (Figure 11d). The mixing process substantially decreased sorting. Clay and silt fill the pore spaces between sand grains.

Faulted silt- and clay-rich sediments locally exhibit a well-developed foliation defined by aligned clay and compositional bands. The margins of these compositional bands can be relatively planar or highly irregular; microstructures suggest that material locally was mixed along these boundaries (Figure 11e). Tiny calcite veins are present within one sample of silt. The majority of these veins are subhorizontal, though a few have irregular orientations and one is roughly parallel to the foliation. Most are not visible in hand specimen; none are more than a millimeter wide.

DISCUSSION

Fault-Zone Width

Previous workers have argued that fault-zone width in rocks correlates with magnitude of displacement [see reviews by *Hull*, 1987 and *Scholz*, 1990]. The Sand Hill fault is a growth fault, thus we might expect architectural elements in the footwall, which experienced greater displacement than the hanging wall, to be wider than elements in the hanging wall. A greater degree of cementation, and thus higher competence, in the footwall sediments should amplify this difference in width. However, hanging wall architectural elements are thicker than footwall elements at both the Shooting Gallery and Waterfall sites, so additional controlling parameters must be considered. Variables that might affect fault width are 1) grain size of parent sediment, 2) bed thickness and distribution of beds of different grain size, and 3) fault geometry. These variables are considered in the following paragraphs.

Variations in fault-zone thickness are accommodated largely by variations in damage zone width. Grain size is the fundamental factor controlling damage-zone width in the Sand Hill fault zone. The damage zone is wide where the fault cuts (a) thickly bedded, coarse-grained sediments (≤80% sand and gravel; hanging wall sediments beneath unconformity at Shooting Gallery site, Figs. 7 and 8), (b) coarse-grained sediments where bedding has been destroyed through bioturbation (hanging wall at Waterfall site, Figs. 9 and 10), or (c) thinly bedded coarse-grained sediments with rare, thin, relatively clay-rich interbeds (footwall at Waterfall site). Where it cuts thickly bedded fine-grained

sediments (>20% clay and silt; hanging wall above unconformity at Shooting Gallery site, Figure 7 and 8), or thin beds that alternate between >20% and ≤20% clay and silt (footwall at Shooting Gallery site), the damage zone is narrow. Bed thickness in itself is not a controlling variable. Where the fault cuts a thick bed, the mean grain size of the bed will dictate damage-zone thickness; however, where it cuts a series of thin beds, the aggregate mean grain size of the sequence of thin beds will prescribe damage-zone thickness. Footwall sediments at the Waterfall site are thinly bedded, most beds are ≤80% sand and gravel, and the damage zone is wide. At the Shooting Gallery site, footwall sediments are thinly bedded, but alternate beds are >20% silt and clay, and the damage zone is narrow.

The mixed zones vary less in width than the damage zones (e.g., Figure 7). Fault geometry appears to exert the most significant control on mixed zone thickness. Where the fault bends, splays are more common and pods of little deformed sand and gravel are more abundant within the mixed zones (compare Figure 9 with Figure 10). Mixed zone thickness is greatest in these bends, where pods of little deformed sediment are most numerous.

Deformation Processes

In the previous section, we noted a correlation between damage-zone width and the grain size of faulted sediments. A number of other observations suggest a correlation between grain size distribution and deformation behavior. Grain-size analyses suggest that clay-rich beds were preferentially incorporated into the mixed and core zones and sandy beds were less thoroughly mixed into the fault zone during the faulting process. For instance, clay-rich sediments were thinned and extended farther within the fault zone than coarser-grained sediments, and blocks and slivers of little deformed sediment are typically composed of sand and gravel. In the following section, we consider how grain size might influence deformation processes. Our long-term goal is to predict the width of a fault zone in poorly lithified sediments based upon the grain size distribution of the adjacent beds and some knowledge of along-strike variations in fault geometry.

Previous studies suggest three ways that fault-zone deformation might be controlled by the grain size distribution of faulted sediments: (1) strain softening of clay and silt-rich sediment and sedimentary rock, (2) strain hardening of sand and sandstone, and (3) incorporating blocks of more competent coarse-grained sediment through the interaction between faulting and flexural slip folding. Each of these processes is discussed below.

Strain softening can occur where platy clays have been rotated into the plane of the fault during slip, forming a clay smear. Clays are mechanically the weakest of the common rock-forming minerals, with very low coefficients of friction [*Byerlee*, 1978], so protoliths with higher clay content should have lower coefficients of friction than those with low clay content. The mechanical alignment of clays should further weaken the fault zone by facilitating frictional grain-boundary sliding. *Maltmann* [1987] demonstrated experimentally that repeated slip in argillaceous sediments can be accommodated within very narrow, discrete shear zones. This observation is consistent with field evidence that suggests greater deformation localization where faults in rock cut shale-rich protoliths [see review by *Caine et al.*, 1996]. *Antonellini and Aydin* [1995] demonstrated that fault zone width correlates inversely with sandstone clay content (Figure 8b); a similar correlation is evident along the Sand Hill fault (Figure8a). Thus, strain softening is expected where the fault cuts clay-rich layers or where clays were incorporated into the fault zone. A throughgoing clay-rich core and / or mixed zone, if sufficiently thick, should strongly localize subsequent deformation.

Strain hardening in sandstone is attributed to an increase in grain contact area, and thus an increase in frictional resistance to slip, as a result of cataclasis [e.g., *Aydin and Johnson*, 1978]. Cataclasis causes an increase in grain contact area through grain-size reduction, greater grain interlocking, and a decrease in porosity with respect to the undeformed protolith. Preferential mineral precipitation along a strain-hardened structure may further increase its strength [*Fowles and Burley*, 1994]. In the faulted sandstones studied by *Aydin and Johnson* [1978], a progression from a single deformation band to a zone of deformation bands to a slip surface is evident. The formation of a zone of numerous closely spaced deformation bands with similar orientations suggests that the strength of each deformation band is greater than the undeformed protolith, indicating strain hardening. In *Aydin and Johnson*'s [1978] model, the zone of deformation bands ultimately gives way to a throughgoing slip surface along which repeated movements may occur. Deformation bands, zones of deformation bands, and slip surfaces are common in sand-rich units within damage and mixed zones of the Sand Hill fault. Furthermore, there is strong evidence for cataclasis (see later section on microstructures). Thus, the strain-hardening processes that control fault-zone development in sandstone probably occurred in the poorly lithified sand-rich sediments cut by the Sand Hill fault, and contributed to the development of both damage and mixed zones.

Folding in fault zones may be accomplished by a number of mechanisms, including flexural slip. Bed-parallel slip, which accommodates flexural-slip folding, is typically partitioned into shale-rich beds sandwiched between more competent sandstone layers [*Gibson*, 1994; *Cooke and Pollard*, 1997; *Watterson et al.*, 1998; *Foxford et al.,* 1996]. Consequently, bending of units during folding may result in the preferential formation of fractures or deformation bands in stiffer or more rigid layers, and bed-parallel slip can create steps between beds, causing asperities to form on a given fault surface [*Watterson et al.*, 1998]. Steps can be

removed by later faulting; weaker material would be smeared within the fault zone, while more competent units, such as sandstone, would be incorporated into the fault zone as intact pods or blocks [*Gibson*, 1994; *Watterson et al.*, 1998; *Foxford et al.,* 1996].

Although we have not conducted a systematic study of folds within the Sand Hill fault zone, it appears that they may have formed through flexural slip. Where clay-rich beds were rotated into the fault zone and thinned, pervasive slickenside striae record bed-parallel slip. Relatively intact pods of sand and gravel are common constituents of mixed zones (e.g., Figure 5c), and could have formed as fold-related asperities. In addition, the coarser grained, more competent beds exhibit a greater density of deformation bands in folded areas. Flexural-slip folding therefore may have played an important role in the development of the Sand Hill fault zone.

Damage and Core Zone Evolution

From the above arguments, as a given fault zone in poorly lithified sediment evolves, it should widen where it cuts sand-rich materials (≤80% sand and gravel) and maintain a relatively constant width where it cuts clay-rich sediment (>20% clay and silt). These observations are the foundation of a simple model that illustrates the development of damage zones and a clay-rich core zone during fault-zone evolution (Figure 12). This model illustrates several critical points about fault-zone evolution in poorly lithified sediments:

(1) Small-displacement faults should initiate in sand and terminate in adjacent clay-rich beds during the incipient stages of deformation. Previous workers have noted that numerous small faults may form during a single seismic event in a sandstone bed [cf. *Aydin*, 1978; *Aydin and Reches*, 1982; *Aydin and Johnson*, 1983]. These faults may propagate into adjacent shaley layers, but they do not initiate in the relatively weak shales. Folds may form at fault tips.

(2) Deformation will be accommodated largely by small faults and deformation bands in sand-rich beds and extension and slip along fault-parallel bedding surfaces in clay-rich beds. The fault zone will widen along sand-sand contacts in response to strain hardening of deformation bands in the damage zones. Widening will continue until a clay smear forms in the fault core, at which point subsequent deformation can be localized in the strain-softened core.

(3) Fault-bounded blocks adjacent to the core can continue to rotate, extend, and undergo mixing with increased displacement and, potentially, flexural slip.

The development of a clay smear in the core of a fault in poorly lithified sediments should be influenced by the clay content of individual beds and the thickness of and spacing between clay-rich beds. Numerous models predicting

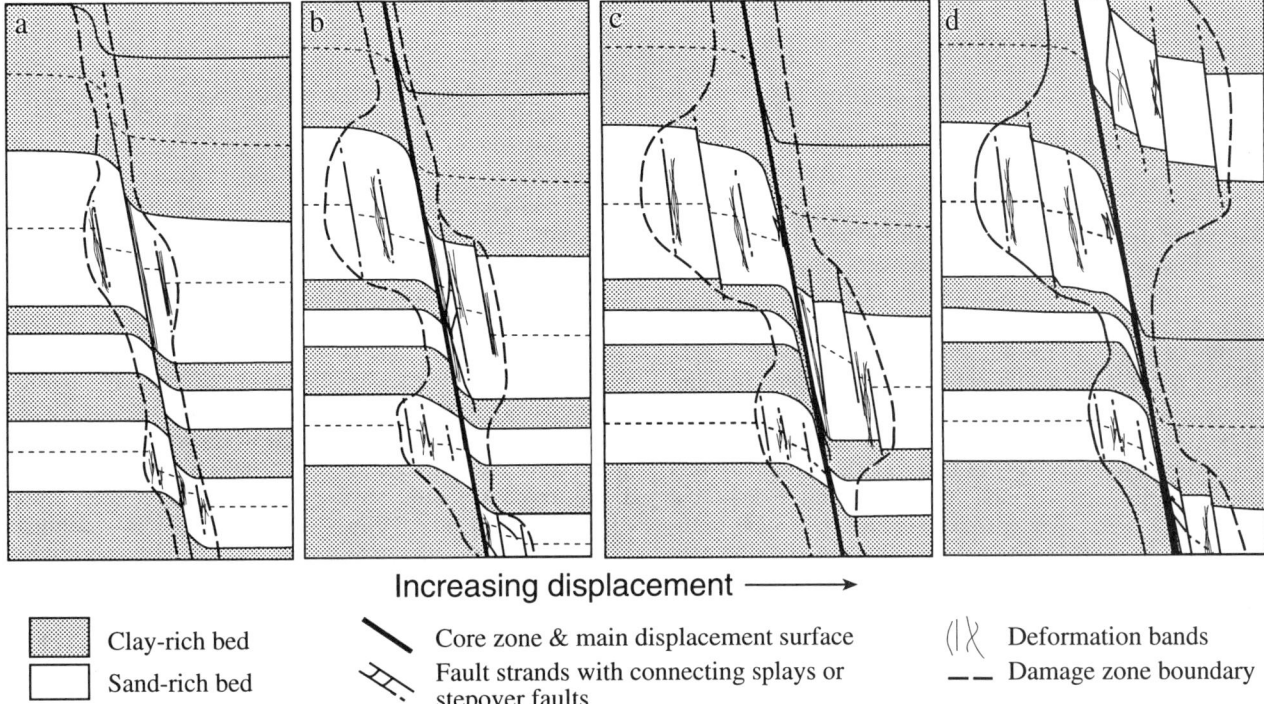

Increasing displacement

Clay-rich bed

Sand-rich bed

Core zone & main displacement surface

Fault strands with connecting splays or stepover faults

Deformation bands

Damage zone boundary

Figure 12. Schematic model of damage and core zone evolution with increasing displacement along a normal fault in poorly lithified sand and clay beds, based on interpretations of field relationships at the Sand Hill fault zone. **(a)** Deformation bands initiate in more competent sand beds. A throughgoing slip surface forms and propagates into clay beds. Strain hardening results in widening of the fault zone where thick sand layers remain in contact during faulting. Strain softening localizes deformation within a narrow zone where the fault cuts thick clay beds and where clay is juxtaposed with sand (thin interbeds of clay and sand). **(b)** The fault zone continues to widen along sand-sand contacts in response to strain hardening of deformation bands. Clay smears form where clay beds are displaced past sandy beds. **(c)** The latter processes continue, resulting in the variations in damage zone width evident in Figs. 7, 9, and 10. Where a continuous clay smear in the core zone separates a sandy bed from the main slip surface, strain is localized within the clay smear and strain-hardening in the sand bed ceases. **(d)** Clay beds have been displaced past all sand beds shown, resulting in a continuous clay smear. The core zone therefore includes both the main slip surface and the clay smear. The mechanically weak core zone will serve to localize subsequent deformation, thus the fault zone will not increase further in width except where the clay-rich core is breached. Rotation and thinning of the clay beds accommodated the differences in fault-zone width that resulted from differential growth of the damage zones in sand and clay beds.

smearing potential relative to bed thickness and spacing in faults in rock have been proposed by previous workers [see review by *Yielding et al.*, 1997]. A great deal of variability exists between models, although the general principles of each are similar. Typically the models assume that the clay smear will thin with increasing distance from its source, eventually becoming thin and patchy. The clay smear at the Sand Hill fault, however, is continuous nearly everywhere with the exception of occasional gaps <30 cm long. These gaps may well be transient features; the relative continuity of the clay smear in the core may indicate that a steady-state system has been attained, in which the low friction core can maintain a relatively consistent thickness. It

may also indicate that the spacing between source beds is sufficiently small, even where few clay beds are visible in outcrop, to maintain a continuous clay smear.

The spacing between clay-rich beds is therefore inferred to have had the greatest influence on the development of the Sand Hill fault zone. Where there are numerous thin, closely spaced, clay- and sand-rich beds with a ratio ≤1, as in the footwall at the Shooting Gallery (Figure 7), the displacement required to create a clay smear would be relatively small and the fault zone would stop widening much earlier in its history. Where there are fewer and more widely spaced clay-rich beds, as at the Waterfall site (Figs. 9 and 10), the displacement required to create a continuous clay smear would be greater and the fault zone would continue to widen for a longer time.

Microstructures: Implications for Permeability

Fault-zone deformation should significantly affect permeability. Petrographic work presented here, by *Mozley and Goodwin* [1995] and by *Goodwin and Haneberg* [1996] indicates that sand within deformation bands and zones of deformation bands exhibits smaller and more angular grains and lower porosity than adjacent undeformed sand. In addition, elongate grains are oriented subparallel to deformation band boundaries rather than subparallel to bedding. Fracturing occurred preferentially in lithic clasts, which are less numerous within deformation bands than in undeformed sands. Brown, clay-size material, interpreted to be pseudomatrix produced by deformation of lithic clasts, fills pore spaces in deformation bands in sand. These observations collectively indicate that cataclasis and mechanical rotation of grains occurred within deformation bands. Cataclasis should decrease permeability.

Processes that should influence the permeability and permeability anisotropy of sediments within the Sand Hill fault zone therefore include: 1) cementation [*Mozley and Goodwin*, 1995], 2) grain and pore-size reduction through cataclasis and grain reorganization, 3) mechanical rotation of grains, and 4) grain-scale mixing of different sediments (e.g., clay and sand). The effectiveness of these processes in occluding pore space and decreasing permeability depends on the grain size of the original sediments, changes in sorting resulting from deformation, and the composition of the faulted sediments. Calcite was preferentially precipitated within sands and gravels, and can effectively destroy porosity in well cemented areas [*Mozley and Goodwin*, 1995]. The magnitude of permeability reduction caused by cataclasis will vary with the composition of the sands. For sands rich in lithic clasts, permeability may be reduced as much as three orders of magnitude as a result of cataclasis [*Sigda et al.*, this volume, provide details on permeability variations]. These data suggest that deformation bands and zones of deformation bands in the damage and mixed zones can significantly reduce the permeability of poorly lithified sands, particularly lithic-rich sands. Mechanical rotation of grains and foliation formation can increase permeability anisotropy within the fault zone [*Goodwin and Haneberg*, 1996]. Grain-scale mixing, and the resulting reduction in sorting, could have the greatest affect on fault-zone permeability (e.g., Figure 11d). Silt and clay-rich sediments have an initially low permeability that may be further reduced by the incorporation of sand. Tiny, subhorizontal veins found locally within fine-grained sediment may be unloading features, or may record transient fracture permeability.

Relative Permeabilities of Architectural Elements

The relative permeabilities of architectural elements may be estimated by applying information gained from other studies. The clay present nearly everywhere in the core zone is inferred to have a very low permeability by virtue of its grain size and fabric [cf. *Knipe*, 1993, 1997; *Gibson* 1994]. This clay-rich core is analogous to clay smears in rock that act as effective barriers to fluid flow [e.g., *Smith*, 1980; *Bouvier et al.*, 1989; *Knipe*, 1993; *Gibson*, 1994]. The influence of the fault core on fluid flow should depend on its thickness, the size and location of holes in the clay, and ambient groundwater flow fields (boundary conditions).

The mixed zones exhibit the greatest density of deformation bands and variation in fabric and sediment characteristics at any given site. It is not uncommon to find a high clay content in mixed zones where the adjacent sediments are sand or, vice-versa, a relatively high sand content where the adjacent sediments are thick clay beds. Consequently, permeabilities of deformed sediments in the mixed zones are expected to fall within a range distinct from that of adjacent sediments. As noted previously, the permeabilities of deformation bands, zones of deformation bands, and mixed zones in small-displacement faults are up to three orders of magnitude lower than adjacent sediment permeabilities [*Sigda et al.*, this volume].

The damage zones exhibit less deformation than any other fault-zone units. Sediments within the damage zones are identical to those immediately adjacent to them, except where they are cut by deformation bands and slip surfaces. Where sediments are silt and clay-rich, slip surfaces are few and deformation bands are absent. Where sediments are sand-rich, deformation bands and zones of deformation bands are common, but typically not laterally and vertically extensive. We therefore expect that permeabilities are somewhat reduced with respect to the parent sediments, but that the damage zones will have permeabilities that are more similar to the parent sediments than other fault-zone units.

The permeability of a fault-zone architectural element should therefore vary along strike and dip as the grain size of the adjacent sediment varies. At a given site, the permeability will depend on: 1) the grain size distribution of the unfaulted sediments, 2) the type and magnitude of deformation, and 3) the degree of cementation. The effects of these variables are illustrated in Figure 13.

We therefore believe that the fault-zone architectural elements identified through field study have hydrologic significance. The influence of a given normal fault in sediment on fluid flow should be a function of the spatial arrangement of these elements and the parent sediments, as well as the ambient flow field. We are currently engaged in a combined laboratory and field study to quantitatively characterize the permeability variation within these units.

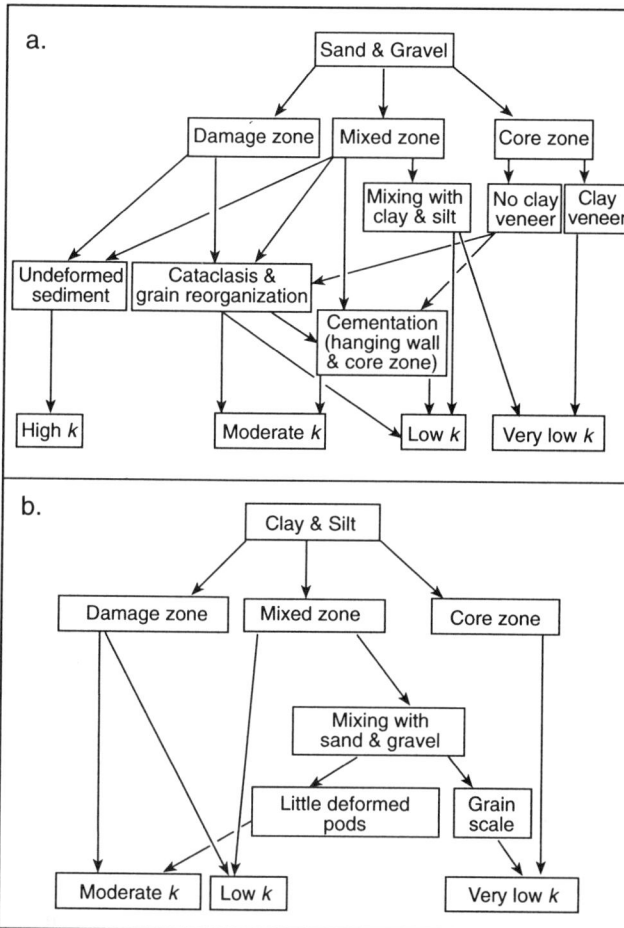

Figure 13. Correlation between grain-size distribution of parent sediment, deformation processes, cementation, and resulting inferred permeability (*k*) within the Sand Hill fault zone. Permeability is expressed in relative terms. **(a)** Parent sediments are dominantly sands and gravels. **(b)** Parent sediments are dominantly silts and clays.

Past Fluid Flow Inferred from Cementation Patterns

If we consider degree of cementation a proxy for fluid flux within and adjacent to the Sand Hill fault zone, we can attempt to evaluate the influence of the fault on fluid flow at the time of cement precipitation. A component of cross-fault fluid flow is indicated by the observations that the cements were deposited on the SE side of the fault in a NW to SE flow system, and that cements are best developed where both footwall and hanging wall sediments are coarse-grained and presumably more permeable than finer-grained sediments (Figure 13). These observations also suggest that at least some cementation occurred late in the faulting history. The magnitude of cross-fault fluid flow should be controlled by the permeabilities of the core and footwall

mixed zones, although a relationship between degree of cementation and gaps in the core zone is not apparent. The hanging wall mixed zone may also have served as a conduit for fluids from different structural levels (elongate concretions typically record subvertical, fault-parallel flow). Mixing of fluids across the fault, and from different structural levels, could have facilitated cement precipitation. The distribution of the different sediments and structures within the fault zone therefore appears to have guided fluid flow.

CONCLUSIONS

Fault-zone deformation in poorly lithified sediments results in mappable fault-zone architectural elements. These elements contain some structures that are similar to those found in porous sandstone, such as deformation bands. However, they can be distinguished from faults in rock in two critical ways: 1) They do not contain open fracture networks. 2) Mixing of parent sediments occurs at the grain scale. Deformation thus acted to decrease, rather than increase, permeability within the fault zone. Because deformation affects permeability, these units have hydraulic significance, and are therefore considered hydrogeologic units.

The magnitude of permeability change within an architectural element depends fundamentally on the grain size of the parent sediment, which controls initial permeability, deformation mechanisms, and the density of structures within the fault zone (expressed as fault-zone width). Deformation in sand is accommodated by cataclasis, grain-boundary sliding, and mechanical rotation of grains resulting in grain reorganization, porosity reduction, and development of fault-zone fabrics. These processes should effect permeability reduction. They also result in strain hardening and fault-zone growth; thus, the fault zone is widest in coarse-grained sediments. Deformation in clay-rich sediments is accommodated by mechanical rotation of grains and sliding, either between grains or along discrete slip surfaces or foliation planes. These processes facilitate thinning of individual beds, mixing of sediments, and strain-softening. Sediment mixing should decrease permeability. Strain-softening will result in a narrow fault zone in fine-grained sediment.

The documented correlation between grain-size distribution of faulted sediments and structures within the fault zone provides a foundation for prediction of fault-zone structure and permeability from hanging wall and footwall stratigraphy. Likely sites of cementation can also be predicted given knowledge of the regional flow system.

Acknowledgements. We are grateful to the New Mexico Bureau of Mines and Mineral Resources, Exxon Production Research, the USGS STATEMAP and EDMAP programs, and the National Science Foundation (EAR-9706482) for support. We also thank the King and Parker families for allowing access to key portions of the study site. Chris Dimeo assisted in fieldwork and developed a method for obtaining oriented samples of poorly lithified material. Discussions with Bill Shea, Jim Holl, Peter Vroljk, and Lee Fairchild helped clarify research directions. Thoughtful reviews by Peter Eichhubl, Casey Moore, and Mark Person improved the manuscript. Last, but not least, we thank Kathy Glesener, Becky *Titus* and Jan Thomas for their exceptional drafting assistance. The views and conclusions contained in this document are those of the authors and should not be interpreted as representing the views of supporting agencies.

REFERENCES

Anderson, L.J., R.H. Osborne, and D.F. Palmer. Cataclastic rocks of the San Gabriel fault - An expression of deformation at deeper crustal levels in the San Andreas fault zone. *Tectonophysics, 98*, 209-251, 1983.

Anderson, T.W., G.E. Welder, G. Lesser, and A. Trujillo. Region 7, Central alluvial basins, in *The Geology of North America (Volume 0-2), Hydrogeology,* edited by W. Back, J.S. Rosenshein, and P.R. Seaber, P.R., pp. 81-86. Geol. Soc. Am., Boulder, CO, 1988.

Antonellini, M. and A. Aydin, 1994, Effect of faulting on fluid flow in porous sandstones: petrophysical properties. *Amer. Assoc. Petrol. Geol. Bull, 78*, 355-377, 1994.

Antonellini, M. and A. Aydin. Effect of faulting on fluid flow in porous sandstones: geometry and spatial distribution. *Amer. Assoc. Petrol. Geol. Bull, 79*, 642-671, 1995.

Aydin, A. Small faults formed as zones of deformation bands and as slip surfaces in sandstone. *Pure Appl. Geophys., 16,* 931-942, 1978.

Aydin, A. and A.M. Johnson. Development of faults as zones of deformation bands and as slip surfaces in sandstone. *Pure Appl. Geophys., 116*, 913-930, 1978.

Aydin, A. and A.M. Johnson, A.M. Analysis of faulting in porous sandstones. *Jour. Struct. Geol., 5*, 19-31.

Aydin, A. and Z. Reches. The number and orientation of fault sets in the field and in experiments. *Geology, 10*, 107-112, 1982.

Beckner, J.R. Cementation processes and sand petrography of the Zia Formation, Albuquerque Basin, New Mexico. M.S. Thesis, New Mexico Institute of Mining and Technology, Socorro, 1996.

Beckner, J., and P.S. Mozley Origin and spatial distribution of early phreatic and vadose calcite cements in the Zia Formation, Albuquerque Basin, New Mexico, USA, in *Carbonate Cements in Sandstones*, edited by S. Morad, pp. 27-51, International Association of Sedimentologists Special Publication 26, 1998.

Bouvier, J.D., C.H. Kaars-Sijpesteijn, D.F. Kluesner, C.C. Onyejekwe, and R.C. van der Pal. Three-dimensional seismic interpretation and fault sealing investigations, Nun River field, Nigeria. *Am. Assoc. Petrol. Geol. Bull., 73.*, 1397-1414, 1989.

Bruhn, R. L., W.T. Parry, W.A. Yonkee, and T. Thompson. 1994, Fracturing and hydrothermal alteration in normal fault zones. *Pure Appl. Geophys., 142*, 139-157, 1994.

Byerlee, J.D. Friction of rocks. *Pure Appl. Geophys., 116*, 615-626, 1978.

Byerlee, J.D. Model for episodic flow of high-pressure water in fault zones before earthquakes. *Geology, 21*, 303-306, 1993.

Caine, J.S., J.P. Evans, and C.B. Forster. Fault-zone architecture and permeability structure. *Geology, 24*, 1023-1028, 1996.

Cather, S.M. Toward a hydrogeologic classification of map units in the Santa Fe Group, Rio Grande rift, New Mexico. *New Mexico Geol., 19*, 15-21, 1997.

Cather, S.M., S.D. Connell, M.R. Heynekamp, and L.B. Goodwin, *Map of the Sky Village SE 7.5-minute quadrangle, Sandoval county, New Mexico.* New Mexico Bureau of Mines and Mineral Resources Open-file report, DGM 9, 1997.

Chester, F. M., and J.M. Logan. Implications for mechanical properties of brittle faults from observations of the Punchbowl fault zone, California. *Pure Appl. Geophys., 124*, 77-106, 1986.

Cooke, M.L. and D.D. Pollard. Bedding-plane slip in initial stages of fault-related folding. *J. Struct. Geol, 19*, 567-581, 1997.

Cowan, D. S. and M.T. Brandon. A symmetry-based method for kinematic analysis of large-slip brittle fault zones. *Am. J. Sci., 294*, 257-306, 1994.

Davis, S.N. and R.J.M. DeWeist, *Hydrogeology.* 604 pp. John Wiley & Sons, New York, NY, 1966.

Foxford, K.A., I.R. Garden, S.C. Guscott, S.D. Burley, J.J.M. Lewis, J.J. Walsh, and J. Watterson. The field geology of the Moab Fault, in *Geology and Resources of the Paradox Basin*, edited by C. Huffman, H.H. Doelling, and G. Willis, Utah Geological Association, Utah, Salt Lake City, 1996.

Fowles, J. and S.D. Burley. Textural and permeability characteristics of faulted, high porosity sandstones. *Mar. Petrol. Geol., 11*, 608-623, 1994.

Gibson, R.G. Fault-zone seals in siliclastic strata of the Columbus Basin, offshore Trinidad. *Am. Assoc. Petr. Geol. Bull., 78*, 1372-1385, 1994.

Goodwin, L.B. and W.C. Haneberg. Deformational fabrics and inferred permeability of faulted sands from the Rio Grande Rift, New Mexico. *Geol. Soc. Am. Abstr. Prog., 28(7)*, A-255, 1996.

Haneberg, W.C. Steady-state groundwater flow across idealized faults. *Water Resour. Res., 31*, 1815-1820, 1995.

Hawley, J.W., and C.S. Haase. *Hydrogeological framework of the northern Albuquerque Basin.* New Mexico Bureau of Mines and Mineral Resources Open File Report 387, 147 pp, 1992.

Hawley, J.W., C.S. Haase, and R.P. Lozinsky. An underground view of the Albuquerque Basin, in *The Water Future of Albuquerque and the Middle Rio Grande Basin (Proceedings of the 39th Annual New Mexico Water Conference)*, pp. 37-55. New Mexico Water Resources Research Institute Report 290, 1995.

Hull, J. Thickness-displacement relationships for deformation zones. *J. Struct. Geol., 10*, 431-435, 1988.

Jameson, W.R., and D.W. Stearns. Tectonic deformation of Wingate Sandstone, Colorado National Monument. *Am. Assoc. Petr. Geol. Bull., 66*, 2584-2606, 1982.

Kelley, V.C. *Geology of the Albuquerque Basin, New Mexico* . New Mexico Bureau of Mines and Mineral Resources Memoir 33, 59 pp., 1977.

Knipe, R. J. The influence of fault zone processes and diagenesis on fluid flow, in *Diagenesis and Basin Development*, edited by A. D. Horbury and A. G. Robinson, pp. 135-148, Am. Assoc. Petrol. Geol. Studies in Geology 36, 1993.

Maltmann, A. Shear zones in argillaceous sediments— an experimental study, in *Deformation of Sediments and Sedimentary Rocks*, pp. 77-90, edited by M.E. Jones and R.M.F. Preston. Geol. Soc. Special Publ. 29, 1987.

Mifflin, M.D., Region 5, Great Basin, in *The Geology of North America (Volume 0-2), Hydrogeology*, edited by W. Back, J.S. Rosenheim, and P.R. Seaber, pp. 69-78. Geol. Soc. Am., Boulder, CO, 1988.

Mozley, P.S., and J.M.D. Davis. Relationship between oriented calcite concretions and permeability correlation structure in an alluvial aquifer, Sierra Ladrones Formation, New Mexico. *J. Sed. Res., A66*, 11-16, 1996.

Mozley, P.S. and L.B. Goodwin, Patterns of cementation along a Cenozoic normal fault: A record of paleoflow orientations. *Geology, 23*, 539-542, 1995.

Mozley, P.S., J. Beckner, and T.M. Whitworth. Spatial distribution of calcite cement in the Santa Fe Group, Albuquerque Basin, NM: Implications for groundwater resources. *New Mexico Geol.,* 17, 88-93, 1995.

Person, M. and G. Garven. A sensitivity study of the driving forces on fluid flow during continental-rift basin evolution. *Geol. Soc. America Bull., 106*, 461-475, 1994.

Person, M., J.P. Raffensperger, S. Ge, and G. Garven. Basin-scale hydrogeologic modeling. *Rev. Geophys., 34*, 61-87, 1996.

Pittman, E. D. Effect of fault-related granulation on porosity and permeability of quartz sandstones, Simpson Group (Ordovician), Oklahoma. *Am. Assoc. Petrol. Geol. Bull., 65*, 2381-2387, 1981.

Scholz, C.H. *The Mechanics of Earthquakes and Faulting.* Cambridge University Press, 1990

Sibson, R.H., 1977, Fault rocks and fault mechanisms. *J. Geol. Soc. London, 133*, 191-231, 1977.

Sibson, R.H. Conditions for fault-valve behavior, in *Deformation Mechanisms, Rheology and Tectonics*, edited by R.J. Knipe and E.H. Rutter, pp. 15-28. Geological Society Special Publication 54, 1990.

Sigda, J.M., L.B. Goodwin, P.S. Mozley, and J.L. Wilson. Permeability alteration in small-displacement faults in poorly lithified sediments: central Rio Grande Rift, New Mexico (this volume).

Smith, D.A. Sealing and nonsealing faults in Louisiana Gulf Coast Salt Basin. *Am. Assoc. Petrol. Geol. Bull., 64*, 145-172, 1980.

Smith, L., C.B. Forster, and J.P. Evans. Interaction of fault zones, fluid flow, and heat transfer at the basin scale, in *Hydrogeology of Low Permeability Environments*, edited by S.P. Neuman and I. Neretnieks, p. 41-67. Verlag Heinz Heise, Hanover, 1990.

Tedford, R. H., 1982, Neogene stratigraphy of the northwestern Albuquerque Basin, in *Albuquerque Country II*, edited by J.F. Callender, J.A. Grambling, and S.G. Wells, p. 271-278. New Mexico Geological Society Guidebook 33, 1982.

Thorn, C.R., D.P. McAda, and J.M. Kernodle, 1993, *Geohydrologic Framework and Hydrologic Conditions in the Albuquerque Basin, Central New Mexico.* U.S. Geological Survey Water Resources Investigations Report 93-4149, 1993.

Titus, F.B., *Geology and Ground-Water Conditions in Eastern Valencia County*, New Mexico: New Mexico Bureau of Mines and Mineral Resources Ground-water Report 7, 1963.

Underhill, J. R., and N. H. Woodcock, Faulting mechanisms in high porosity sandstones: New Red Sandstone, Arran, Scotland, in *Deformation of Sediments and Sedimentary Rocks*, edited by M. E. Jones and R. M. F. Preston, p. 91-105. *Geol. Soc. London Spec. Publ. 29*, 1987.

Watterson, J., C. Childs, and J. J. Walsh. Widening of fault zones by erosion of asperities formed by bed-parallel slip. *Geology, 26*, 71-74, 1998.

Wright, H. E. Tertiary and Quaternary geology of the lower Rio Puerco area, New Mexico. *Geol. Soc. Am. Bull., 57*, 383-456, 1946.

Yielding, G., B. Freeman, and D. T. Needham. Quantitative fault seal prediction. *Amer. Assoc. Petrol. Geol. Bull., 81*, 897-917, 1997.

Laurel B. Goodwin, Department of Earth and Environmental Science, New Mexico Tech, 801 Leroy Place, Socorro NM 87801 (e-mail: lgoodwin@nmt.edu)

William C. Haneberg, Haneberg Geoscience, 10411 SE Olympiad Drive, Port Orchard WA 98366, (e-mail: bill@haneberg.com)

Michiel R. Heynekamp, New Mexico Bureau of Mines and Mineral Resources, New Mexico Tech, 801 Leroy Place, Socorro NM 87801 (e-mail: heynekam@gis.nmt.edu)

Permeability Alteration in Small-Displacement Faults in Poorly Lithified Sediments: Rio Grande Rift, Central New Mexico

John M. Sigda, Laurel B. Goodwin, Peter S. Mozley, and John L. Wilson

Department of Earth and Environmental Science, New Mexico Institute of Mining and Technology, Socorro, NM

Faults in clastic rocks influence fluid flow by juxtaposition of different lithologic units and by localized alteration of petrophysical properties through cataclasis, cementation, or other deformational and diagenetic processes. Extensional tectonic settings, such as the Basin and Range Province and the Rio Grande rift, are characterized by numerous faults in both sedimentary rocks and poorly lithified basin-fill sediments. Faults in poorly lithified sediments have received little attention; our study is the first to examine their permeability. We tested whether faulting of poorly lithified sediments significantly affects permeability by comparing two uncemented, small-displacement, normal faults in New Mexico's central Rio Grande rift: one with a clay-rich core (displacement > bed thickness) and one without a clay-rich core (displacement < bed thickness). Using *in situ* permeametry and thin section analysis, permeability, porosity, and clay size fraction were mapped across undeformed and faulted sediments. Both fault zones display permeability values two to three orders of magnitude lower than those for undeformed sand. Clay size fraction increased four to five-fold over undeformed sand, even in the fault without a clay-rich core, and is inversely correlated with permeability. Our results indicate that small-displacement faults are much less permeable than their poorly lithified parent sediments and that permeability reduction is associated with an increase in clay size fraction, but does not depend solely on formation of a clay-rich core. Under saturated conditions these faults impede fluid flow, but may act as preferential flow paths through thick, dry vadose zones common in the arid Southwest. Numerous in extensional basins but typically not included on most geologic maps, such faults could significantly influence flow through basin-fill sediments.

INTRODUCTION

Within the Rio Grande rift and the Basin and Range Province, poorly lithified basin-fill sediments were repeat-edly faulted as basins subsided relative to mountain blocks. Varying from kilometers to tens of kilometers in length and millimeters to hundreds of meters in displacement, the faults cross-cut the basin aquifers on which most of the region's rapidly growing population depends. Yet the hydrologic impacts from faulting of basin-fill sediments - which depend on fault areal density, geometry, and permeability - remain poorly understood.

Although geologists typically do not map small-displacement faults (their displacement is not significant at

Faults and Subsurface Fluid Flow in the Shallow Crust
Geophysical Monograph 113

the scale of most geologic maps), they appear to be common features in areas of crustal extension such as the Rio Grande rift. Data on power law relationships between fault number and displacement for the Rio Grande rift's Española Basin, just north of the Albuquerque Basin, suggest small-displacement faults, 0.01 - 1.0 m, may number more than 3000 within a roughly 400 km^2 area [*Carter and Winter*, 1995]. We have observed them throughout the Rio Grande rift in central New Mexico. The areal density of large-displacement faults appears to be greatest near the margins of the Rio Grande rift's central basins and least near basin centers [e.g., *Hawley and Haase*, 1992]. The spatial distribution of small-displacement faults may follow a similar pattern.

Faulting of clastic rocks can induce large- and small-scale changes in the spatial distribution of petrophysical characteristics, particularly porosity and permeability. Large-scale changes in permeability spatial distribution are typified by juxtaposition of lithologic units with different permeabilities following large fault displacements (>10^2 m), and are often revealed by sizable head drops across fault zones [*Tolman*, 1937; *Dutcher and Garrett*, 1963; *Haneberg*, 1995]. Deformational and diagenetic processes elicit small-scale changes in permeability which are confined to the fault zone. However, both large and small-scale changes in permeability spatial distributions are sufficiently variable that a single fault in clastic rocks seldom behaves strictly as either a permeability barrier or a capillary barrier across its entirety [see *Knipe*, 1992, 1993; *Caine et al.*, 1996; *Yielding et al.*, 1997].

A number of fault-zone deformational and diagenetic processes can alter the permeability and porosity of clastic sedimentary rocks. Incorporation of fine-grained (particularly clay-rich) beds into a fault zone can significantly reduce its permeability relative to that of the parent rocks. Clay can smear along the fault or mix with fault-zone materials during slip, creating zones with reduced pore throat and pore sizes that retard fluid flow and commonly support build-up of non-wetting phases such as hydrocarbons [*Gibson*, 1994; *Berg and Avery*, 1995; *Yielding et al.*, 1997]. Cataclasis (grain-size reduction through distributed microfracturing) decreases grain and pore sizes in clastic sedimentary rocks; displacements of a centimeter or less can reduce grains to clay-sized fragments, leading to a poorly sorted grain-size distribution, and in many cases almost entirely eliminating porosity [*Aydin*, 1978; *Pittman*, 1981; *Edwards et al.*, 1993]. Elongate grains and layer silicates can be rotated from a bed-parallel to a roughly fault-parallel orientation with a concomitant increase in permeability anisotropy within the fault zone [e.g., *Davis and DeWiest*, 1966; *Knipe*, 1992, 1993; *Mozley and Goodwin*, 1995]. Cementation and compaction can also reduce or eliminate porosity [*Edwards et al.*, 1993; *Antonellini et al.*, 1994]. Recently published intrinsic permeability measurements of faulted, well indurated (i.e., hard or well lithified), porous

sandstones indicate that fault zones that show cataclasis and compaction, but no cementation, are 10^3 to 10^7 less permeable than the parent material [*Antonellini and Aydin*, 1994; *Fowles and Burley*, 1994].

Although faults in poorly lithified sediments have received little attention until recently, deformational and diagenetic features of the Sand Hill fault, a major fault that cuts the poorly lithified sediments of the Albuquerque Basin of New Mexico, have been studied [*Heynekamp et al.*, this volume]. Extending for nearly 50 km, the Sand Hill fault accommodates ~600 m of displacement and serves as the western boundary for the Albuquerque Basin [*Hawley and Haase*, 1992]. Features described include deformation bands and zones of deformation bands within which porosity and grain size were significantly decreased, zones of tectonically mixed bedding up to 3 m wide within the fault zone (including pods of little deformed material), fracturing and/or preferred grain orientation within and adjacent to deformation bands, and extensive cementation [*Mozley and Goodwin*, 1995; *Goodwin and Haneberg*, 1996; *Heynekamp et al.*, this volume]. The deformational and diagenetic processes responsible for these features may also locally alter the petrophysical properties of the more numerous, typically unmapped, small displacement (≤ 10 m) faults in poorly lithified sediments.

We assess the hypothesis that small-displacement faulting can significantly alter the permeability of poorly lithified sand by comparing differences in permeability and petrologic characteristics of two faults. A fault with a clay-rich core (displacement > bed thickness), where we expect alteration of fault-zone permeability [cf. *Caine et al.*, 1996], is contrasted with a fault that lacks a clay-rich core (displacement < bed thickness). (We define "bed" as a sedimentary layer with consistent lithology and texture, e.g., cross-stratified fine-grained sandstone.) Focusing on small-displacement faults minimizes confounding effects from numerous slip events and repeated juxtapositions of beds with differing lithologies. We use *in situ* mini-permeametry and thin section analysis to map changes in permeability and petrophysical characteristics (e.g., clay size fraction, porosity) across undeformed and faulted sediments at each site.

GEOLOGIC SETTING

The Rio Grande rift is expressed within central New Mexico as several west-stepping, *en echelon* basins (Figure 1). Crustal extension, most of which occurred during the Miocene but which began in the late Oligocene, and rotation of the Colorado Plateau formed the basins [*Chapin and Cather*, 1994]. Although the majority of the basin fill was laid down in the Miocene, deposition of the Santa Fe Group continued until the Rio Grande became a through-flowing river during the Plio-Pleistocene [*Kelley*, 1977; *Hawley and Haase*, 1992; *Cather et al.*, 1994; *Chapin*

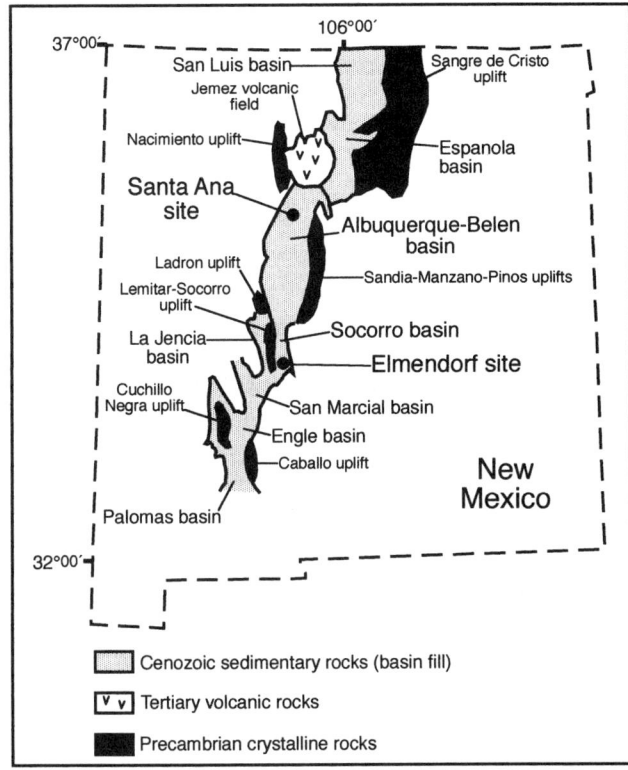

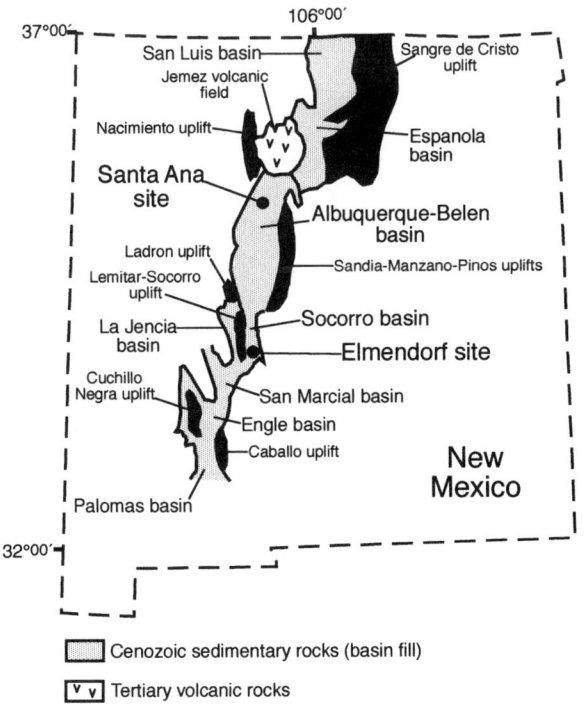

Figure 1. Rio Grande rift basins of New Mexico. Study sites shown with filled circles. Modified from *Russell and Snelson* [1994].

and Cather, 1994; *Lozinsky*, 1994]. Santa Fe Group sediments include gravels deposited in alluvial fan and fluvial environments, sands from alluvial fan, fluvial, and eolian environments; and muds from fluvial and playa lake environments [*Kelley*, 1977; *Hawley et al.*, 1978; *Chapin et al.*, 1978; *Hawley and Haase*, 1992; *Hawley et al.*, 1995]. The units of the Santa Fe Group vary in degree of lithification from unlithified sediment to well indurated rock. Faulting continued throughout the Quaternary after regional sedimentation had ceased, as demonstrated by the large number of faults displacing late and post-Santa Fe deposits [e.g., *Kelley*, 1977; *Machette*, 1982; *Chapin and Cather*, 1994].

The Albuquerque Basin, one of the largest and deepest of the rift's New Mexico basins, is estimated to have undergone 17% extension across its northern sub-basin and nearly 30% across the southern sub-basin [*Russell and Snelson*, 1990 and 1994]. Syn-rift Santa Fe Group thicknesses vary from less than 600 m to more than 4000 m within the basin [*May and Russell*, 1994]. Large, mappable faults are numerous throughout the basin, though they are better exposed in more indurated rock than in the poorly lithified Santa Fe Group sediments. The fault site with a clay core is located near the center of the basin, on the bank of an active arroyo within the city of Rio Rancho, ~32 km north of the city of Albuquerque (Figure 1). The sediments exposed in the outcrop are interpreted to be part of the upper Santa Fe Group (J. Hawley, personal communication, 1994).

The Socorro Basin possesses a more complicated structure because the Popatosa Basin, precursor to the modern-day La Jencia and Socorro Basins, underwent 50% extension during the Oligocene and Miocene [*Chamberlin and Osborn*, 1984; *Chapin and Cather*, 1994]. The Popatosa Basin split into the west-tilting La Jencia and Socorro half-grabens with the uplift of the Lemitar-Socorro Mountains during the Miocene [*Cather et al.*, 1994]. Faulting continued throughout the Quaternary and many faults are considered still active [*Machette*, 1982; *Cather et al.*, 1994; *Lewis and Baldridge*, 1994]. Basin-fill thicknesses are believed to vary from 300 to 3000 m [*Chapin et al.*, 1978]. The second study site is located near the southern end of the Socorro Basin, where it opens into the Jornada del Muerto and San Marcial Basins. Located 200 km south of Albuquerque, NM, the second fault site is part of the Elmendorf fault zone found near the southern boundary of the Bosque del Apache Wildlife Refuge (Figure 1). The outcrop is part of the Popatosa Formation within the Santa Fe Group (D. Love, personal communication, 1995).

STUDY METHODS

We identified lithologic, sedimentary, and structural features at each outcrop and constructed geologic maps on photo-mosaics. Mean grain sizes were estimated using comparators. Transect lines provided a local two dimensional coordinate system on each outcrop face. Representative undisturbed samples were collected from both undeformed and deformed zones at each site. Samples of undeformed sediments are considered representative of their respective units because the sands are well sorted and consistent in both their bedding characteristics and average grain size. After drying, the samples were impregnated with blue-dyed epoxy, and thin sections were prepared.

Point counts (300 points per area of interest) were taken to estimate the abundance of quartz, potassium feldspar, plagioclase, rock fragments, pores, and clay size fraction within deformed and undeformed zones. Repeated counts were made on two thin sections from the Elmendorf site and five thin sections from the Santa Ana site. Porosity was classified as either macroporosity or microporosity. Macroporosity was defined as an area of wholly blue epoxy with little or no clay-sized material apparent within an intergranular void. Microporosity was defined as an area of small amounts of blue-dyed epoxy interspersed within clay-sized material or within an altered grain. When an area contained equal amounts of both clay-sized material and blue-dyed epoxy the count was assigned to microporosity and clay size fraction categories in an alternating fashion. We also qualitatively evaluated changes in grain size, angularity, and fabric on the seven thin sections used for point counting and on an additional four thin sections from the Elmendorf site. All observations were made using a petrographic microscope equipped with a mechanical stage driven by an electronic counter.

Air and gas mini-permeameters were used to measure the permeability at each site. Employing a modified form of Darcy's Law, permeability is computed as a function of the differential between absolute atmospheric pressure and the pressure behind the tip seal, gas flow rate, temperature, gas viscosity, tip seal geometry, and the sample or sample location geometry [Goggin et al., 1988; Davis et al., 1994]. The observed permeability is also a function of the force applied to hold the tip seal against the outcrop face [Tidwell and Wilson, 1997]. The rock volume interrogated by the mini-permeameter is a function of the tip seal size: typical sample volumes range from 1 to 7 cc. Initial experimentation indicated two mini-permeameters were required to adequately capture the wide range in permeabilities at each study site.

Permeability of undeformed sand was measured with a syringe air mini-permeameter (SAMP) designed and constructed at the New Mexico Institute of Mining and Technology [Davis, 1994; Davis et al., 1994]. It measures the amount of time taken to deliver a constant volume of air from a glass syringe and was specifically developed for measuring the relatively high permeability (1 to 200 darcies) of poorly lithified sands and gravels of the Upper Santa Fe Group [Davis et al., 1993]. The SAMP tip seal has inner and outer radii of r_i = 2.5 mm and r_o = 13.0 mm. A recently improved version of the original SAMP

(SAMP2) was used to measure the permeability of undeformed footwall sands at the Santa Ana site; SAMP2 tip seal radii are r_i = 3.3 mm and r_o = 19.9 mm. The relatively low permeability of the deformed sands was measured with a continuous flow N_2 gas mini-permeameter (CFMP) developed at Sandia National Laboratories similar to the one described by Tidwell and Wilson [1997]. This instrument directly measures the pressure behind the tip seal and the gas flow rate. It has a permeability range of roughly between 0.001 and 5 darcies and a tip seal with r_i = 1.25 mm and r_o = 3.25 mm.

Measurements were taken along vertical and horizontal transects with standard spacing of 10 cm. More closely spaced measurements were required within the fault zones to adequately detail abrupt changes in permeability. Nearly all of the measurements were made with the mini-permeameter tip seal oriented perpendicular to the outcrop face (i.e., roughly parallel to fault strike) and are referred to as "parallel-to-fault" permeabilities. A limited number of measurements, collected on an exposed portion of Elmendorf fault A, were oriented perpendicular to the fault plane and are referred to as "normal-to-fault" permeabilities. Measurement error was quantified by recording a minimum of three replicates for each sampling location and by repeating measurements at selected locations one or more times on each sample date. Selected sample locations were also measured on several different sampling dates.

All sampling locations were scraped with a putty knife or trowel and gently brushed to remove weathering effects, such as clay or silt wash, prior to permeability measurement. Measurements were collected at least several days after a rainfall to minimize the confounding effects of moisture content on permeability estimation. Ambient air temperature was measured several times during each sampling date to help correct for changes in air (or gas) viscosity.

TERMINOLOGY

Antonellini and Aydin [1995] offer a conceptual model of fault-zone hydrology based on detailed field study of faults in porous sandstone, material more similar to the sediments we are studying than any other rock studied to date. Their model is founded on the observation that faults in porous sandstone consist of a hierarchy of structures, from isolated deformation bands, to zones of deformation bands and adjacent slip surfaces [Aydin, 1978; Aydin and Johnson, 1978]. This conceptual model is distinctly asymmetric, with zones of deformation bands on the footwall and less deformation in the hanging wall. We have observed structures similar to deformation bands and zones of deformation bands in the fault zones we have studied. We therefore retain the terminology outlined by Aydin [1978] and Aydin and Johnson [1978].

Caine et al. [1996] have included Antonellini and Aydin's [1995] observations and numerous other field, laboratory, and numerical models of flow within and near faults in a general model for fault zone architecture and permeability of faults in rock. They divide fault zones into three basic architectural elements: a low permeability fault core, where most of the deformation is accommodated; a less deformed damage zone, within which permeability is generally enhanced by a relatively high fracture density; and the protolith or host rock, characterized by intermediate permeability values. Two end-member fault types can be distinguished in this model. In one, the core is poorly developed or absent and the damage zone acts as a conduit for fluid flow. In the other, the core is well developed but the damage zone is poorly developed or absent, and the fault acts as a barrier. *Caine et al.* [1996] place *Antonellini and Aydin*'s [1995] data into the latter category, contending that deformation bands, zones of deformation bands, and slip surfaces associated with a given fault can be collectively considered a low permeability fault core. Different combinations of core and damage zone may result in combined conduit-barrier systems.

Heynekamp et al. [this volume] describe a somewhat different conceptual model for fault-zone architecture in poorly lithified sediments, which is based on study of the large-displacement (~600 m) Sand Hill fault in central New Mexico. In faults in poorly lithified sediments, the fault core is bracketed by footwall and hanging wall mixed zones, within which sediments from different structural levels have been tectonically mixed. The mixed zones are highly heterogeneous, and include lens-shaped pods of intact sediments, highly attenuated beds elongate subparallel to the fault, and sediments that have been so thoroughly mixed that evidence of bedding has been destroyed. Sediments in the mixed zone are typically characterized by a tectonic foliation, defined by compositional banding and/or aligned elongate sand grains and clay plates. They are also cut by deformation bands, zones of deformation bands, and slip surfaces.

On either side of the mixed zones are damage zones [*Heynekamp et al.*, this volume]. The boundaries between the mixed and damage zones are defined by minor faults across which displacement exceeded bed thickness. The damage zones are the least deformed of the fault-zone architectural elements, and include broad folds, deformation bands, zones of deformation bands, and minor slip surfaces. The fractures and fracture networks described by *Caine et al.* [1996] in damage zones in rock are absent from these poorly lithified sediments.

FAULT ZONE DESCRIPTIONS

Fault with a clay core: Santa Ana site

The fault with a clay core appears to be the southern tip or a splay of the major intrabasinal Santa Ana normal fault (J. Hawley, personal communication, 1996). At the study site, the Santa Ana fault is estimated to have approximately 10 m of vertical displacement based on the difference in elevation between a sequence of clay and sand beds located in the footwall north of the outcrop studied and a similar sequence in the hanging wall at the outcrop itself. Deformation is accommodated within a zone rather than a single, discrete slip surface (Figure 2a). The fault zone juxtaposes a sequence of fine- to medium-grained sands (footwall) against fine-grained sands and a meter-thick clay bed (hanging wall).

The fault zone includes architectural elements similar to those described in the Sand Hill fault zone [*Heynekamp et al.*, this volume], so our description is framed in terms of these elements. A 1 - 15 cm clay-rich core zone (Fig. 2b) accommodated most of the fault slip. Narrow footwall and hanging wall mixed zones flank the fault core. The width of the combined mixed and core zones is greatest (1.5 m) at the top of the outcrop, then narrows to 15 - 20 cm along the lowest half meter. The clay core steepens in dip where the mixed and core zones narrow. The mixed zones comprise a number of domains that are delineated on the basis of variations in width, color, induration, grain size, and structure (Figure 2b). These sandy domains are roughly lens-shaped, are elongate parallel to the fault zone, and lack evidence of bedding. The orientations of the domain boundaries vary: dips range from 68 to 80°, with an average of ~70°. Isolated deformation bands cut the mixed zones and a zone of deformation bands < 1 cm wide is present within the footwall mixed zone (Figure 2b). The outermost domains can be traced to source beds in the outcrop. The domains therefore are interpreted to have formed as beds were incorporated into the zone, extended parallel to the fault, and disrupted by deformation bands during faulting.

Two white, quasi-tabular zones of closely spaced deformation bands, which vary from 1 to 3 cm wide, separate the mixed from the damage zones (Figure 2b). These narrow zones of deformation bands exhibit smaller average grain sizes and better induration than the sediments to either side. Though relatively homogeneous in appearance for much of their length, the deformation bands locally anastomose around centimeter-scale, lens-shaped pods of macroscopically undeformed sediment.

The footwall and hanging wall sands shown in Fig. 2 lie within the damage zones. Deformation within the damage zones is confined to minor folds (Fig. 2a) and isolated deformation bands (Fig. 2b). The footwall and hanging wall buff-colored, fine-grained sands are similar in both grain size and bedding characteristics. They are well sorted and massively bedded, and grains are sub-rounded. Both are interpreted to be eolian. The red fine- to medium-grained sand in the footwall displays planar bedding as well as small scours and clay envelopes along the erosional contact with the buff fine-grained sand. The red-brown clay in the hanging wall consists primarily of clay with thin

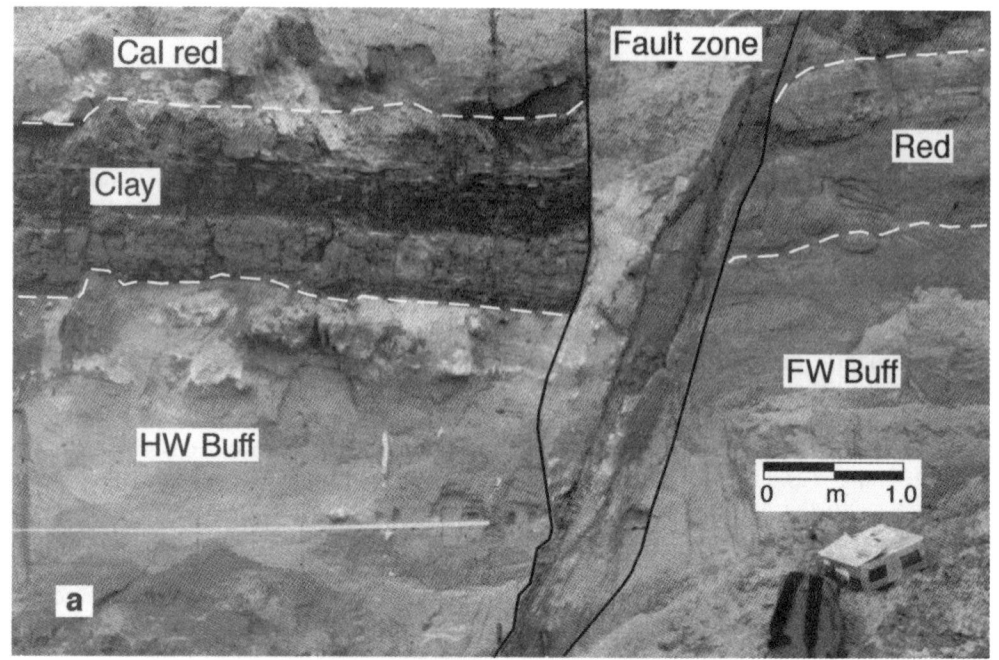

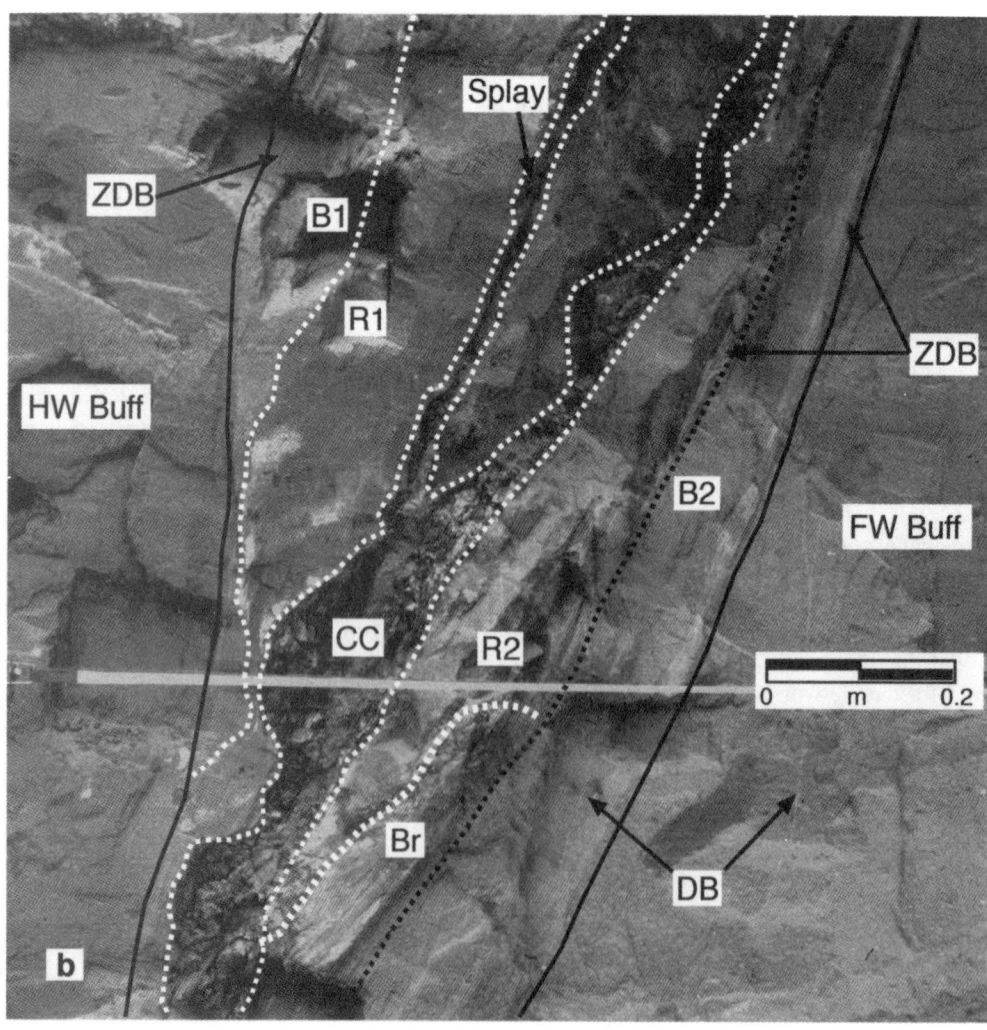

Figure 2. Santa Ana site: (a) Black lines mark fault-zone boundaries; dashed white lines mark bedding planes. Eolian buff fine-grained sand (HW buff) overlain by fluvial red-brown clay bed (Clay) and cemented fine-grained red sand (Cal red) units comprise the hanging wall, left (east) of the fault zone. Footwall includes another eolian buff fine-grained sand (FW Buff) overlain by a fluvial red fine- to medium-grained sand (Red). (b) Close-up of a portion of (a), showing detailed fault structure. Two white, narrow zones of deformation bands (ZDB) bound the fault zone. The clay core (CC), which splays (Splay), separates the footwall mixed zone (B2, R2, and Br domains) from the hanging wall mixed zone (B1 and R1 domains). A third very narrow zone of deformation bands (dashed black line) divides the Br and B2 domains. Isolated deformation bands are evident as thin white lines both within and outside the fault zone (DB). See text for further explanation.

interbedded sands; sandy mottled areas, which may indicate bioturbation, are common. The latter two units are fluvial.

In order to correlate changes in permeability with changes in petrophysical characteristics and structural position, we have named the mixed zone domains according to their color and location within the fault zone. Domains B1 and R1 are located in the hanging wall mixed zone and share a sharp, stair-stepped contact (Figure 2b). Domains R2, Br, and B2 lie in the footwall mixed zone. The B1 and B2 domains are slightly better indurated but similar in color and grain size to the hanging wall and footwall undeformed buff sands from which they were derived. The domains are best indurated adjacent to the core zone (Br, R1, R2) and least indurated at the fault-zone margins (domains B1 and B2). Degree of induration is inversely correlated with grain size; average grain size is least near the core zone (domains Br, R1, R2) and greatest near the fault-zone margins (domains B1 and B2).

Our petrographic data are limited to a few thin sections, but these provide important constraints on interpretations of the permeability data presented in a later section. Table 1 lists mineral abundances, clay size fraction, and both macro- and micro-porosity obtained from point counts of thin sections from the hanging wall undeformed buff sand (Figure 3a) and the sandy domains within the fault zone (e.g., Figure 3b). The most important observations are: 1) The undeformed sand is 5% clay size fraction, whereas the clay size fraction varies from 6 - 27% within the fault zone. 2) Macroporosity varies inversely from clay size fraction. It is 23% in the undeformed sand, but only 3 - 19% in the sands in the fault zone. 3) Where macroporosity is highest, microporosity is lowest, and vice-versa, suggesting the operation of a process that decreased macroporosity and increased microporosity within the fault zone.

We have looked at microstructures within and along the boundaries between the hanging wall damage and mixed zones, B1 and R1, R2 and Br, and Br and B2. Within the zones of deformation bands that bound the mixed zones, elongate grains were rotated from sub-horizontal to roughly parallel to the diffuse deformation band margins. Grains are also aligned subparallel to all of the domain boundaries examined. Sub-horizontal grains within B1 were rotated into parallelism with the B1 / R1 domain boundary.

Elongate grains in the Br domain are particularly strongly aligned, forming a well developed foliation subparallel to the fault (Figure 3b). Sand grains appear to be more angular in domains (such as Br and R2) with more clay-sized material (Figure 3, compare a and b). The low porosity, tight packing, angularity of grains, and foliation collectively indicate significant deformation within the Br domain. Nearly all of the pore space is occluded by clay-sized materials in the footwall domains closest to the core zone (R2 and Br), whereas the domains at the margins (B1 and B2) and the R1 domain display less pore occlusion (Figure 3b). Pore occlusion by the clay size fraction is much less in the hanging wall's undeformed buff sand than in any of the fault-zone domains.

If we consider that the B1 domain of the fault zone was derived from the adjacent buff sand, other patterns emerge. The percentages of feldspar and rock fragments in B1 have decreased relative to the parent sand, the clay size fraction has increased slightly, and macroporosity has decreased (Table 1). We explore these patterns further at the second study site.

In summary, faulting at the Santa Ana site produced damage zones that exhibit little deformation, mixed zones, and a clay core. We anticipate that the clay core would act as a low permeability barrier to fully saturated flow [cf. *Caine et al.*, 1996]. We would also expect that the sandy domains within the mixed zones, which have lower porosity and higher clay size fraction than the adjacent hanging wall undeformed sand, would have lower permeabilities than the adjacent undeformed sands.

Fault without clay core: Elmendorf site

Two salient differences between the Santa Ana and Elmendorf sites are immediately evident (cf. Figures 2 and 4). First, displacement at the Elmendorf site is less than the thickness of beds, so we do not see juxtaposition of lithologically distinct layers. Second (and probably related to the first point), the Elmendorf site does not exhibit either mixed zones or a clay-rich core. The Elmendorf outcrop's primary facies consists of a fine-grained, well sorted, poorly lithified, buff-colored sand. Calcareous cemented beds and concretions form several horizontal planes that serve as markers for estimating vertical displacement.

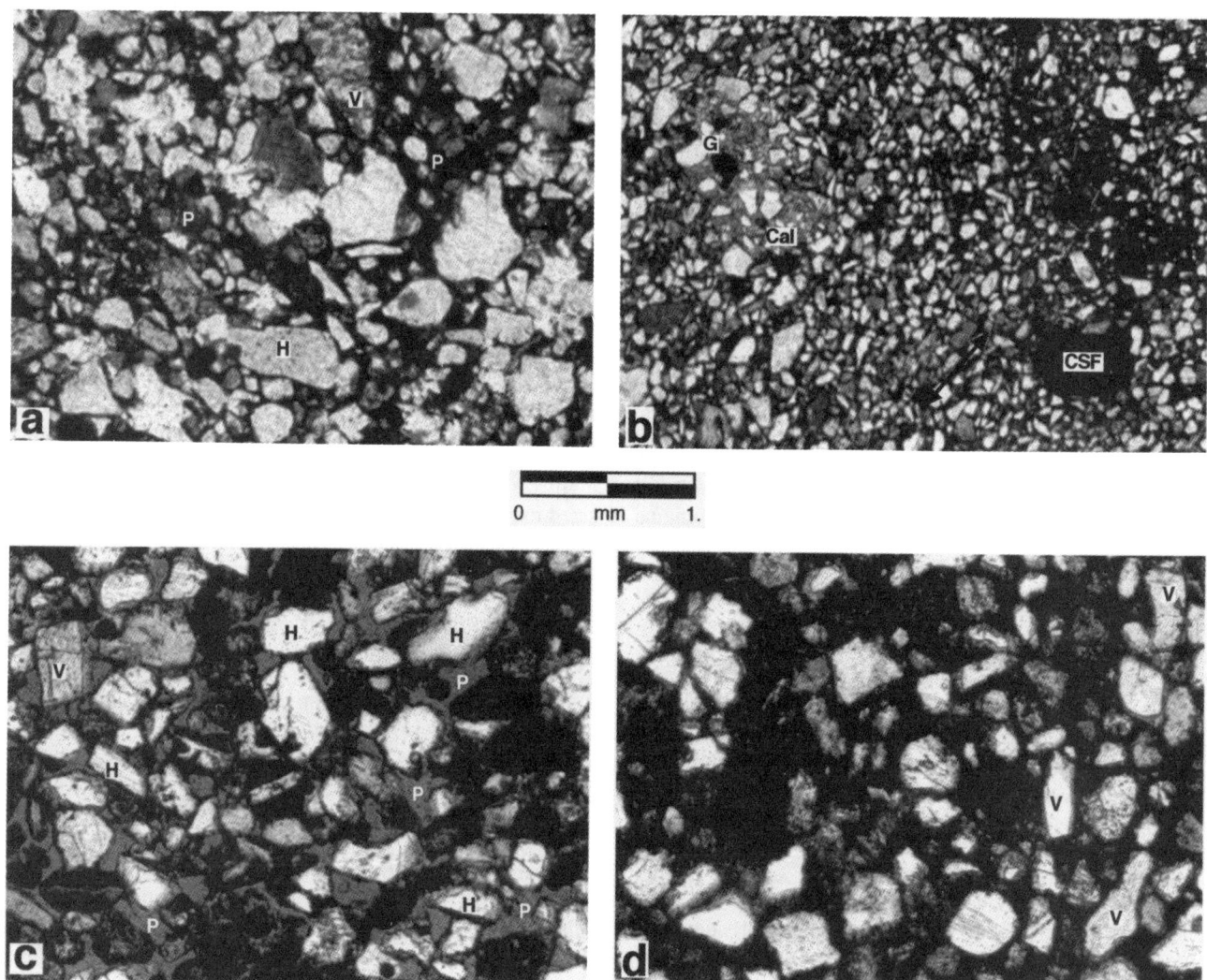

Figure 3. Photomicrographs of undeformed and deformed sand transformed to grayscale images. (a) Santa Ana site: undeformed buff sand from hanging wall. Most elongate grains are subhorizontal (H), but some are nearly vertical (V). Relatively large pores (P) are filled with blue-dyed epoxy with very little clay-sized material present (partially obscured by grayscale transform). (b) Santa Ana site: Br domain from footwall mixed zone. There is a strong grain preferred orientation of elongate grains from upper right to lower left in this image (arrow). This foliation is roughly fault-parallel. The grain size of this domain is much finer than in undeformed sand shown in (a). Macroporosity is just 3%, compared to 23% in undeformed sand. Note significant clay size fraction occluding pores. Rare calcite cement (Cal) is visible surrounding a dissolution ghost grain (G). (c) Elmendorf site: hanging wall undeformed sand. Elongate grains are subhorizontal (H), roughly parallel to tangential cross beds (Fig. 4b). The large, blue-epoxy filled pores contain very little clay-sized material (the grayscale transform obscures some of the small-scale details). (d) Elmendorf site: fault A. Elongate grains are nearly vertical (V), subparallel to the fault. Pores are interconnected, but are partly to mostly occluded with clay-sized material (nearly all of the dark intergranular material represents clay-sized material under the grayscale transform, which obscures the very small amounts of blue-dyed epoxy present between grains). Grain and pore sizes are small relative to parent sand shown in (a).

Table 1. Mineral Abundance, Clay Size Fraction, and Porosity: Santa Ana Site

Constituent	Hanging wall undeformed buff sand	Fault zone domains				
		B1	R1	Br	R2	B2
Detrital quartz	12	20	24	18	13	21
Potassium feldspar	9	12	10	7	10	9
Plagioclase	25	20	18	18	22	18
Total rock fragments (RF)	21	18	16	13	13	21
Volcanic RF	5	<1	0	2	1	1
Metamorphic RF	0	0	<1	0	0	0
Sedimentary RF	2	2	2	3	1	3
Undifferentiated RF	14	15	14	8	11	17
Clay-size fraction	5	6	8	27	22	9
Porosity						
Macroporosity	23	19	16	3	7	14
Microporosity	5	5	8	14	9	9
Total porosity	28	24	24	17	16	23

All values are abundance (% of sample volume occupied by the constituent) and are calculated by dividing the total number of observations for each zone (or domain) into the number of observations made for each constituent in that zone (or domain).

Three high-angle (65-75°) normal faults -- A, B, and C — delineate blocks of relatively undeformed sand: domains A-B and B-C (Figure 4a). These domains display little deformation beyond minor folds and a few deformation bands, and therefore constitute damage zones associated with the faults.

The core zone of each fault consists of a narrow, white, quasi-tabular zone of closely spaced deformation bands (Figure 4b). These zones vary in width from 5 to roughly 15 mm and, like analogous structures at the Santa Ana site, display reduced grain size but increased induration relative to the surrounding sand. We refer to these structures as faults for simplicity, but it is important to recall that they are zones of deformation bands. A zone of more widely spaced deformation bands, also better indurated than the parent sand, fans out from fault B roughly half way up the outcrop (Figure 4b). This zone reaches its greatest width of ~25 cm near the bottom of the outcrop. The 10 - 15 deformation bands within the zone, each 2 - 3 mm wide, accommodated a total vertical displacement of 20 - 25 cm. Faults A and B accommodated 54 cm and 35 - 40 cm of vertical slip, respectively. Vertical displacement across fault C could not be estimated because no marker beds or other correlative features were observed.

Petrographic examination indicates that the undeformed sand is characterized by a sub-horizontal orientation of elongate clasts, roughly parallel to bedding (Figure 3c). In contrast, elongate grains within fault A are typically sub-vertical in orientation, and both grain and pore size are reduced relative to undeformed sand (Figure 3d). The trends suggested by point-count data from the Santa Ana site are more evident here: volcanic rock fragments and potassium feldspar each constitute a much smaller proportion of the fault than of the parent undeformed sand (Table 2, columns 2 and 3); quartz and plagioclase abundances exhibit negligible change. Clay-sized materials form only a minor portion (6%) of undeformed sand in the hanging wall, yet represent nearly one quarter (23%) of fault A (Table 2, columns 2 and 3). Macroporosity constitutes 20% of the undeformed sand versus only 7% of fault A, whereas microporosity constitutes twice as much of fault A as of the undeformed sand (Table 2, columns 2 and 3). Larger pores are almost entirely occluded and larger grains are surrounded by clay-sized materials within fault A (Figure 3d). Total porosity (macro + microporosity) represents 25% of undeformed sand but only 17% of fault A (Table 2, columns 2 and 3). Recalculating mineral and lithic clast abundances of undeformed and deformed sand by excluding porosity eliminates the confounding influence of porosity differences, assuming mass is conserved. These calculations further support the conclusion that there are large differences in the relative frequencies of potassium feldspar, volcanic lithic fragments, and clay size fraction as well as plagioclase, whereas the relative frequency of quartz in fault A differs little from that observed in the undeformed sand (Table 2, columns 4 and 5).

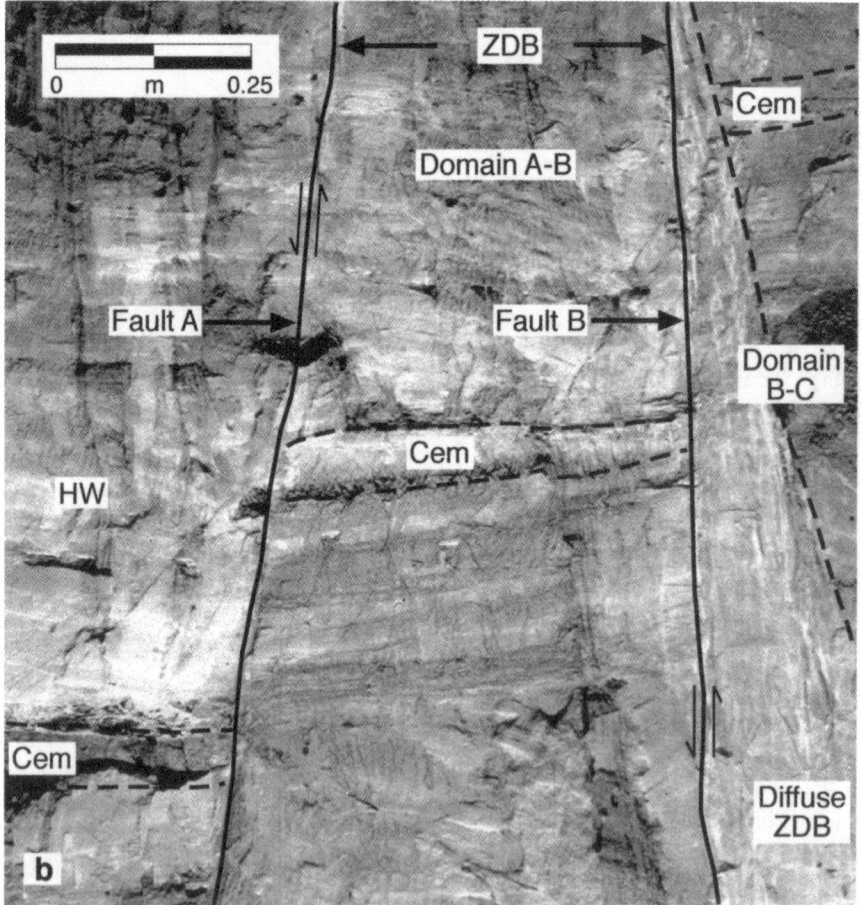

Figure 4. Elmendorf site: (a) Outcrop of fine-grained sand, locally with large trough cross bedding (X-beds). Faults A, B, and C separate domains of relatively undeformed sand. (b) Close-up of faults A and B shows details of structure. Faults A and B comprise 0.5-1.3 cm wide zones of deformation bands (ZDB). Cemented layers (Cem) record vertical displacement across zones of deformation bands. Numerous white sub-vertical deformation bands are visible in the wide zone of deformation bands (Diffuse ZDB), located in the footwall of fault B. In situ air miniper- meameter measurements were made in sand in the hanging wall of fault A (HW). Shallowly dipping cross beds are evident in this area.

Table 2. Mineral Abundance, Clay Size Fraction, and Porosity: Elmendorf Site

Constituent	Abundance (%) [a]		Abundance without porosity (%) [b]	
	Hanging wall undeformed sand	Fault A	Hanging wall undeformed sand	Fault A
Detrital quartz	14	15	19	18
Potassium feldspar	8	4	10	5
Plagioclase	20	19	27	22
Total rock fragments (RF)	27	21	36	26
Volcanic RF	10	5	13	6
Metamorphic RF	0	<1	0	0
Sedimentary RF	4	3	5	4
Undifferentiated RF	13	13	17	16
Clay size fraction	6	23	8	28
Porosity				
Macroporosity	20	7		
Microporosity	5	10		
Total porosity	25	17		

[a] Abundance is defined as the percentage of sample volume occupied by the constituent and is calculated by dividing the total number of observations for each zone (i.e., undeformed or fault) into the number of observations made for each constituent in that zone.
[b] Abundance without porosity is calculated as above except that the total number of observations for each zone is reduced by the number of observations for porosity (both micro- and macroporosity).

PERMEABILITY

Descriptive statistics

More than 1000 replicate permeameter measurements were collected at 195 sampling locations across the Santa Ana site and at 67 locations across the Elmendorf site. Nearly all measurements were made parallel to the faults (i.e., normal to the outcrop face), with the majority taken within the undeformed zones at each site, carefully avoiding isolated deformation bands. A total of 23 normal-to-fault replicate measurements at five locations in the Elmendorf fault zone were also collected. Data reduction yielded a mean permeability for each of the 145 undeformed and 48 fault zone sampling locations at the Santa Ana site and each of the 47 undeformed zone and 20 fault zone sampling locations at the Elmendorf site (Table 3).

Log permeability distributions for the undeformed sands display very narrow ranges and very small coefficients of variation relative to the log permeability distributions for the fault zone (Table 3 and Figure 5). Sample cumulative frequency curves for undeformed log permeability closely match cumulative normal distribution frequency curves computed using the observed sample means and variances (Figure 6) and form straight lines when plotted on probability-scale paper, indicating the permeability of undeformed sands is log normal. Kolmogorov-Smirnov test statistics further support the hypothesis of log normal per-

meability in the undeformed zones at a 95% confidence level [*Sigda*, 1997].

Fault-zone permeability measurements do not follow log normal distributions and are strongly bimodal (Figures 5 and 6). The sample cumulative frequency curves for fault-zone log permeability do not resemble curves expected from log normal distributions with the same moments (Figure 6) nor do they form straight-line plots on probability-scale paper. The two modes correspond to 1) the range of low permeability values associated with fault-zone deformation structures (e.g., domains within mixed zones and both broad and narrow zones of deformation bands) and 2) the range of higher permeability values associated with pods of less deformed sediments. Permeability values for the macroscopically undeformed fault zone sediments differ little from the mean permeability values for undeformed sands (Figure 6).

Permeability and deformation: Santa Ana site

Santa Ana site parallel-to-fault permeability measurements show a two order of magnitude span between the lowest mixed zone permeability and the highest undeformed sand permeability (Table 3). Permeability in the footwall and hanging wall mixed zones ranges from 0.1 to 7.6 darcies whereas undeformed sand permeability ranges from 7 to 13 darcies in the hanging wall buff sand, 3 to 14 darcies in the footwall buff sand, and 16 to 46 darcies in the

Table 3. Summary Statistics for *in situ* Permeability

Outcrop/Zone	Mean (darcy)	CV [a]	Min (darcy)	Median (darcy)	Max (darcy)	n	Mean grain size (mm)	Sorting
Santa Ana site								
Undeformed sands								
Hanging wall buff	10.4	0.1	7.7	10.4	13.6	96	0.125 - 0.25	well
Foot wall buff	9.4	0.1	3.2	9.8	14.6	40	0.125 - 0.25	well
Foot wall red	26.5	0.1	16.2	27.9	46.1	9	0.50 - 1.00	moderate
Fault zone domains								
B1	6.8	0.1	5.3	6.6	7.6	9	0.125 - 0.25	well
R1	3.2	0.4	0.1	3.5	4.8	19	NM	moderate
Br	0.4	0.6	0.1	0.4	1.0	9	NM	poor
R2	1.2	1.0	0.3	0.8	3.9	9	NM	poor
B2	2.0	NC [b]	1.9	NC	2.2	2	0.125 - 0.25	well
Elmendorf site								
Hanging wall undeformed sand	7.9	0.3	4.2	7.7	12.3	47	0.125 - 0.35	well
Fault A	0.3	0.3	0.2	0.3	0.5	4	NM	poor
Normal to fault A	0.4	1.0	0.04	0.3	1.0	5	NM	poor
Fault B	0.2	NC	0.2	NC	0.3	2	NM	poor
Wide zone of deformation bands	5.8	0.4	4.4	4.8	8.4	3	NM	moderate
Domain A-B	6.7	0.5	0.6	7.3	9.8	6	0.125 - 0.35	well

All permeability values were measured parallel-to-fault unless stated otherwise.

[a] CV = coefficient of variation = (standard deviation)(mean)$^{-1}$

[b] NC = Not calculated   NM = Not measured

footwall red sand (Table 3). No normal-to-fault measurements were collected.

The mixed zones vary widely in parallel-to-fault permeability, but average permeability is greatest in domains at the fault-zone margin and least in domains adjacent to the core zone. Average domain-specific permeability is lowest for the Br domain, coincident with the greatest induration (Table 3) and smallest grain size.   The footwall mixed zones (R2 and B2) are less permeable than the hanging wall mixed zones (R1 and B1), yet the measured permeabilities for footwall undeformed sands are either just as permeable (footwall buff mean permeability = 9.4 darcies) or more permeable (footwall red mean permeability = 26.5 darcies) than the hanging wall undeformed buff sand (mean permeability = 10.4 darcies).   The *in situ* permeameter measurements support our expectations from field and petrographic observations that permeabilities within the mixed zones would be less than those within the undeformed sands.

Permeability and deformation: Elmendorf site

At the Elmendorf site, parallel-to-fault permeability ranges from 4 - 12 darcies within the undeformed sand to 0.2 - 10 darcies in the fault zones, yielding a span of roughly two orders of magnitude between the lowest fault-zone permeability and the highest undeformed permeability (Table 3).   Normal-to-fault measurements at fault A vary from 0.04 - 1.0 darcies, increasing the span between undeformed and deformed sand permeability to roughly three orders of magnitude.

The lowest permeabilities are found within the narrow zones of deformation bands that constitute the fault cores: fault A, which accommodated the greatest amount of vertical displacement, has a parallel-to-fault permeability range of 0.2 to 0.5 darcies whereas the two measurements made on fault B, which is thinner and less well defined than fault A, were both on the order of 0.2 darcies (Table 3). Measurements made normal to fault A are lower still.   Perme-

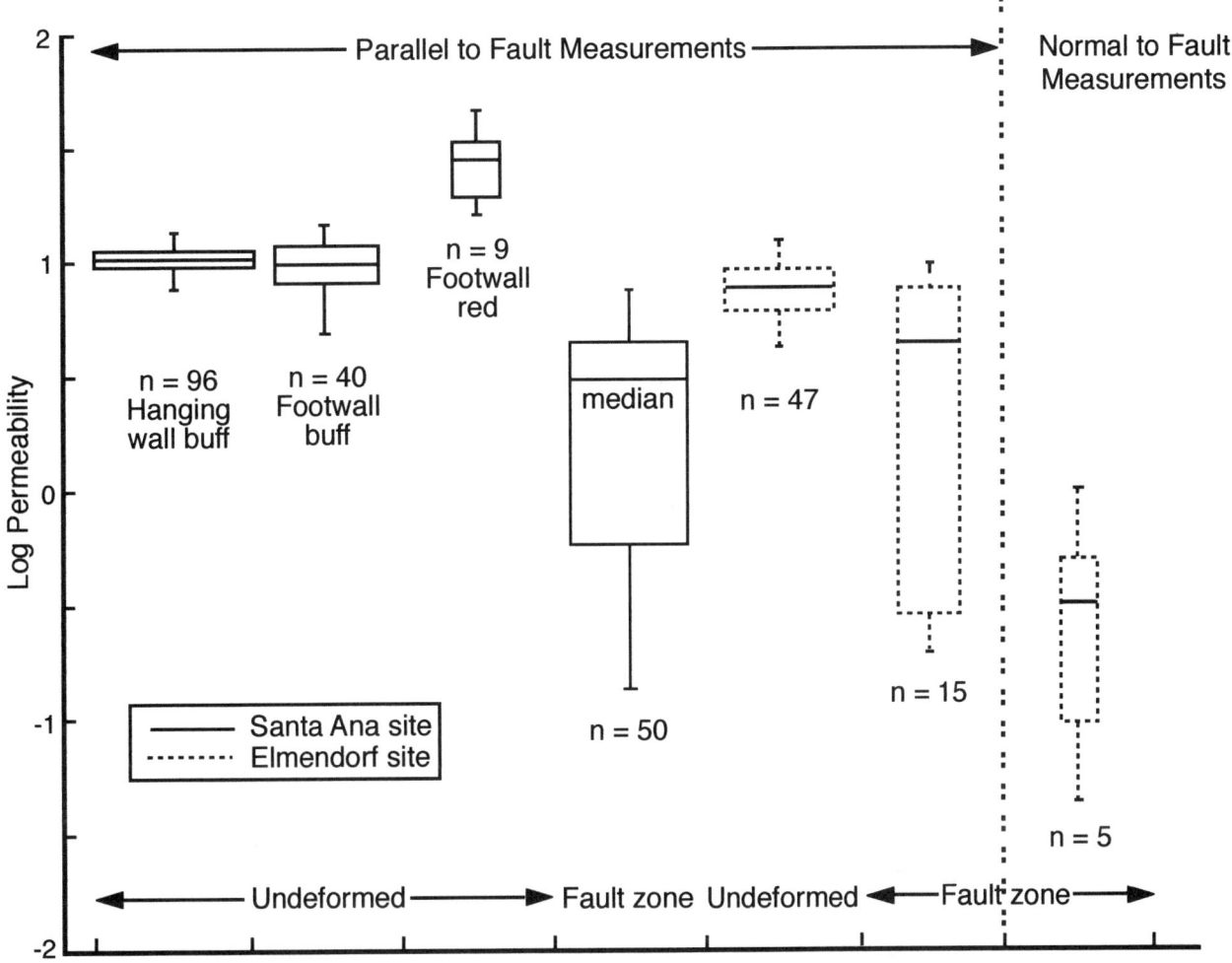

Figure 5. Box plots of log permeability. Boxes encompass 50% of data; bars show extrema. Box widths are proportional to the square root of n for each group.

ability measurements fell between 4 and 8 darcies (Table 3) within the widely spaced zone of deformation bands, which accommodated less vertical displacement than the two faults (Figure 4).

DISCUSSION

Faulting and petrophysical changes: Santa Ana site

Displacement exceeding bed thickness at the Santa Ana site created outcrop-scale changes in the spatial distribution of permeability by juxtaposing units with different permeabilities. This juxtaposition took two forms: 1) a classic juxtaposition seal was created where the thick hanging wall clay bed was faulted against the coarse-grained fluvial sands in the footwall, and 2) sandy units were incorporated into and deformed within the fault zone, forming tabular to lens-shaped domains within narrow mixed zones [cf. *Heynekamp et al.*, this volume]. The clay core, which likely formed by extension of clay bed(s) within the fault zone, probably has a permeability many orders of magnitude lower than any of the mixed zone domains under water-saturated conditions. As such, it is a small-scale analog to the clay smear seals created by faults in the interbedded shales and sandstones found in many hydrocarbon reservoirs [*Smith*, 1966 and 1980; *Berg and Avery*, 1995].

Deformation also produced smaller-scale changes in the spatial distribution of permeability as evidenced by the permeability variability between and within the mixed-zone domains. Petrographic data indicating a higher clay size fraction and lower porosity within these domains than in the hanging wall undeformed buff sand suggest that some of this variability may be related to processes resulting in grain-size reduction and grain reorganization. This inter-

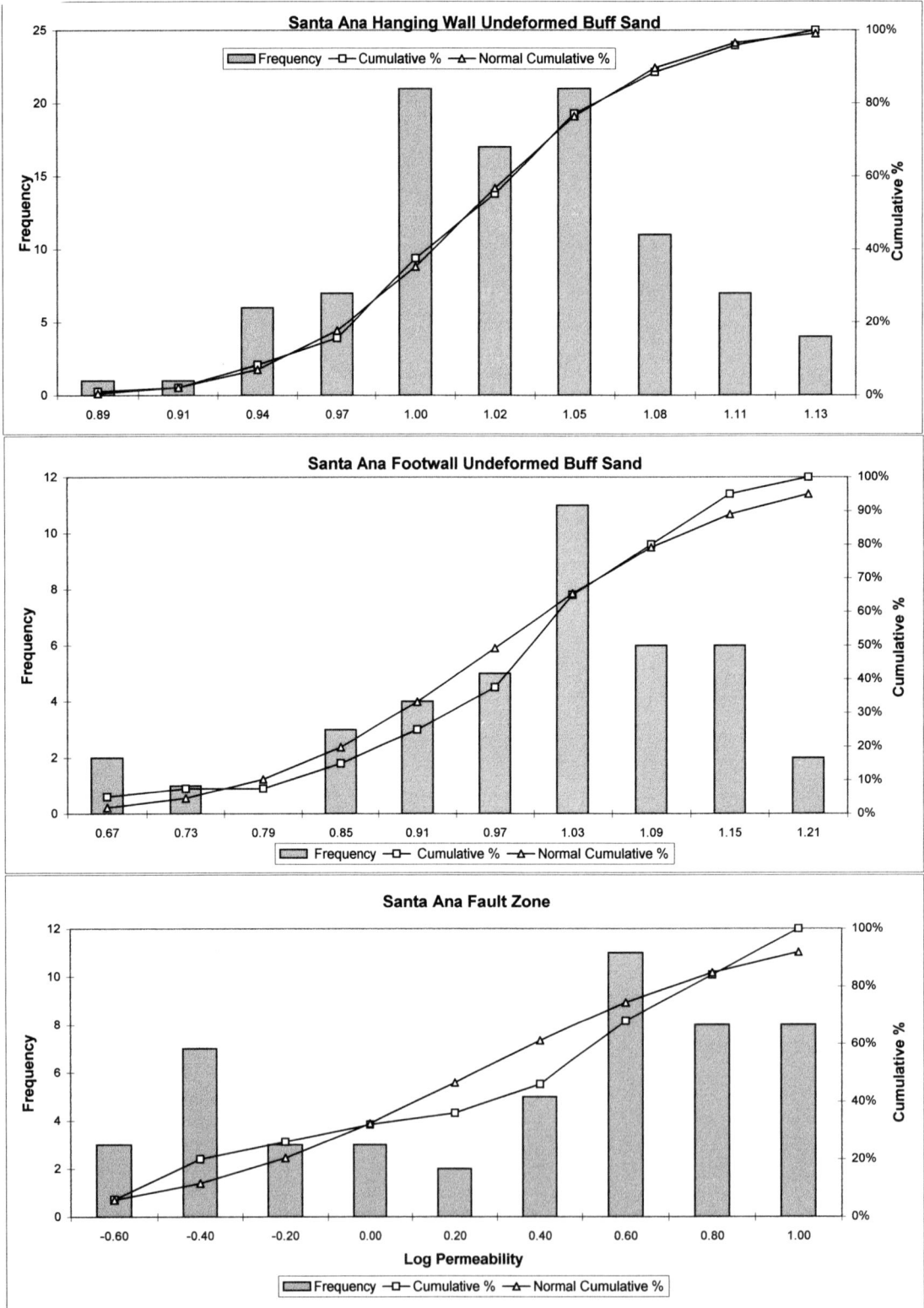

Figure 6. Histogram plots of log permeability. All measurements were made parallel to the fault. Normal cumulative % shows the cumulative frequency (in %) for a normal distribution with the same mean and variance as the observed log permeability data.

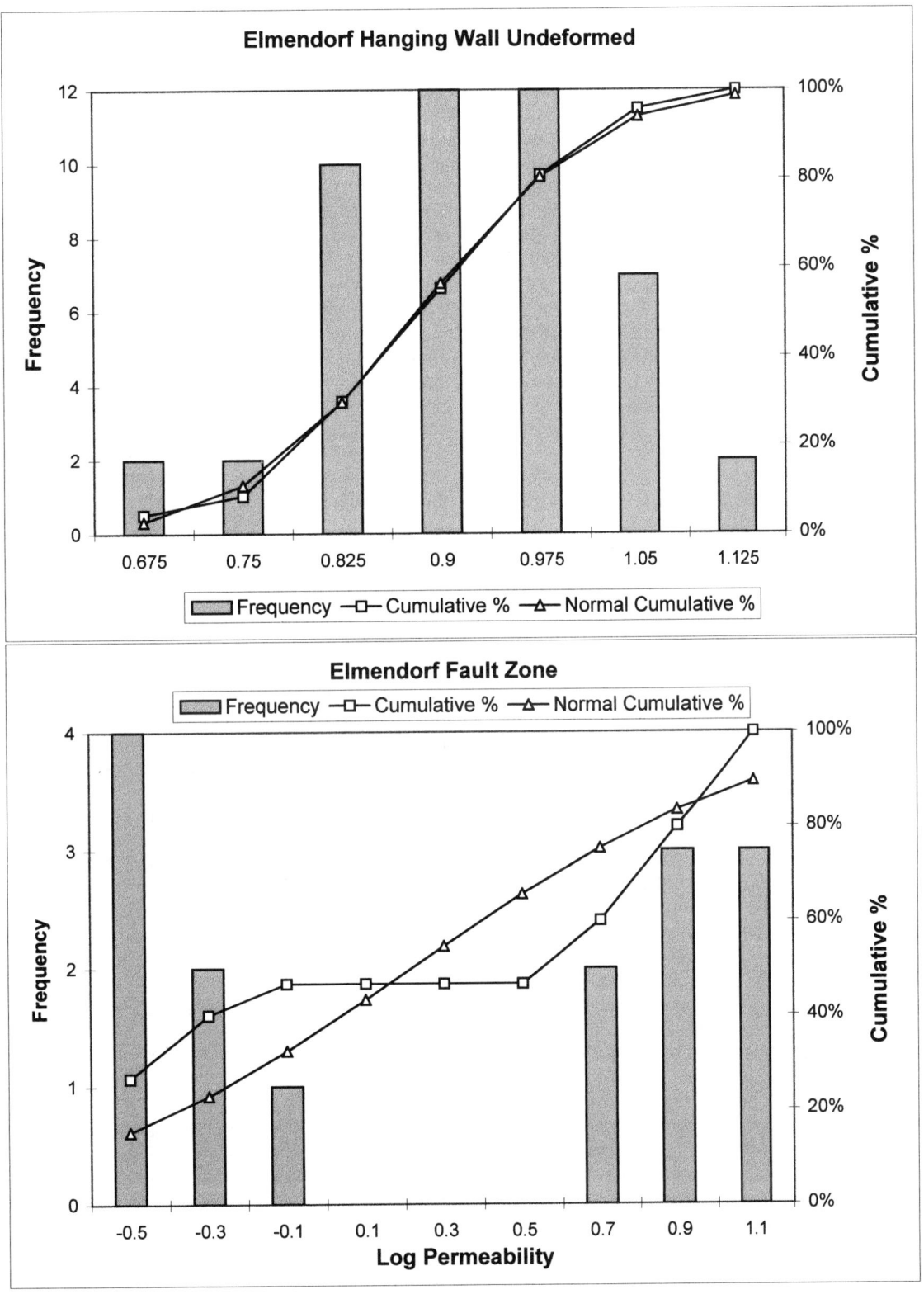

Figure 6. Continued.

pretation is supported by the presence of a roughly fault-parallel foliation along domain boundaries and throughout much of the domains. We note, however, that a conclusive determination of deformation mechanisms at the Santa Ana site is not possible with our limited petrographic data and lack of information about source material for some of the mixed-zone domains. It is, for example, possible that low permeability undeformed sediments not visible in outcrop were mixed into the fault zone and thereby contributed to the low permeability observed in the Br and R2 domains. Whether the clay size fraction within the mixed-zone domains was developed by grain-size reduction or by tectonic mixing of different sediments within the fault zone, the net result is an inverse association between clay size fraction and permeability. An effective permeability for one-dimensional flow across the entire fault zone is controlled by the thickness of those domains with large clay size fraction and perhaps reduced even more by formation of the vertically continuous clay core.

Faulting and petrophysical changes: Elmendorf site

At the Elmendorf site, the largest reduction in permeability and macroporosity relative to the undeformed parent sand is associated with the narrow zones of deformation bands that constitute the macroscopically visible faults. Incorporation and mixing of clay-rich beds into the fault zone could not have been significant because clay laminae are rare and clay beds are absent.

Petrographic data provide evidence of deformation mechanisms at this site, where we can directly relate deformed sands to their undeformed protoliths. We observe 1) elongate grains were rotated from sub-horizontal to roughly fault-parallel through deformation, 2) faulting resulted in a substantial increase in clay size fraction and concomitant reduction in macroporosity, and 3) point-count data indicate a reduction in abundance of feldspar and (especially) lithic fragments following deformation. These data collectively indicate selective grain-size reduction of lithic fragments and feldspar grains relative to quartz grains, resulting in an increase in clay size fraction within deformed zones. Pore-size reduction is indicated by the reduction in both macroporosity and total porosity and by grain reorientation. We therefore interpret the deformation bands to have formed through cataclasis. To summarize, cataclasis and associated grain reorganization of weaker framework grains (e.g., feldspars and volcaniclastic rock fragments) during faulting produced large increases in clay size fraction and microporosity as well as large decreases in average grain and pore sizes and macroporosity. These, in turn, caused the several order of magnitude reduction in fault permeability. It is possible that diagenesis may also have contributed to permeability reduction through alteration of the clay size fraction subsequent to deformation.

Along fault A, normal-to-fault permeability values are significantly lower than nearby parallel-to-fault measure-ments. We identify two factors which are responsible for this disparity. First, and most important, paired parallel-to-fault and normal-to-fault measurements were collected along a length of Fault A where it is bounded by air on the hanging wall side (hanging wall sands were eroded away). Any parallel-to-fault measurement will be influenced by the negligible resistance to flow (equivalent to a very high permeability) from the fault core's open air boundary just a few millimeters from the tip seal (CFMP tip seal inner orifice diameter = 2.5 mm, fault core width = 7 - 13 mm). We expect the measured parallel-to-fault permeability would be lower if the fault core were bounded on both sides by sand. Second, we expect that fault-induced foliation would lead to permeability anisotropy: permeameter measurements taken normal to the foliation should be lower than those taken parallel to the foliation.

Normal-to-fault measurements may still underestimate the permeability of fault A because CFMP depth of investigation may have exceeded the thickness of the fault cores. Depth of investigation for mini-permeameters has been defined as some multiple of some part of the tip seal: as equaling $4r_i$ based on theory [*Goggin et al.*, 1988] or as equaling $1.7r_o$ for $r_o/r_i = 2$ from empirical observations [*Suboor and Heller*, 1995]. If the CFMP depth of investigation falls between the estimates (5 mm and 11 mm) based on these two relationships, then the CFMP normal-to-fault measurements are mostly sensitive to permeability in the 7 - 13 mm thick zone of deformation bands. If, however, the CFMP interrogates a significant thickness of the high permeability (10 darcies) footwall sands adjacent to the fault, then the fault-core permeability may be less than that indicated by the normal-to-fault measurements shown in Table 3.

The mini-permeameter measurements demonstrate that fault-zone deformation can reduce permeability by up to three orders of magnitude even where vertical displacement is less than bed thickness and there is no potential for development of a clay-rich core or mixed zones.

IMPLICATIONS

The petrophysical changes resulting from small-displacement faulting of poorly lithified sediments could engender potentially significant hydrologic impacts. Given sufficient lengths and spatial density, such small-displacement faults could act as permeability barriers to saturated single-phase flow, perhaps even compartmentalizing aquifers or hydrocarbon reservoirs at small scales. Furthermore, pore size distributions altered by cataclasis, grain boundary sliding, grain reorganization, and diagenesis may cause fault zones to act as capillary barriers to non-wetting fluids, like oil, and capillary conduits to wetting fluids, like water. In the latter case, small-displacement faults could act as fast paths for contaminant transport through the vadose zone to the water table under the high matric potentials (water tensions) common in the arid Southwest.

CONCLUSIONS

Faults in poorly lithified Santa Fe Group sediments exhibit reduced macroporosity and average grain sizes, significantly increased clay size fraction relative to undeformed sediments, and realignment of grains from bedding-parallel to roughly fault-parallel orientations. We have demonstrated that these changes result from cataclasis and grain reorganization, and can significantly reduce saturated permeability, even in faults where displacement is less than bed thickness. Where displacement is greater than bed thickness (e.g., the Santa Ana site), the potential exists for developing both mixed and clay-rich core zones, which can further reduce fault-zone permeability. In general, the fault-zone elements with the greatest abundance of clay-sized material and microporosity have the lowest permeability, macroporosity, and average grain-size values.

The clay size fraction within the Elmendorf fault zone, where the total vertical displacement across three faults is much less than bed thickness, increased nearly six-fold relative to that of undeformed sand, despite the absence of clay units. The large increases in clay size fraction and microporosity and large decrease in macroporosity and average grain size were caused by cataclasis of weaker framework grains (feldspars and volcanic rock fragments) and possibly diagenesis. The range of individual parallel-to-fault permeability measurements for undeformed sand and the Elmendorf fault zone spans two orders of magnitude; normal-to-fault permeability measurements increase the span to roughly three orders of magnitude.

The impact of small-displacement faults on fluid flow through poorly lithified basin-fill sediments has not been previously considered by hydrogeologists, perhaps because these faults are too small to be included on geologic maps at most scales. Our results demonstrate that such faults impede cross-fault fluid flow under saturated conditions. The deformational and diagenetic processes responsible for altering pore-size distributions within such faults may also cause them to form preferential flow paths for wetting fluids under unsaturated conditions, as well as capillary barriers to the flow of non-wetting fluids.

Acknowledgments. This research was funded in part by National Science Foundation grant EAR-9614385 as well as the U.S. Department of Energy and Sandia National Laboratories under the guidance of James T. McCord. We gratefully acknowledge Dave Love and John Hawley of the New Mexico Bureau of Mines and Mineral Resources for their enthusiastic assistance; Vince Tidwell (Sandia National Laboratories) for providing his field mini-permeameter; William Haneberg (New Mexico Bureau of Mines and Mineral Resources) for many helpful discussions; and the staff at the Bosque del Apache National Wildlife Refuge (San Antonio, NM) for access to the Elmendorf site. We are also grateful for very helpful reviews by J. Matthew Davis and Larry W. Lake.

REFERENCES

Antonellini, M.A., A. Aydin, and D.D. Pollard, Microstructure of deformation bands in porous sandstones at Arches National Park, Utah, *J. Struct. Geol.*, 16, 941-959, 1994.

Antonellini, M. and A. Aydin, Effect of faulting on fluid flow in porous sandstones: petrophysical properties, *AAPG Bull.*, 78, 355-377, 1994.

Aydin, A. and A.M. Johnson, Development of faults as zones of deformation bands and as slip surfaces in sandstone, *Pure Applied Geophys.*, 116, 931-942, 1978.

Aydin, A., Small faults formed as deformation bands in sandstone, *Pure Applied Geophys.*, 11, 913-930, 1978.

Berg, R.R. and A.H. Avery, Sealing properties of Tertiary growth faults, Texas Gulf Coast, *AAPG Bull.*, 79, 375-393, 1995.

Caine, J.S., J.P. Evans, and C.B. Forster, Fault zone architecture and permeability structure, *Geology*, 24, 1025-1028, 1996.

Carter, K.E. and C.L. Winter, Fractal nature and scaling of normal faults in the Española Basin, Rio Grande rift, New Mexico: Implications for fault growth and brittle strain, *J. Struct. Geol.*, 17, 863-873, 1995.

Cather, S.M., R.M. Chamberlin, C.E. Chapin, and W.C. McIntosh, Stratigraphic consequences of episodic uplift in the Lemitar Mountains, central Rio Grande rift, in *Basins of the Rio Grande rift: Structure, stratigraphy, and tectonic setting*, edited by G.R. Keller and S.M. Cather, *Geol. Soc. Amer. Spec. Paper 291*, pp. 157-170, 1994.

Chapin, C.E. and S.M. Cather, Tectonic setting of the axial basins of the northern and central Rio Grande rift, in *Basins of the Rio Grande rift: Structure, stratigraphy, and tectonic setting*, edited by G.R. Keller and S.M. Cather, *Geol. Soc. Amer. Spec. Paper 291*, pp. 5-26, 1994.

Chapin, C., R. Chamberlin and J. Hawley, Socorro to Rio Salado, in *Guidebook to Rio Grande rift in New Mexico and Colorado*, edited by J. Hawley, New Mexico Bureau of Mines and Mineral Resources Circular 163, pp. 121-134, 1978.

Davis, J.M., A conceptual sedimentological-geostatistical model of aquifer heterogeneity based on outcrop studies, Ph.D. thesis, New Mexico Tech, Socorro, New Mexico, 135 pp., 1994.

Davis, J.M., J.L. Wilson, and F.M. Phillips, A portable air-minipermeameter for rapid in situ field measurements, *Ground Water*, 32, 258-266, 1994.

Davis, J.M., R.C. Lohman, F.M. Phillips, J.L. Wilson, and D.W. Love, Architecture of the Sierra Ladrones Formation, central New Mexico: Depositional controls on the permeability correlation structure, *Geol.Soc. Amer. Bull.*, 105, 998-1007, 1993.

Davis, S.N. and R.J.M. DeWiest, *Hydrogeology*, New York, John Wiley, 463 pp., 1966.

Dutcher, L.C. and A.A. Garrett, Geologic and hydrologic features of the San Bernardino area, California, *U.S. Geol. Surv. Water-Supply Paper 1419*, 114 pp., 1963.

Edwards, H.E., A.D. Becker, and J.A. Howell, Compartmentalization of an aeolian sandstone by structural heterogeneities: Permo-Triassic Hopeman Sandstone, Moray Firth, Scotland, in *Characterization of fluvial and aeolian reservoirs*, edited by C.P. North and D.J. Prosser, *Geol. Soc. (London) Spec. Publ. 73*, pp. 339-365, 1993.

Fowles, J. and S. Burley, Textural and permeability characteristics of faulted, high porosity sandstones, *Mar. Petrol. Geol.*, 11, 608-623, 1994.

Gibson, R.G., Fault-zone seals in siliciclastic strata of the Columbus Basin, offshore Trinidad, *AAPG Bull.*, 78, 1372-1385, 1994.

Goggin, D.J., R. Thrasher, and L.W. Lake, A theoretical and experimental analysis of minipermeameter response including gas slippage and high velocity flow effects, *In Situ*, 12, 79-116, 1988.

Goodwin, L. and W.C. Haneberg, Deformational fabrics and inferred permeability of faulted sands from the Rio Grande rift, New Mexico, *Geol. Soc. Amer. Abstracts with Programs, 28*, A-255, 1996.

Haneberg, W.C., Steady state groundwater flow across idealized faults: *Water Resour. Res.*, 31, 1815-1820, 1995.

Hawley, J., C. Chapin and G.R. Osburn, Elephant Butte Reservoir to Socorro, in *Guidebook to Rio Grande rift in New Mexico and Colorado*, edited by J. Hawley, New Mexico Bureau of Mines and Mineral Resources Circular 163, pp. 91-110, 1978.

Hawley, J. and C.S. Haase, Hydrogeologic framework of the northern Albuquerque Basin, *Open-File Report 387*, 165 pp., New Mexico Bureau of Mines and Mineral Resources, 1992.

Hawley, J., C. Haase, and R. Lozinsky, An underground view of the Albuquerque Basin, in *The water future of Albuquerque and middle Rio Grande basin*, edited by C. Ortega-Klett, New Mexico Water Resources Research Institute Technical Report 290, pp. 37-55, 1995.

Heynekamp, M.R., Controls on fault-zone architecture and fluid flow in poorly consolidated sediments: the Sand Hill fault, central New Mexico, Masters thesis, New Mexico Tech, Socorro, New Mexico, 1997.

Heynekamp, M.R., L.B. Goodwin, and P.S. Mozley, Structural and lithologic controls on patterns of cementation along a Cenozoic normal fault, *Geol. Soc. Amer. Abstracts with Programs*, 27, no. 6, A-218, 1995.

Heynekamp, M.R., L.B. Goodwin, P.S. Mozley, and W.C. Haneberg, Controls on fault-zone architecture and fluid flow in poorly consolidated sediments: the Sand Hill fault, central New Mexico, this volume, 1999.

Kelley, V., *Geology of Albuquerque Basin, New Mexico*, Memoir 33, 59 pp., New Mexico Bureau of Mines and Mineral Resources, 1977.

Knipe, R.J., Faulting processes and fault seal, in *Structural and tectonic modeling and its application to petroleum geology*, edited by R.M. Larsen, Stavanger, NPF, 325-342, 1992.

Knipe, R.J., The influence of fault zone processes and diagenesis on fluid flow, in *Diagenesis and basin development*, edited by A. Horbury and A.G. Robinson, *AAPG Studies in Geology no. 36*, 135-151, 1993.

Lewis, C.J. and W.S. Baldridge, Crustal extension in the Rio Grande rift, New Mexico: half-grabens, accommodation zones, and shoulder uplifts in the Ladron Peak-Sierra Lucero area, in *Basins of the Rio Grande rift: Structure, stratigraphy, and tectonic setting*, edited by G.R. Keller and S.M. Cather, *Geol. Soc. Amer. Spec. Paper 291*, pp. 135-156, 1994.

Lozinsky, R.P., Cenozoic stratigraphy, sandstone petrology, and depositional history of the Albuquerque Basin, central New Mexico, in *Basins of the Rio Grande rift: Structure, stratigraphy, and tectonic setting*, edited by G.R. Keller and S.M. Cather, *Geol. Soc. Amer. Spec. Paper 291*, pp. 73-82, 1994.

Machette, M., Quaternary and Pliocene faults in the La Jencia and southern part of the Albuquerque -Belen Basins, New Mexico: evidence of fault history from fault-scarp morphology and Quaternary geology, in *New Mexico Geological Society Guidebook - Albuquerque Country II 33rd field conference*, edited by J.A. Grambling and S.G. Wells, pp. 161-169, 1982.

May, S.J. and L.R. Russell, Thickness of the syn-rift Santa Fe Group in the Albuquerque Basin and its relation to structural style, in *Basins of the Rio Grande rift: Structure, stratigraphy, and tectonic setting*, edited by G.R. Keller and S.M. Cather, *Geol. Soc. Amer. Spec. Paper 291*, pp. 113-124, 1994.

Mozley, P. and L. Goodwin, Patterns of cementation along a Tertiary normal fault: A record of paleoflow orientations, *Geology*, 23, 539-542, 1995.

Pittman, E.D., Effect of fault-related granulation on porosity and permeability of quartz sandstones, Simpson Group (Ordovician), Oklahoma, *AAPG Bull.*, 65, 2381-2387, 1981.

Russell, L.R and S. Snelson, Structure and tectonics of the Albuquerque Basin segment of the Rio Grande rift: insights from reflection seismic data, in *Basins of the Rio Grande rift: Structure, stratigraphy, and tectonic setting*, edited by G.R. Keller and S.M. Cather, *Geol. Soc. Amer. Spec. Paper 291*, pp. 83-112, 1994.

Russell, L.R. and S. Snelson, Structural style and tectonic evolution of the Albuquerque Basin segment of the Rio Grande rift, in *The potential of deep seismic profiling for hydrocarbon exploration*, edited by B. Pinet and C. Bois, Institut Francais Petrole Research Conference Proceedings, Paris, Editions Technip, pp. 175-207, 1990.

Sigda, J., Effects of small-displacement faults on the permeability distribution of poorly consolidated Santa Fe Group sands, Rio Grande rift New Mexico, Masters thesis, New Mexico Tech, Socorro, New Mexico, 111 pp., 1997.

Smith, D.A., Theoretical considerations of sealing and non-sealing faults, *AAPG Bull.*, 50, 363-374, 1966.

Smith, D.A., Sealing and non-sealing faults in Louisiana Gulf Coast salt basin, *AAPG Bull.*, 64, 145-172, 1980.

Suboor, M. and J. Heller, Minipermeameter characteristics critical to its use, *In Situ*, 19, 225-248, 1995.

Tidwell, V.C. and Wilson, J.L., Laboratory method for investigating permeability upscaling, *Water Resour. Res.*, 33, 1607-1616, 1997.

Tolman, C.F., *Ground Water*, New York, McGraw-Hill, 593 pp., 1937.

Weber, K., G. Mandl, W. Pilaar, and R. Precious, The role of faults in hydrocarbon migration and trapping in Nigerian growth fault structures, Proceedings of Offshore Technology Conference 10, paper OTC 3356, pp. 2643-2653, 1978.

Yielding, G., B. Freeman, and D. Needham, Quantitative fault seal prediction, *AAPG Bull.*, 81, 897-917, 1997.

J. Sigda, L. Goodwin, P. Mozley, and J. Wilson, Department of Earth and Environmental Science, New Mexico Institute of Mining and Technology, Campus Station, Socorro, NM 87801

Fault-Fracture Networks and Related Fluid Flow and Sealing, Brushy Canyon Formation, West Texas

Eric P. Nelson, Aaron J. Kullman, Michael H. Gardner

Department of Geology and Geological Engineering, Colorado School of Mines, Golden, Colorado

Michael Batzle

Department of Geophysics, Colorado School of Mines, Golden, Colorado

We describe fault-fracture networks and associated cementation in excellent three-dimensional exposures of the Permian Brushy Canyon Formation in the central Delaware Mountains of west Texas. Faults and fractures are present in two main, nearly orthogonal sets (NNW and NE trending). A third set is present adjacent to some faults, or between two closely-spaced faults. Spacing of fault-parallel fractures decreases near some faults, in some cases parabolically. Fault core zones contain banded carbonate veins, complex breccias with banded vein clasts, and hydrocarbon material. Fault damage zones contain carbonate veins, and iron oxide and carbonate matrix alteration. These features are interpreted to record a complex history of slip, multiple fluid flow events and multiple cementation events along the fault-fracture network. Outcrop velocity probe data and porosity data suggest that matrix cementation halos have formed around the faults. The width of such halos controls the effectiveness of fault seals and the volume of reservoir rock that may be damaged by porosity occlusion. Fracture orientation and spacing are dissimilar in the hanging wall vs. footwall of some faults, suggesting that cementation-related fault sealing and reservoir damage may be asymmetric around such faults. Fracture-fault networks thus focused fluid flow at times, and were seals at times. In analogous reservoirs, open fractures would allow fracture-assisted production, and sealed fractures would compartmentalize the reservoir into semi-independent production zones.

INTRODUCTION

Fault and fracture networks are important structural features in hydrocarbon systems. Fault-fracture networks can act as fluid flow conduits during hydrocarbon migration over geological time, and/or during reservoir production over short time spans. Such fracture networks can evolve from high permeability zones into low permeability flow barriers (seals) due to cementation and/or pore collapse [e.g., *Antonellini and Aydin*, 1995]. Also, permeability may fluctuate as a consequence of deformation during seismic cycles [*Sibson*, 1990; *Knipe*, 1993].

Fault seals can be of various types, including juxtaposition seals in which reservoir rocks are faulted against

Faults and Subsurface Fluid Flow in the Shallow Crust
Geophysical Monograph 113

seal rocks [*Knipe*, 1997], fault rock seals in which fault core deformation forms impermeable fault rocks, and diagenetic halo seals where fluids flowing through fault-fracture networks precipitate cements that occlude permeability [*Knipe*, 1992]. Fracture networks are most important in development of diagenetic halo seals.

Characterization of fault and fracture networks is important for assessing reservoir quality and fault seal potential. Knowledge of how lithology, bed thickness, fault characteristics, and fault proximity control fracture spacing and orientation can enhance the predictability of fractures in the subsurface. One method of characterizing these relationships is through outcrop analog studies in areas where excellent three-dimensional exposures exist.

Outcrops of the Permian Brushy Canyon Formation in the Delaware Mountains of west Texas provide an excellent analog to many large, deep-water sandstone reservoirs. The Brushy Canyon Formation is interpreted to consist of slope and basin low-stand deposits of a third order depositional sequence [*Gardner and Sonnenfeld*, 1996]. Studies of sedimentology and stratigraphic architecture have shown this section to be stratigraphically analogous to reservoirs in sandstone-rich slope and basin deposits around the world, including offshore west Africa, the North Sea, and the Gulf of Mexico. However, today the rocks are much different. Whereas typical deep-water sandstone reservoir rocks may have porosity greater than 30%, the Brushy Canyon sandstones have an average porosity of about 10%. Thus, the Brushy Canyon Formation may be more analogous to fractured sandstone reservoirs. Nonetheless, prior to cementation these rocks may have been more porous.

We studied faults and fractures, recording spacing, orientation, and sets, both as a function of sedimentological and stratigraphic parameters and as a function of fault proximity. We also examined the nature and effectiveness of diagenetic fault sealing by describing fault and fracture fill and by measuring petrophysical parameters closely tied to permeability variations.

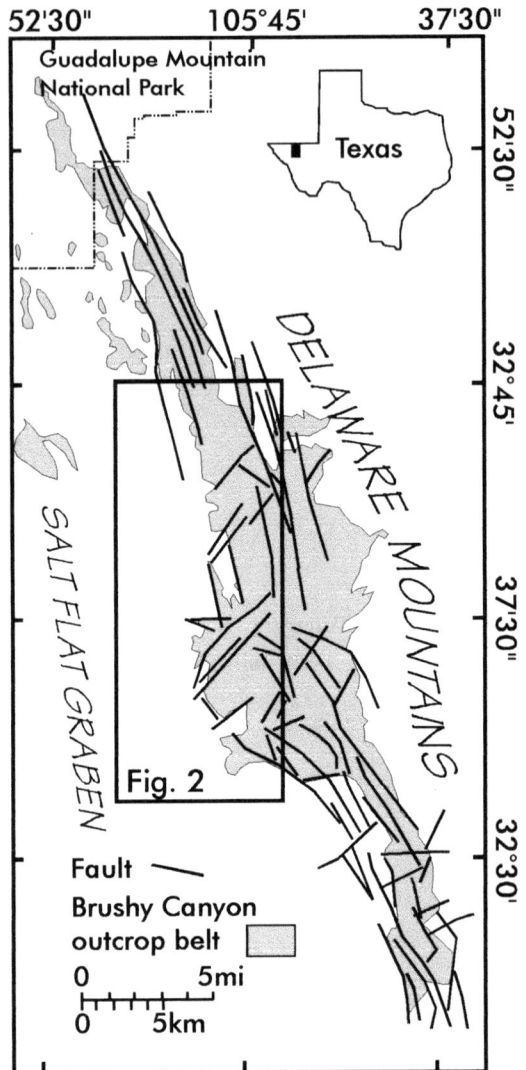

Figure 1. Location map showing outcrop belt of Brushy Canyon Formation in the Delaware Mountains; after *Gardner* [1996]. Location of Figure 2 is shown by rectangle.

GEOLOGIC SETTING AND BRUSHY CANYON STRATIGRAPHY

The north-northwest-trending Delaware Mountains constitute an east-tilted horst block southeast of Guadalupe National Park and approximately 50 km north of Van Horn, Texas (Figure 1). The range is bounded on the west by the Salt Flat graben. Regional dip is 4 degrees east-northeast into the Delaware basin, forming a gentle dip slope topography on the east flank of the range. Topography along the steeper, western escarpment is controlled by east- and west-down normal faults with a cumulative down-to-

west throw of over 1000m [*King*, 1948; *Adams*, 1988; *Gardner and Sonnenfeld*, 1996]. Northeast-trending transverse faults control NE-trending topographic elements in the range, particularly at Bitterwell Mountain. Here a transverse horst block forms a salient which extends west into the Salt Flat graben.

Stratigraphically, the Brushy Canyon Formation is above clastic and carbonate strata of the Cutoff and Bone Spring Formations and below clastic strata of the Cherry Canyon Formation [*Zelt and Rossen*, 1995]. Litho-logically, the Brushy Canyon Formation is dominantly composed of very fine to medium sandstone and subordinate siltstone. The Brushy Canyon Formation was deposited in a slope and basin environment and constitutes a complex array of sediment bodies, with architectural elements deposited in various canyon, channel and fan environments. Reservoir-scale (10s x 100s meters) architectural elements include amalgamated sandstone channels, broadly tabular sandstone sheets and convex-up lobes, interbedded sandstone and siltstone successions, and condensed sections of organic-rich siltstone forming thin tabular sheets.

FIELD METHODS AND DATA ANALYSIS

Three main field areas were studied (Figure 2). The 6-Bar Ranch area north of the Bitterwell salient contains sites at Chinaman Hat, Plane Crash Canyon, the Popo-Pollo fault blocks, and Colleen Canyon. The other sites are south of the Bitterwell salient in the Cordoniz Canyon area.

Most fracture spacing measurements were made on scan lines along cliff or arroyo exposures. Fracture spacing was measured as the perpendicular distance between fractures of a particular fracture set. A one-dimensional average fracture spacing was derived, in units of number of fractures per unit length. However, to describe fracture length, truncation relationships, and two-dimensional fracture density, two other procedures were used. In the first procedure (here called the Titley method) fracture density was determined on pavement exposures using the method described by *Haynes and Titley* [1980]. In this method, the cumulative length of all fractures exposed in a one-meter square area was measured, and the integrated fracture density was derived, in units of cm^{-1}. Three pavement exposures were measured at varying distances from a given fault. In the second procedure, one bedding-plane pavement was photographed, traced, and studied to determine the characteristics of fracture spacing, length, and truncation relationships in plan view. This pavement was actually a mini-pavement (approximately 50 cm x 50 cm), but shows the same overall fracture pattern seen at the megascopic scale.

Fracture orientations were determined at sites where spacing and other data were acquired, and were plotted on lower hemisphere equal area nets for statistical analysis. All orientation data are given in right-hand rule convention (facing the strike azimuth, dip is to the right).

Reconnaissance porosity, permeability, and compressional velocity measurements were made on 1-inch diameter core plugs collected in the field. These measurements were made by *Batzle et al.* [1996] along measured sections throughout the Delaware Mountains to characterize the petrophysical nature of various architectural elements in the Brushy Canyon Formation. In the present study we compared the statistical distribution of

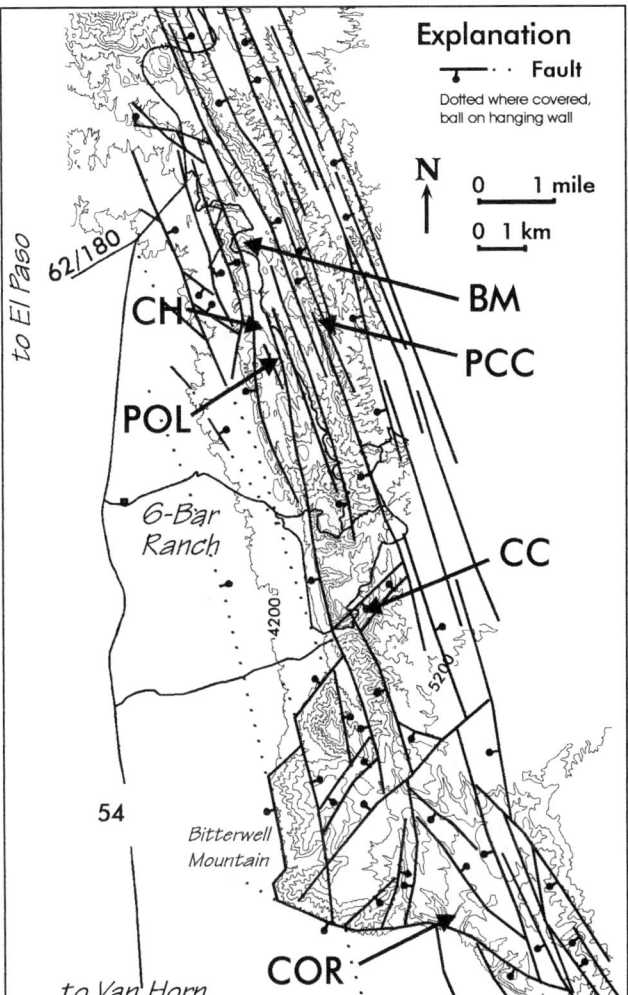

Figure 2. Location map showing field sites in central Delaware Mountains; modified from *Gardner and Sonnenfeld* [1996]. BM = Brushy Mesa, CH = Chinaman Hat; PCC = Plane Crash Canyon; POL = Pollo site; COR = Cordoniz Canyon; CC = Colleen Canyon.

values between sections distant from faults and those near faults.

Compressional velocities were also measured in the field using a portable outcrop velocity probe [*Batzle and Smith*, 1992]. *Batzle and Smith* [1992] showed the probe to be useful in examining relative velocity changes in outcrop. In tests, they found the standard deviation of measurements to be within 1-3% of the average measurements. The depth of penetration is usually only a few millimeters, thus measurements are sensitive to weathering and surface roughness. Therefore, we were careful to choose outcrops with little or no weathering rinds, such as in the bottom of arroyos. We measured velocity in the same sandstone bed

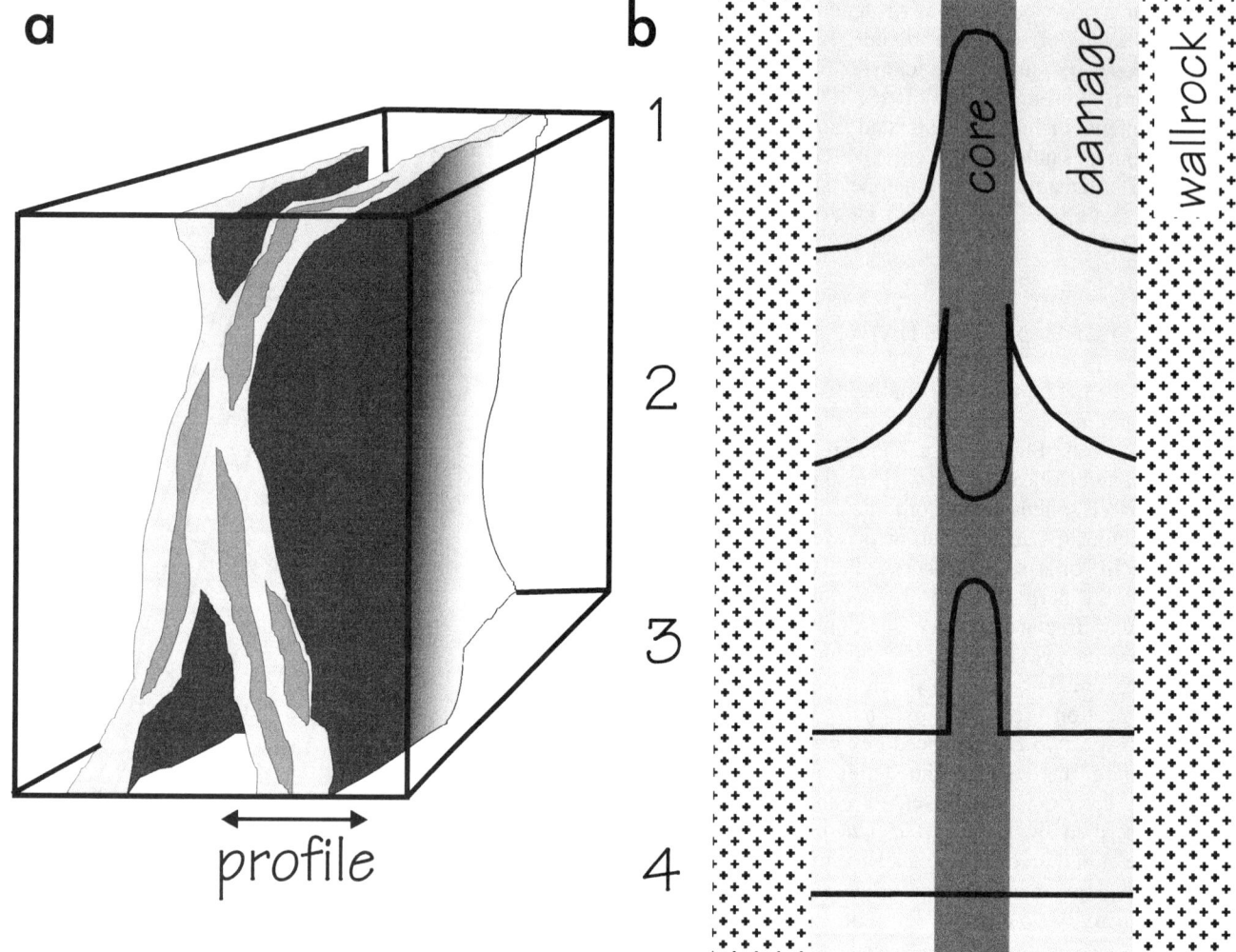

Figure 3. Fault zone structure and permeability models.   a. Anastomosing fault zone with core zone (dark) and surrounding damage zone (light); after *Caine et al.* [1996]. b. Models of permeability (*k*) structure across fault zone; after *Scholz and Anders* [1994]. Model permeability profiles are: 1. high *k* in core zone, high *k* across fault; 2. low *k* in core, high *k* in damage zone; 3. high *k* only in core zone; 4. constant *k* across damage and core zones.

as a function of distance from faults and veins. In this way compressional velocity is used as an indirect measure of porosity and cement content, and not of variations in lithology.

FAULT ZONE PERMEABILITY MODELS

Primary components of upper-crustal fault zones, although variable, are fault core, damage zone, and undeformed protolith in adjacent fault blocks (Figure 3) [*Caine et al.*, 1996]. The core zone, which may contain anastomosing slip surfaces, clay-rich gouge, cataclasite, and fault breccia, accommodates most of the fault displacement.   The damage zone is the network of

subsidiary structures between the fault core and the protolith, and contains much lower bulk shear strain than the fault core. Typically, the damage zone is characterized by heterogeneously distributed strain, generally represented by a mesh of shear and extension fractures. These features cause heterogeneity and anisotropy in the permeability structure of the fault zone [*Bruhn et al.*, 1994]. Wide damage zones may indicate multiple episodes of slip.

Permeability can vary across fault zones, and may be higher in core or damage zone [*Scholz and Anders*, 1994]. Generally, grain-size reduction and/or mineral precipitation produce fault cores with lower porosity and permeability than adjacent protolith. However, renewed fault move-ment, particularly under high fluid pressures, can cause extension fracturing and increased permeability in such core zones [*Scholz and Anders*, 1994; *Sibson*, 1996]. For example, banded vein fill in many fault zones records multiple fracturing and transient fluid flow events [e.g., *Logan and Decker*, 1994]. *Sibson et al.* [1988] and *Cox et al.* [1991] have shown that seismic faulting, dilation, and fault-controlled fluid flow are cyclic and are controlled by fluid pressure cycles. This process, generally known as the fault valve model, involves fluid pressure buildup prior to seismic faulting. Seismic faulting may occur when fluid pressure exceeds lithostatic stress, or in the case of wrench or normal faults, at pressures less than lithostatic stress [*Sibson*, 1981; *Behrmann*, 1991].

Permeability of protolith outside the fault zone, or in blocks between faults and fractures in the damage zone, is much lower than that in open fractures and includes a combination of matrix and microfracture permeability. Fracture permeability should be considered in assessing the total permeability of a fractured or faulted reservoir system.

FAULT CHARACTERISTICS

Faults in the Delaware Mountains form a belt approximately 16 kilometers wide which follows the crest and west flank of the range (Figure 1) [*King*, 1948]. Two main sets of faults are recognized in this portion of the Delaware Mountains. Set 1 faults, with an average orientation of 347/87 (Figure 4a), are the primary normal faults in the region and control the NNW geomorphic trend of the range. The western flank of the range is formed by the Border fault zone [*King*, 1948], which comprises a series of discontinuous fault segments, and which has between 600-1200m of throw. Faults we studied show much less throw (generally < 100m). Most set 1 faults dip steeply to the west, although east-dipping faults also are present, forming horst and graben structure in the central corridor of the range. Set 2 faults, with an average orientation of 233/83 (Figure 4a), are transverse faults and control NE-trending topographic elements in the range. These faults are generally shorter than those of set 1, and can also form grabens such as those at Colleen Canyon and Cordoniz Canyon. These grabens form the northern and southern boundaries of the Bitterwell Mountain horst. Most faults have slickenlines with rakes of 80-85°, and are thus primarily normal faults. Some lower rakes were recorded, indicating oblique and/or strike slip components of displacement.

Fault core zones are characterized by fault-parallel veins, some up to 2 meters thick. These veins contain open-space fill calcite with two mesoscopic morphologies. The older morphology is coarse calcite spar (crystals up to 4 cm across); the younger morphology is banded fibrous calcite. Some calcite bands are white or translucent,

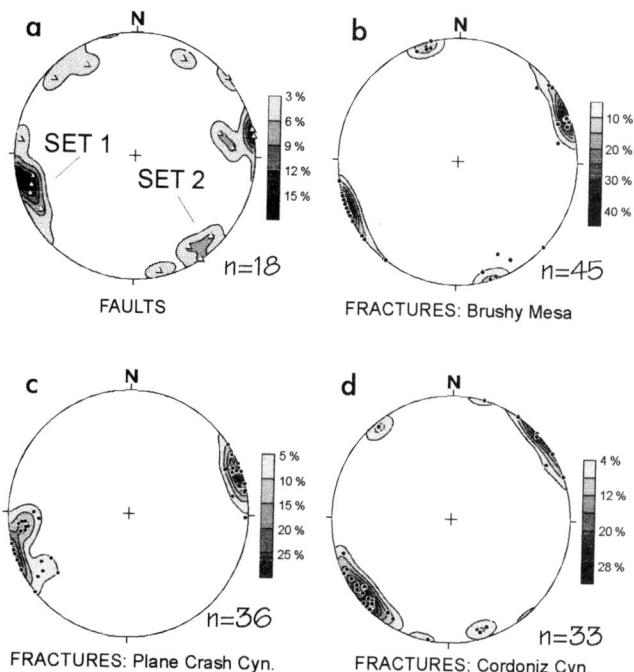

Figure 4. Poles to faults and fractures plotted on lower hemisphere equal area nets. Contours are percent per one percent area; n = number of data points. a. faults in all study areas; b. fractures from Brushy Mesa site; c. fractures from Plane Crash Canyon site (set 1 only); d. fractures from Cordoniz Canyon site.

whereas other, generally younger bands, are brown and petroliferous. Younger bands may also contain dark brown hydrocarbon. Fault core zones also contain breccia, which can be older or younger than the calcite fill. Older breccia contains blocks of the wall rock, whereas younger breccia may also contain blocks of banded veins (Figure 5).

A damage zone is present around the core zone of faults in both sets. The damage zone is characterized by fractures that are commonly filled with open-space calcite fill and, in places, by iron oxide alteration of the host rock. These fractures are typically extension fractures, although shear fractures and subsidiary faults also were documented. Vein aperture is greatest in and near the core zone and decreases with distance from the core. Vein networks in and near the core zone may evolve to calcite-cemented breccia with increased displacement on the fault. Also, calcite-cemented breccia may form at fault stepovers and jogs. Transverse faults, such as in Colleen Canyon and Cordoniz Canyon, commonly have very thick breccia zones, variable calcite fill, and relatively narrow alteration zones.

The spacing of set 1 faults is either bimodal or clustered. A bimodal distribution represents fault doublets, in which two faults are closely spaced (10's to 100's m) in a population of faults spaced far apart. An example occurs

Figure 5. a. Photograph of banded carbonate vein; note symmetry of bands indicating syntaxial growth (growth from margins toward middle of vein). Dark bands are hydrocarbon rich; vein is ~10cm wide. b. Photograph of fault breccia with blocks of banded carbonate vein. From Colleen Canyon site; fault zone is ~50 cm wide.

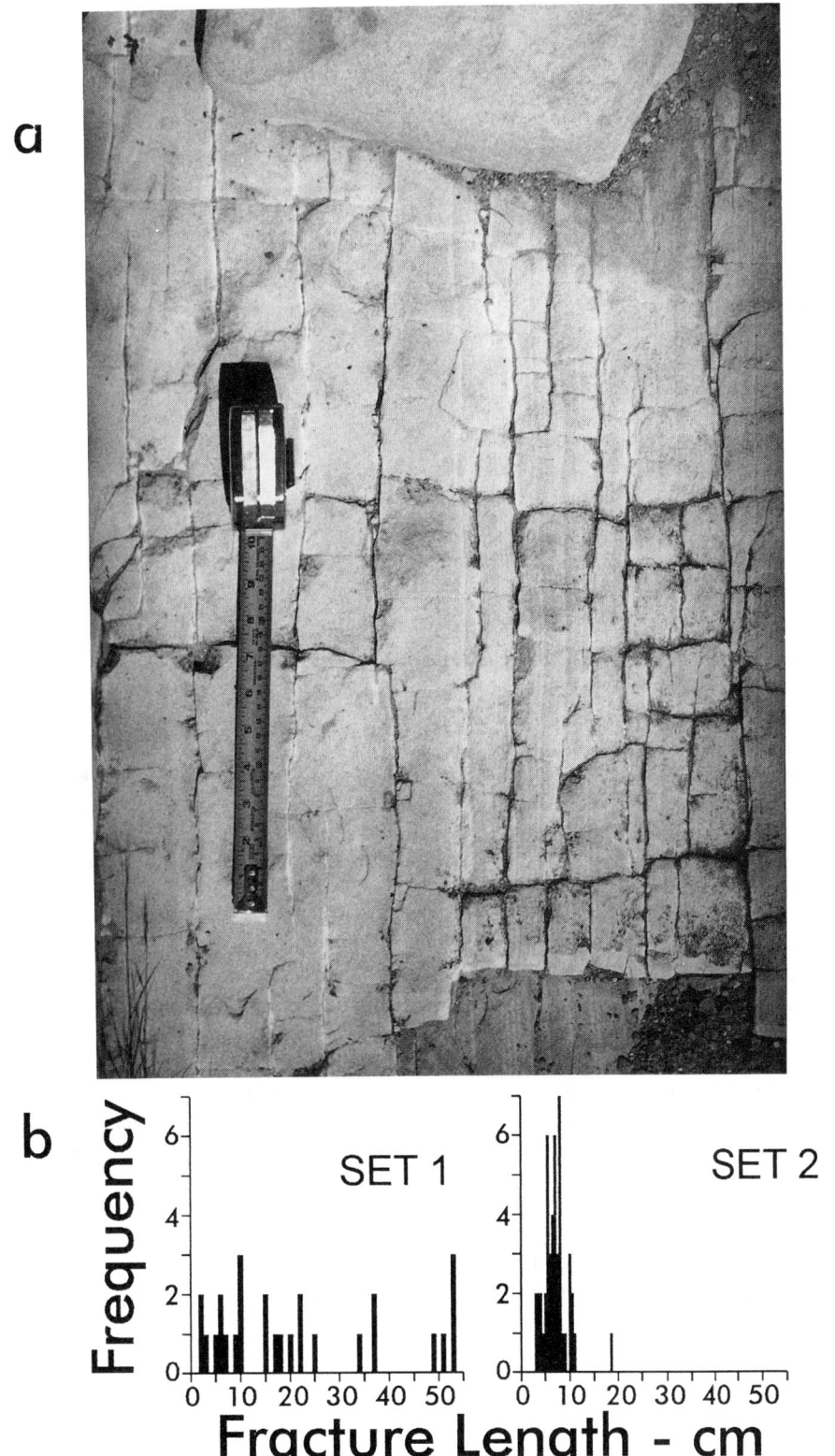

Figure 6. Mini-pavement at Pollo site. a. Photograph of horizontal bedding plane showing set 1 fractures parallel to tape and set 2 fractures perpendicular to tape (tape approximately 25 cm long). b. Histograms of fracture length.

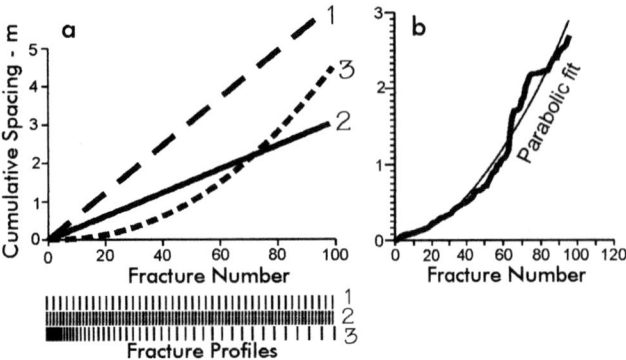

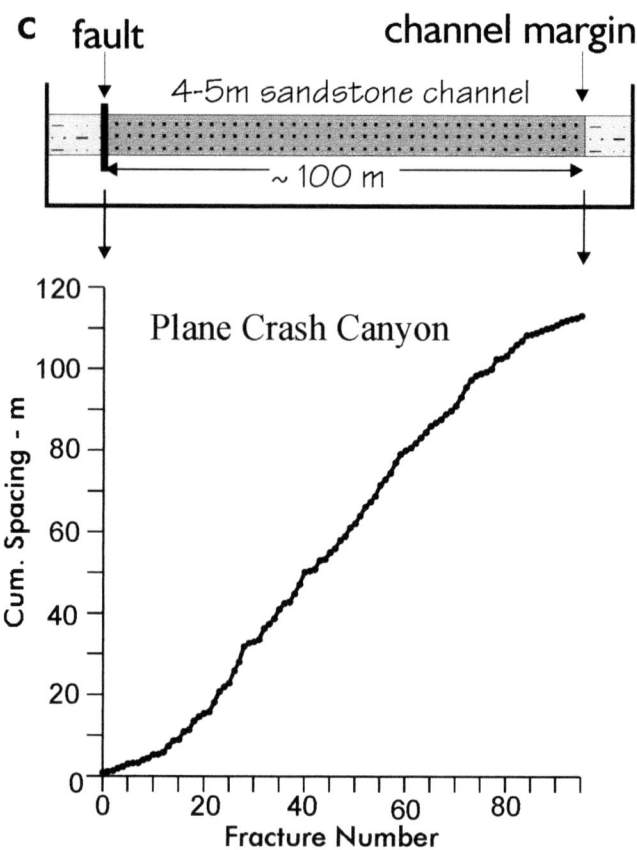

Figure 7. Cumulative fracture spacing curves. a. Model cumulative spacing curves also showing synthetic fracture profiles below. Patterns 1 and 2 are for constant spacing, and pattern 3 is for parabolic spacing pattern. b. Cumulative spacing curve from Chinaman Hat site; fracture number 0 = fault. Model curve fit (dashed) is second order polynomial with R-squared = .976. c. Cumulative spacing curve from Plane Crash Canyon site; cross section above shows position of fault and channel margin.

in the Pollo area, where two faults, spaced 10-20m, form at a fault stepover; a relay ramp is developed between the faults. Where spacing is clustered, numerous faults are closely spaced in one area, whereas the spacing is generally much greater. An example of clustering is present in the Plane Crash Canyon area, and also may be related to a fault stepover.

FRACTURE CHARACTERISTICS

Fractures in the region are mostly extension fractures and are present in most outcrops, both distal and proximal to faults. Two main fracture sets are recognized regionally [*King*, 1948], and constitute a regional fracture system. Fractures in set 1 have an average orientation of 160/88 (trend 340, Figures 4b-4d), and generally are the longest fractures. Fractures in set 2 have an average orientation of 070/88 (Figures 4c-4d), and are generally shorter than, and transverse to, set 1 fractures. The typical spacing, length, and crosscutting relationships between set 1 and set 2 fractures are illustrated on a mini-pavement exposed on a sub-horizontal bedding surface (Figure 6a). Set 1 fractures are longer and have a more dispersed length distribution than set 2 fractures (Figure 6b). Set 2 fractures usually truncate against set 1 fractures, indicating that they formed later than those of set 1.

In areas distant from faults, fracture spacing is controlled by a combination of lithology, bed thickness, bed stacking patterns, and sandstone body architecture. However, in areas near faults, fracture spacing parallel to faults is generally less and, in some cases, decreases as the

fault is approached; *King* [1948] first recognized this pattern. Also, fracture orientation and spacing are dissimilar in the hanging wall vs. footwall of some faults.

To illustrate the variation of fracture spacing with distance from a fault, fracture spacing is plotted as cumulative spacing versus fracture number, where fracture number one is at the fault margin. Figure 7a shows model cumulative curves and fracture spacing profiles for three synthetic fracture spacing patterns. Curves one and two are for constant spacing pattern, and curve three shows a parabolic spacing pattern.

Fault-proximal spacing was measured in the Chinaman Hat and Plane Crash Canyon areas (Figure 2). In these areas, spacing of set 1 fractures decreases parabolically with distance from the fault (Figure7b). Variations from the model fit may be related to sampling problems such as inaccurate spacing measurements where cliff faces trend at a low angle to the fractures. Variations may also be related to fracture clustering. For example, because the bending moment imposed by fault shear on the adjacent wallrock is greatest at a distance from the fault [see *Turcotte and Schubert*, 1982, p. 129], a cluster of fractures might form at or near this distance from the fault. In the Plane Crash Canyon site, spacing of set 1 fractures in a sandstone channel also decreases parabolically with distance from the steep channel margin (Figure 7c). This suggests that, during fracturing, the sandstone near the channel margin acted mechanically similar to sandstone near a fault.

The Titley method was also used to determine fracture density variations with distance from faults. Three sites were chosen in the Cordoniz Canyon area and all show a pattern of initial decrease in fracture density with distance from faults (Figure 8). Two of the sites also show a partial increase following the initial decrease.

In some areas a third fracture set was observed near faults. Fractures in this set generally dip less than those in sets one or two. Examples are at the Plane Crash Canyon site within about 10 meters of a fault, and in the Cordoniz Canyon area, where a third set is present between two faults (Figure 9). Also, set 3 fractures and small displacement faults are present in the Popo fault block, possibly related to a major fault step-over, and/or oblique slip on the bounding faults. These qualitative observations also suggest that fracture orientation and spacing are dissimilar in the hanging wall vs. footwall of some faults.

FAULT-PROXIMAL DIAGENESIS AND SEALING

Evidence for fluid flow and sealing along fault and fracture networks comes from two sources. First, fractures in fault core and damage zones are filled with carbonate cements, commonly with a banded pattern (Figure 5). Carbonate-filled veins are concentrated near faults, and some veins are over 1 meter thick. In addition, iron oxide alteration is present along most faults and many veins. For example, in the Plane Crash Canyon site, two faults spaced about 15 m apart have an extensive zone of iron oxide alteration between them. Also, Liesegang banding is present around many fractures and veins.

Second, petrophysical measurements vary as a function of proximity to faults and veins. Laboratory measurements were taken on cores from intact rock between fractures and

Figure 9. Tracing of photograph of two faults; thin vertical lines are set 1 fractures and thick lines are inclined set 3 fractures between faults; Cordoniz Canyon site. Note, faults are normal faults but appear as reverse faults due to perspective of photograph.

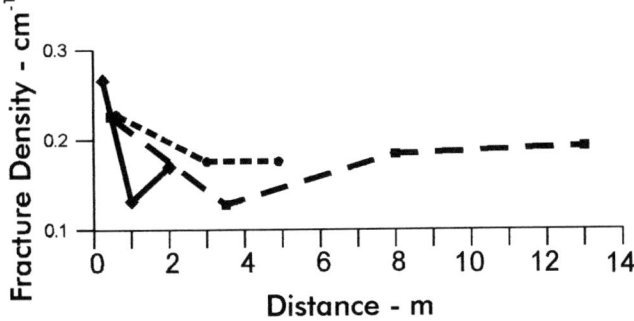

Figure 8. Plots of fracture density (cm^{-1}) with distance from faults; fault at 0 distance. Data from Cordoniz Canyon site.

faults, and thus do not account for the effects of mesoscopic fractures. Permeability, porosity, and seismic velocity were measured by *Batzle et al.* [1996] along stratigraphic sections to characterize facies-related variations in the Brushy Canyon Formation. The distribution of these properties was compared between two sections in the Colleen Canyon site. One section is located near a fault zone and the other section is located distant from faults. Average porosity is systematically reduced in the section near the fault (Figure 10). Permeability varies from near zero to over 20 md, but shows no systematic shift between the two sections. Velocity data also show no systematic shift in the distribution of S- or P-velocities between the two sections. Further petrophysical studies are in progress.

Outcrop velocity probe measurements also vary as a function of proximity to faults. Two profiles were measured in the Cordoniz Canyon area (Figure 11). In one profile, velocity increases for about 3 meters from the fault, then decreases to a background level (Figure 11a). In the other profile, velocity is high within about 3 meters of the fault then decreases to a background level (Figure 11b). The higher velocities near the faults are interpreted to represent the effect of carbonate cement in the matrix and

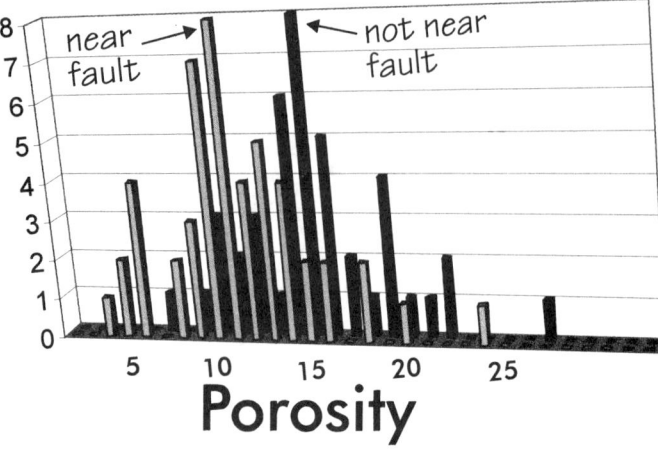

Figure 10. Frequency plot of porosity (%) distributions for stratigraphic sections measured near fault and distant from fault. Data from Colleen Canyon site. F-test shows sample variances to be not statistically different; Student's t-test shows means to be statistically different.

decreased porosity, and suggest that a diagenetic alteration halo is present in the core and damage zones of these faults.

A short velocity profile (2 meters long) was measured across a calcite vein exposed on a horizontal bedding plane. The velocity profile across the vein correlates with the topographic profile (Figure 12), suggesting that lateral variations in carbonate cement content are present in the matrix, forming a halo around the vein. This is consistent with the observation that many similar veins and fault zones have halos of calcite cement and/or iron oxide alteration, in some cases developing Liesegang banding.

DISCUSSION

In the central Delaware Mountains, *King* [1948] first showed that faults and fractures occur in two main, nearly orthogonal sets and noted that fracture spacing is smaller near faults. We have shown that spacing of fault-parallel fractures decreases parabolically near some faults. A pattern of decreased fracture spacing near faults also has been documented by other workers for both macrofractures [e.g., *Brock and Engelder*, 1977; *Chernyshev and Dearman*, 1991] and microfractures [*Anders and Wiltschko*, 1994]. We also have documented a third set of fractures near some faults, particularly in areas between two closely-spaced faults. Commonly, the closely spaced faults are in the overlap region of a fault step-over. Development of a third fracture set is consistent with the observation of *Rawnsley et al.* [1992], who noted that joint patterns can be perturbed in the vicinity of faults.

In fault core zones we have documented banded carbonate veins, complex breccias with clasts of banded

veins, and hydrocarbon material. In fault damage zones we have documented halos of carbonate and iron oxide alteration around faults and veins using outcrop velocity probe data and preliminary porosity data. These fault zone features are interpreted to record a complex history of slip, accompanied by multiple fluid flow events and multiple cementation events along the fault-fracture network. Fractures were clearly open at times allowing fluid flow along the fault zone. Subsequent cementation closed the fracture-controlled flow paths, and younger faulting events re-opened the fractures allowing younger flow events. Thus, fault-controlled fluid flow and sealing were episodic.

The hydrocarbon material is confined to the central portion of syntaxial fault-zone veins and has not been observed in sandstone adjacent to the faults. This is interpreted to indicate that, in this area, hydrocarbon migration was late, and flow into adjacent sandstone beds was prohibited by earlier vein fill. Based on studies in the Delaware basin, the source rock for the Brushy Canyon Formation reservoirs is believed to be organic-rich siltstones within the Delaware Mountain Group [*Hayes et al.*, 1996], but may also include the organic-rich Bone Spring Formation below. In the field area, the source of hydrocarbons is unknown. On the basis of Lopatin thermal maturation modeling, *Hays et al.* [1996] suggest that oil maturation occurred in the Brushy Canyon Formation between about 210-235 Ma.

Whereas maturation initially occurred in the Permian, normal faulting probably occurred during Tertiary Basin and Range development. Thus, fault-controlled migration

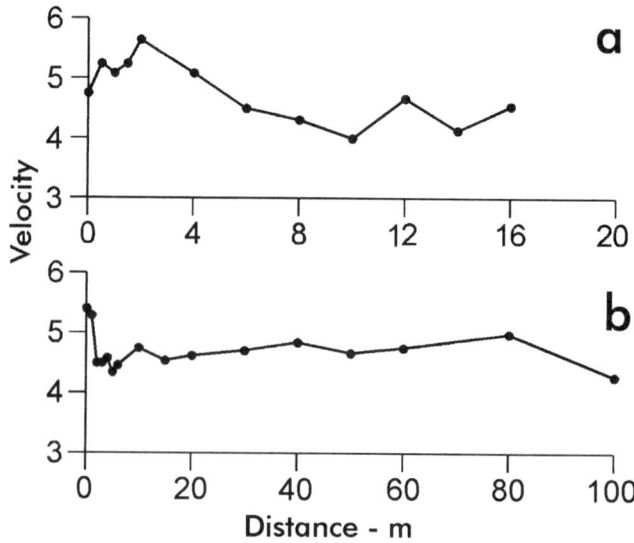

Figure 11. Seismic velocity (V_p) plotted versus distance from fault (at 0). Data from two faults in Cordoniz Canyon site. Note different scales on x-axis.

of carbonate- and hydrocarbon-bearing fluids probably occurred when the strata were well lithified. However, sands in relatively young, deep-water reservoir systems commonly are only partially lithified. Application of the results of this outcrop analog study to these younger systems may thus be questionable. Nonetheless, there is evidence that fluid flow can be focused along faults in poorly lithified sediments. For example, *Sample et al.* [1993] showed that fluid flow can be channelized along active faults on the seafloor and can cause carbonate cementation in poorly lithified sediments. Also, *Mozley and Goodwin* [1995] documented fault-controlled fluid flow and cementation in poorly consolidated continental sediments along a Tertiary-Quaternary fault. Thus, it appears that fluid flow through fault-fracture networks in well-lithified strata may be focused in fault zones at higher stratigraphic levels where sediments are less lithified. If such focusing occurs, then it is possible for reservoir damage and fault zone seals to form in poorly lithified sediments.

It is unclear how this focusing occurs, although it may be related to feedback between fault zone damage and syn-

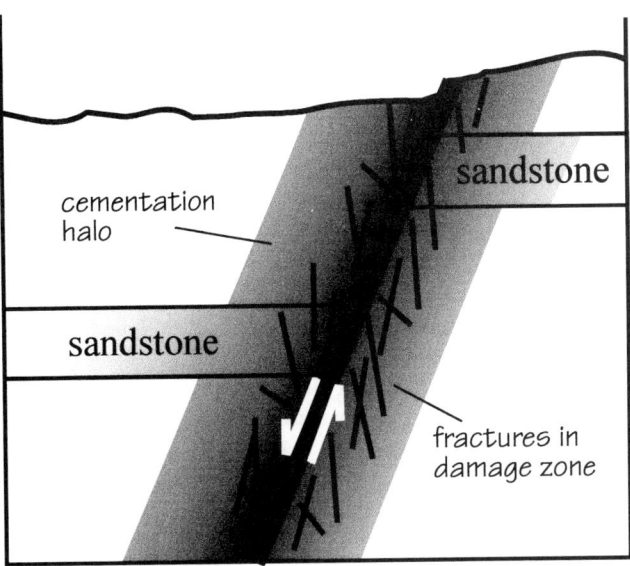

Figure 13. Cross section model of diagenetic cementation halo formed around fault zone. The fault zone contains an initially high permeability fracture network that focuses fluid flow and controls cementation. Reservoir damage might occur in under-pressured reservoir rocks where the cementation halo widens.

faulting cementation in poorly lithified sediments. It is speculated that faulting in poorly lithified sediments may focus fluid flow by incipient fracturing and alignment of clay particles. Focused fluid flow would lead to localized cementation, more brittle behavior, and subsequent development of fracture networks and enhancement of flow.

Figure 13 presents a model showing development of a cementation halo around a fault-fracture network in well-lithified rock. In this model, the fault-proximal fracture network forms in a process zone near the propagating fault tip [*Anders and Wiltschko*, 1994], although some fractures may also form during shear along the fault. It is proposed, but not confirmed, that higher permeability sandstone beds may develop a wider cementation halo through flow or diffusion away from the fault. The width of the halo is important because it will control not only the effectiveness of fault sealing, but also the volume of reservoir rock that is potentially damaged by porosity occlusion. Because fracture orientation and spacing are dissimilar in the hanging wall vs. footwall of some faults, cementation-related fault sealing and reservoir damage may be asymmetric around such faults. Further study is underway to quantify this relationship.

We have used physical and petrophysical observations to document the nature of fault and fracture networks and related cementation in well-exposed outcrops of reservoir analog rocks. In subsurface reservoirs, borehole or remote

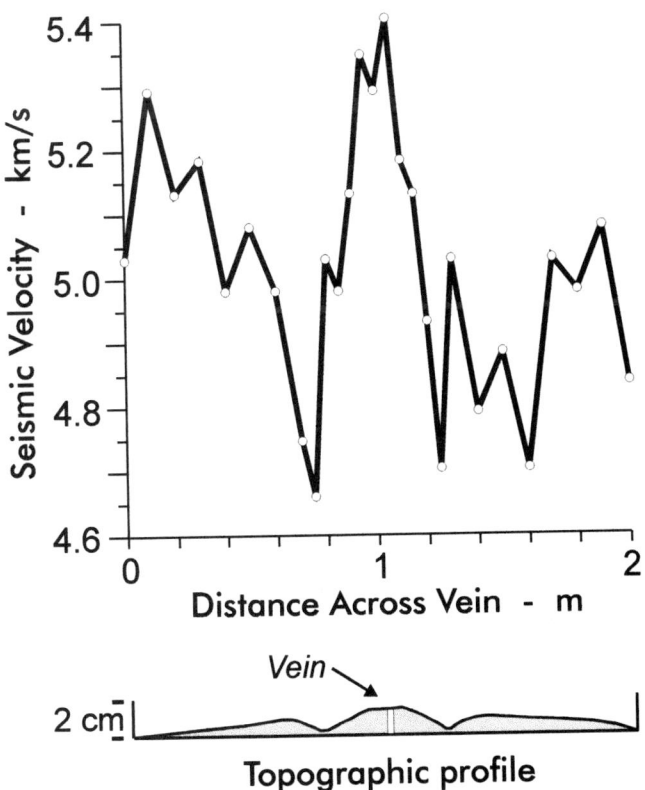

Figure 12. a. Seismic velocity (V_p) plotted versus distance across vein. b. Approximate topographic profile of vein and adjacent wallrock.

sensing techniques (e.g., seismic reflection) may be used to infer the presence of fault and fracture networks and related cementation at various scales. Patterns such as parabolic decrease of fracture spacing may indicate the presence of nearby faults, and when associated with anomalous cementation may indicate the presence of cementation halo fault seals. If fractures are open, fracture assisted production may be possible [*Nelson*, 1985]; if fractures are sealed, then the reservoir may be compartmentalized into semi-independent production zones. In the field area, such flow and/or compartmentalization would have been dominantly controlled by the NNW trending set 1 faults and fractures, and secondarily by NE trending set 2 faults and fractures. It is thus recommended that, in analyzing fault seal potential, diagenetic studies be combined with fracture network studies using subsurface data. Such studies should strive to determine spacing and orientation of faults and fractures, and amount and type of cementation associated with fractures.

Acknowledgements. Financial support was supplied by the following companies: Amoco EPTG, Conoco Inc., Exxon Production Research Company, Marathon Oil Company, Statoil, and Texaco Inc. Support for access to the field area was supplied from the Slope and Basin Consortium at Colorado School of Mines (Amoco Production Company, Arco Exploration & Production Technology, British Petroleum, Conoco Inc., Elf Exploration Production, Exxon Production Research Company, Marathon Oil Company, Mobil Exploration and Production Technical Center, ORYX Energy Company, Phillips Petroleum Company, Shell Offshore Inc., Statoil, Texaco Exploration and Production, Inc., Unocal Corporation, Landmark Corporation, RC^2, and Smedvig Technologies). We thank Tony Kunitz and Anne and Mike Capron of the 6-Bar Ranch, and the Sibley Ranch for granting access and helping with logistics. Brad Sinex, Jill Savage, Laura Sweezy, and Paul Burger are thanked for field assistance, and Christie Callens helped with laboratory data collection.

REFERENCES

Adams, J.W., Comments on the structure of the area, Road log second day from Carlsbad, New Mexico, via White City and Pine Springs, to west face overlook Williams Ranch and Shumard Canyon, in *West Texas Geological Society 1988 Field Seminar to Guadalupe Mountains*, 19-21, 1988.

Anders, M.H., and D.V. Wiltschko, Microfracturing, paleostress, and the growth of faults, *J. Structural Geol., 16*, 795-815, 1994.

Antonellini M., and A. Aydin, Effect of faulting on fluid flow in porous sandstones: geometry and spatial distribution, Amer. *AAPG Bull., 79*, 642-671, 1995.

Batzle, M.L., and B.J. Smith, Hand-held velocity probe for outcrop and core characterization, in *Proceedings of the 33rd Symposium on Rock Mechanics 1992, Santa Fe, New Mexico,* edited by J.R. Tillerson, and W.R. Wawersik, pp.949-958, Balkema, Rotterdam, 1992.

Batzle, M., M. Gardner, and M. Sonnenfeld, Geophysical signature of slope fan facies architecture, *Society of Exploration Geophysicists 66th Annual Technical Conference, Denver, Colorado*, pp.380-383, 1996.

Behrmann, J.H., Conditions for hydrofracture and the fluid permeability of accretionary wedges, *Earth Planet. Sci. Lett., 107*, 550-558, 1991.

Bruhn, R. L., W. T. Parry, W. A Yonkee, and T. Thompson, Fracturing and hydrothermal alteration in normal fault zones, *Pageoph, 142*, 609-644, 1994.

Brock, W. G., and T. Engelder, Deformation associated with the movement of the Muddy Mountain overthrust in the Buffington window, southeastern Nevada, *Geol. Soc. Amer. Bull., 88*, 1667-1677, 1977.

Caine, J.S., J.P Evans, and C.B Forster, Fault zone architecture and permeability structure, *Geology, 24*, 1025-1028, 1996.

Chernyshev, S.N., and W. R. Dearman, *Rock Fractures*, Butterworth-Heinemann, London, 1991.

Cox, S.F., V.J. Wall, M.A. Etheridge, and T.F., Potter, Deformational and metamorphic processes in the formation of mesothermal vein-hosted gold deposits – examples from the Lachlan Fold Belt in central Victoria, Australia, *Ore Geology Review, 6*, 391-423, 1991.

Gardner, M.H., and M.D Sonnenfeld, Stratigraphic changes in facies architecture of the Permian Brushy Canyon Formation in Guadalupe Mountains National Park, West Texas: in *The Brushy Canyon Play in Outcrop and Subsurface: Concepts and Examples*, edited by W.D. DeMis, and A.G Cole, Permian Basin Section, *Soc. of Econ. Paleontol. Mineral. Pub. 96-38*, 17-40, 1996.

Hays, P.D., S.D. Walling, and T.T.Tieh, Organic and authigenic mineral geochemistry of the Permian Delaware Mountain Group, west Texas: Implications for the chemical evolution of pore fluids: in *Siliciclastic Diagenesis and Fluid Flow: Concepts and Applications*, edited by L.J. Crossey, R. Loucks, and M.W. Totten, Soc. of Econ. Paleontol. Mineral. Spec. Pub. 55, 163-186, 1996.

Haynes, F.M., and S.R.Titley, The evolution of fracture-related permeability within the Ruby Star granodiorite, Sierrita porphyry copper deposit, Pima County, Arizona: *Econ. Geol., 75*, 673-683, 1980.

King, P.B., Geology of the southern Guadalupe Mountains, Texas: *USGS Prof. Paper 215*, 183pp., 1948.

Knipe, R.J., Faulting processes and fault seal, in *Structural and tectonic modeling and its application to petroleum geology*, edited by R.M. Larsen, Norweigan Petroleum Society, Stavanger, 325-342, 1992.

Knipe, R. J., The influence of fault zone processes and diagenesis on fluid flow: in *Diagenesis and Basin Development*, edited by A.D. Horbury and A.G. Robinson, AAPG Studies in Geology 36, 135-151, 1993.

Knipe, R.J., Juxtaposition and seal diagrams to help analyze fault seals in hydrocarbon reservoirs, *AAPG Bull. 81*, 187-195, 1997.

Logan, J.M., and C.L. Decker, Cyclic fluid flow along faults, in *The Mechanical Involvement of Fluids in Faulting*, edited by S. Hickman, R. Sibson, and R. Bruhn, USGS Workshop LXIII, 190-203, 1994.

Mozley, P.S., and L.B. Goodwin, Patterns of cementation along a Cenozoic normal fault: A record of paleoflow orientations, *Geology, 23*, 539-542, 1995.

Nelson, R.A., *Geologic Analysis of Naturally Fractured Reservoirs*, Gulf Publishing Co., Houston, 320pp., 1995.

Rawnsley, K.D., T. Rives, J.-P. Petit, S. R. Hencher, and A. C. Lumsden, Joint development in perturbed stress fields near faults: *J. Structural Geol., 18*, 939-951, 1992.

Sample, J.C., M.R. Reid, H.J. Tobin, and J.C. Moore, Carbonate cements indicate channeled fluid flow along a zone of vertical faults at the deformation front of the Cascadia accretionary wedge (Northwest U.S. coast). *Geology, 21*, 507-510, 1993.

Scholz, C.H., and M.H. Anders, The permeability of faults: in *The Mechanical Involvement of Fluids in Faulting*, edited by S. Hickman, R. Sibson, and R. Bruhn, USGS Workshop LXIII, 247-253, 1994.

Sibson, R.H., Control on low-stress hydrofracture dilatancy in thrust, wrench and normal fault terrains, *Nature, 289*, 665-667, 1981.

Sibson, R.H., F. Robert, and K.H. Poulsen, High-angle reverse faults, fluid-pressure cycling, and meso-thermal gold-quartz deposits, *Geology, 16*, 551-555, 1988.

Sibson, R.H., Conditions for fault-valve behavior, in *Deformation Mechanisms, Rheology and Tectonics*, edited by R.J. Knipe, and E.H. Rutter, Geol. Soc. Spec.Pub. 54, 15-28, 1990.

Sibson, R.H., Structural permeability of fluid-driven fault-fracture meshes, *Jour. Structural Geol., 18,* 1031-1042, 1996.

Turcotte, D.L. and G. Schubert, *Geodynamics: Applications of Continuum Physics to Geological Problems*, John Wiley & Sons, New York, 450 pp., 1982.

Zelt, F.B. and C. Rossen, Geometry and continuity of deep-water sandstones and siltstones, Brushy Canyon Formation (Permian) Delaware Mountains, Texas, in *Atlas of Deep Water Environments: Architectural Style in Turbidite Systems*, edited by K.T. Pickering, R.N. Hiscott, N.H. Kenyon, E. Ricci-Luchi, and R.D.A. Smith, Chapman and Hill, London, 167-183, 1995.

Nelson, Eric P., Department of Geology and Geological Engineering, Colorado School of Mines, Golden, CO 80401, enelson@mines.edu

Kullman, Aaron J., Department of Geology and Geological Engineering, Colorado School of Mines, Golden, CO 80401

Gardner, Michael H., Department of Geology and Geological Engineering, Colorado School of Mines, Golden, CO 80401

Batzle, Michael, Department of Geophysics, Colorado School of Mines, Golden, CO 80401

Brittle Faulting and Permeability Evolution: Hydromechanical Measurement, Microstructural Observation, and Network Modeling

Teng-fong Wong

Department of Geosciences, State University of New York at Stony Brook, Stony Brook, New York

Wenlu Zhu

Department of Geology and Geophysics, Woods Hole Oceanographic Institute, Woods Hole, Massachusetts

Dilatancy is universally observed as a precursor of brittle faulting in geomaterials. Laboratory data on permeability of triaxially compressed rocks indicate that its evolution during deformation can be very complex. In low-porosity crystalline rocks, permeability enhancement is generally observed during dilatancy and a positive correlation exists between permeability and porosity. In porous siliciclastic rocks, experimental data show that the permeability would actually decrease while the pore space dilates as a sample is stressed to brittle failure. With constraints from microstructural observations, percolation network models provide physical insights into the interplay of stress-induced cracking and pre-existing pore structure that induces such negative correlation between porosity and permeability changes. The laboratory data suggest that tectonic deformation in a seismogenic system located in crystalline rock can potentially induce massive fluid discharge along conduits, which may be localized in a high-permeability heterogeneity such as the damage zone. However, it is unlikely that faulting in porous rock formations and shear displacement along gouge zones can provide conduits for fluid along fault zones. To emphasize the fundamental differences between a compact and a porous rock, separate deformation-permeability maps in the stress space are proposed as conceptual models for the permeability evolution. In a region of active faulting, the loading path typically involves variations in the deviatoric stresses, as well as the mean stress and pore pressure. Laboratory data provide important insight into the interplay of deviatoric stresses, mean stress, and pore pressure in relation to permeability evolution and fluid transport. Hydromechanical parameters that control the generation and maintenance of pore pressure excess are also identified.

1. INTRODUCTION

The transport of fluid is dynamically linked to many sedimentary, metamorphic, and tectonic processes in the crust. It exerts significant mechanical and chemical effects

Faults and Subsurface Fluid Flow in the Shallow Crust
Geophysical Monograph 113

on numerous processes of importance in seismotectonics, economic geology, petroleum geology, aqueous geochemistry, and geotechnical engineering [*Bredehoeft and Norton*, 1990; *Torgersen*, 1991; *Parnell*, 1994; *Jamtveit and Yardley*, 1997]. The key physical parameter that governs fluid transport is permeability, a quantity that varies by >10 orders of magnitude in geomaterials [*Freeze and Cherry*, 1979; *Brace*, 1980; *Hanson*, 1995]. Considerable permeability contrast may arise from lithological difference, for example, an argillaceous formation may act as a hydraulic barrier with permeability down to 10^{-23} m^2 [*Neuzil*, 1994].

Since it is difficult to even come up with an order-of-magnitude estimate, the permeability in a given formation is often idealized in models as not changing temporally or spatially. However, there has been a growing appreciation that the permeability structure and consequently the fluid flow of many active geologic settings involve significant spatio-temporal heterogeneity. As a discontinuous structure that has sustained localized shear deformation, a fault system is characterized by a spatial distribution of permeability that is highly complex [*Pittman*, 1981; *Antonellini and Aydin*, 1994; *Fowles and Burley*, 1994; *Caine et al.*, 1996; *Evans et al.*, 1997; *Seront et al.*, 1998]. Furthermore, the permeability of a seismogenic system may evolve with time in response to tectonic loading and hydrothermal activity [*Byerlee*, 1993; *Chester et al.*, 1993; *Muir Wood*, 1994; *Hickman et al.*, 1995]. At different stages of the earthquake cycle, a fault system may provide conduits or barriers for fluid transport. A localized zone that has undergone dilatancy in the inter-seismic stage may act as a suction pump in the post-seismic stage. Relatively impermeable barriers may be established by healing and sealing processes that can maintain pore pressure excesses in hydraulically isolated compartments during the inter-seismic stage. Significant discharges through fault-valve actions may subsequently occur if the hydraulic barriers are ruptured during an earthquake [*Sibson*, 1994].

Permeability is intimately related to pore geometry, and since the phenomenon of dilatancy is universally observed as a precursor to brittle faulting in geomaterials [*Paterson*, 1978], the developments of dilatancy and permeability are expected to be coupled. For mathematical convenience, permeability is commonly prescribed as a function of the porosity in theoretical models [*Person et al.*, 1996; *Roberts et al.*, 1996] with two implicit assumptions. First, the permeability and porosity changes are positively correlated, and permeability enhancement is commonly inferred to arise from dilatancy and faulting. Second, the functional relation is independent of loading path and applicable to both elastic and inelastic deformation.

To what extent are such relations applicable to rocks of different lithology and porosity? As elaborated upon by *Sibson* [1994], dilatancy during an earthquake cycle may be induced by variations in the deviatoric stresses, mean stress or pore pressure. Whether the dilation is elastic or inelastic hinges on the loading path, which can influence the permeability significantly [*Zhu and Wong*, 1997]. Laboratory data also indicate that the stresses at the onset of dilatancy and the amount of dilation both vary significant with rock type [*Brace*, 1978]. In porous siliciclastic rocks, recent data show that the permeability would actually decrease while the pore space dilates as a sample is stressed to brittle failure [*Zhu and Wong*, 1996]. A deeper understanding of such complex interplay of dilatancy, brittle faulting and permeability evolution would require the synthesis of data from controlled experiments on deformation and fluid flow and from theoretical modeling.

This paper explores the mechanical effects of brittle faulting on permeability evolution. We will first review recent hydromechanical measurements in the brittle faulting regime to highlight the fundamentally different paths for permeability evolution that have been observed in compact and porous rocks. We next summarize relevant data from quantitative characterization of the deformation-induced microstructure that provide important insight into the contrasts in permeability evolution. Theoretical modeling based on percolation theory is then employed to connect the hydromechanical and microstructural measurements. Implications of the laboratory and theoretical results on seismotectonics will also be discussed.

2. BRITTLE FAULTING AND FLUID FLOW: CONTRASTS IN HYDROMECHANICAL BEHAVIOR BETWEEN COMPACT AND POROUS ROCKS

A wide range of pressure, temperature, and stress state are encountered in tectonic, metamorphic and diagenetic processes. Laboratory measurements of permeability under crustal conditions of pressure, temperature and stress can therefore provide important insight into the coupling of deformation and fluid transport. Although preliminary attempts have been made (by e.g. *Fischer and Paterson*, [1992]) to conduct permeability measurements under elevated temperatures and nonhydrostatic loading, most experimental data on the effect of stress on permeability have been obtained under room-temperature conditions.

Such laboratory studies have provided a fairly complete picture of how permeability depends on porosity and deformation mechanism. In a compact rock, dilatancy (defined as the inelastic increase in volume during deformation under applied differential stress) is generally observed whether it fails by brittle faulting or cataclastic flow. Since the volumetric strain and differential stress are linearly related in the elastic regime, the onset of dilatancy is defined to be at that stress level (commonly designated C') at which the relation between the volumetric strain and stress becomes nonlinear [*Brace*, 1978; *Paterson*, 1978]. In the brittle faulting regime, the development of dilatancy is illustrated in Figure 1a by the triaxial compression data from a recent study of *Kiyama et al.* [1996]. The Inada granite sample attained a peak stress and failed by shear localization while undergoing strain softening. The onset of dilatancy was at a stress level C' that was ~ 1/2 of the peak value. In the ductile regime, the development of dilatancy is illustrated in Figure 1b by the data for Carrara marble obtained by *Fredrich et al.* [1989]. The differential stress did not attain a maximum, and the sample failed by cataclastic flow while undergoing strain hardening. The onset of dilatancy C' is as indicated.

In a rock with a relatively high porosity (>5% or so), although the mechanical behavior is qualitatively similar to that of a compact rock in the brittle regime, fundamental differences were observed in the ductile regime. The similarity and difference are illustrated in Figure 2 with data for Berea sandstone (that undergoes the brittle-ductile transition at an effective pressure of ~ 100 MPa). In such a material with appreciable porosity, the relation between elastic volumetric strain and stress is typically nonlinear. In this case, the onset of dilatancy is defined to be at the stress level C' at which the (compactive) volumetric strain at a given effective mean stress becomes less than that observed in hydrostatic loading to the corresponding effective pressure, and similarly the onset of shear-enhanced compaction is defined to be at the stress level C^* at which the volumetric strain at a given mean stress becomes greater [*Brace*, 1978; *Paterson*, 1978; *Wong et al.*, 1997a]. In the brittle faulting regime, the Berea sandstone sample showed strain softening and failed by shear localization (Figure 2a). The onset of dilatancy C' was at a stress level that was ~ 1/2 of the peak value, and the inelastic volume change is small relative to the initial porosity. In the ductile regime, dilatancy was inhibited and the sample failed by cataclastic flow while undergoing appreciable inelastic compaction (Figure 2b). This phenomenon of compactive yield initiates at the stress level C^* that corresponds to the onset of shear-enhanced compaction.

To investigate the concomitant evolution of porosity and permeability, it is important to measure both these quantities during the deformation. As indicated in Figure 1a, the loading ram was locked at different stages of deformation, and the *in situ* permeability measured as a function of the stress state. If the permeability at a certain stress state is relatively high ($k > 10^{-16}$ m^2), it is measured by the steady-state flow technique. For $k < 10^{-16}$ m^2 or so, thermal fluctuation would render it difficult to achieve steady-state flow, and instead the pulse transient technique pioneered

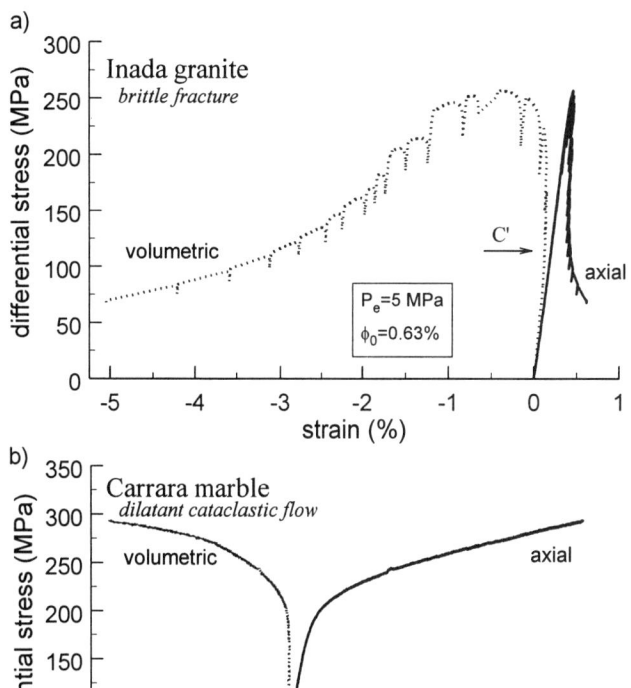

Figure 1. Stress-strain curves of a) Inada granite in the brittle faulting regime [*Kiyama et al.*, 1996]; and b) Carrara marble in the dilatant cataclastic flow regime [*Fredrich et al.*, 1989]. The solid and dotted curves represent axial and volumetric strain, respectively. The effective pressure (P_e) and initial porosity (ϕ_0) are as indicated, and the arrows mark the onset of dilatancy C' corresponding to the onset of nonlinearity in the stress-volumetric strain curves.

by *Brace et al.* [1968] may be used for the relatively impermeable sample. Alternatively the pore pressure oscillation method [*Kranz et al.*, 1990; *Fischer and Paterson*, 1992] can be employed.

During the development of dilatancy and brittle faulting, the permeability evolves along fundamentally different paths in compact and porous rocks. The contrast is illustrated in Figure 3. A triaxial compression experiment is performed in two sequential stages. First the confining pressure is increased, and then the differential stress is applied while the confining pressure is maintained at a fixed value. We are not aware of any permeability and porosity measurements for both stages from a single experiment on a compact rock, and therefore experimental data from several studies are incorporated in Figure 3a to illustrate the

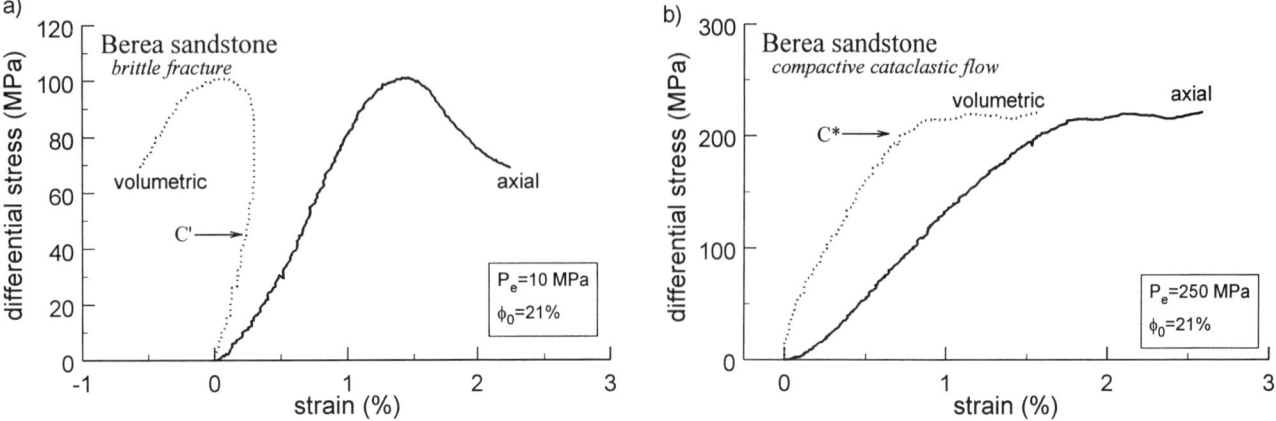

Figure 2. Stress-strain curves of Berea sandstone in the a) brittle faulting and b) cataclastic flow regimes. The solid and dotted curves represent axial and volumetric strain, respectively. The effective pressure (P_e) and initial porosity (ϕ_0) are as indicated. The arrows mark the onset of dilatancy C' in a) and the onset of shear-enhanced compaction C^* in b), respectively. Criteria for picking out these critical stresses are explained in the text.

permeability evolution in a compact granite rock. As shown by the Westerly granite data of *Brace et al.* [1968], hydrostatic loading reduces the porosity and permeability (Figure 3a), due to elastic closure of pre-existing microcracks. The permeability change due to the differential stress is illustrated by *Zoback and Byerlee's* [1975] data for Westerly granite at an effective pressure of 39 MPa. Initially the porosity and permeability both decrease due to elastic compression of the pore space. However, at the onset of dilatancy this trend is reversed with concomitant increases in porosity and permeability. The Westerly granite data in Figure 3a were for deformation up to ~75-95% of the peak stress. The study of *Kiyama et al.* [1996] is more comprehensive in that the deformation was carried through the post-peak stage (Figure 1a). As shown in Figure 3a, the permeability change was considerably higher in the Inada granite. Nevertheless, the three sets of data for compact granite are qualitatively similar in that the permeability and porosity changes are positively correlated.

In contrast, the porosity and permeability changes in the Berea sandstone sample (Figure 3b) did not track one another during the development of dilatancy. While the porosity and permeability changes were positively correlated for hydrostatic loading and for triaxial loading during the inital elastic compression stage before C', permeability continued to decrease even though the pore space dilated in response to the differential stress. Although such a negative correlation between porosity and permeability changes are somewhat counter-intuitive and contradict most empirical models for correlating these two quantities, *Zhu and Wong* [1997] presented data for several sandstones with a wide range of porosities that showed similar behavior.

To characterize the fundamentally different evolution paths for permeability in compact and porous rocks during the development of dilatancy and brittle failure, *Zhu and Wong* [1997] introduced the following parameter

$$\xi = \frac{k(peak)}{k(C')} - 1 \qquad (1)$$

where $k(C')$ and k(peak) denote the permeabilities at the onset of dilatancy and peak stress, respectively. Since dilatancy is universally observed in geomaterials during brittle failure, ξ is a parameter that describes the correlation between permeability and porosity changes. We compile in Table 1 values of ξ for several rock types and unconsolidated materials that were triaxially compressed and failed by brittle faulting. The correlation parameter is also plotted as a function of porosity in Figure 4. There is an overall trend for the correlation between permeability and porosity to change from positive to negative with increasing porosity. In compact crystalline rocks, permeability is positively correlated with porosity (with $\xi > 0$) as is commonly expected. In contrast, permeability is negatively correlated with porosity (with $\xi < 0$) in a sandstone with porosity >15% or so. *Jones'* [1981] data for crushed Westerly granite suggest that a negative correlation may persist to porosities down to 11% in unconsolidated materials (Figure 4). Although it is difficult to extract values of ξ from their data, results for crushed Westerly granite and Ottawa sand of *Zoback and Byerlee* [1976a, b] and for quartz/clay mixture of *Brown et al.* [1994], indicate qualitatively similar trends.

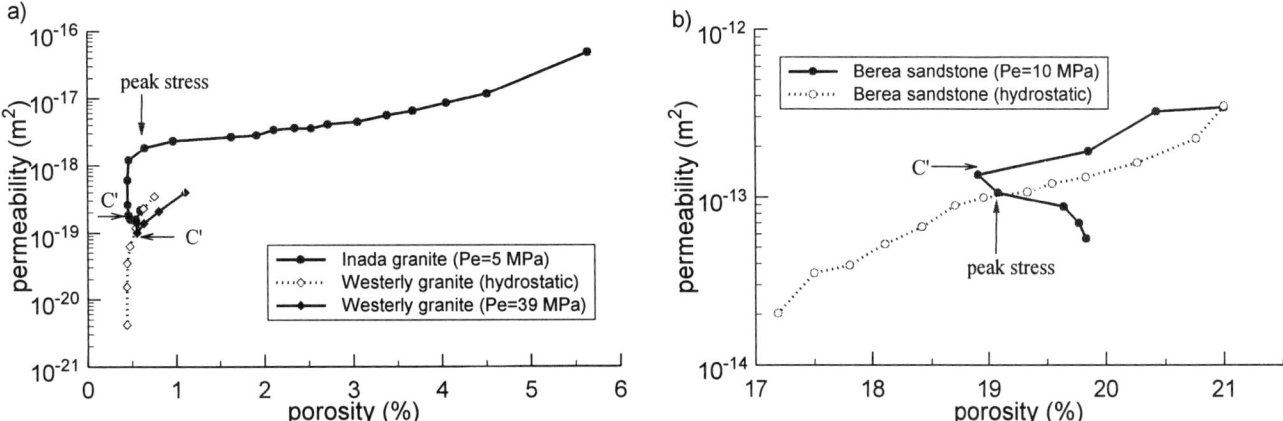

Figure 3. a) Permeability as a function of porosity for Inada granite and Westerly granite. The solid circles are data of *Kiyama et al.* [1996] for Inada granite triaxially compressed at an effective pressure of 5 MPa. The permeability first decreased due to elastic compression, increased drastically from the onset of dilatancy (marked by *C'*) to the peak stress, and then increased gradually in the post-peak stage. The solid diamonds are *Zoback and Byerlee's* [1975] data for Westerly granite triaxially compressed at an effective pressure of 39 MPa. The experiment was terminated before the peak stress had been attained, so only two stages of permeability evolution (initial decrease due to elastic compression and subsequent increase due to dilatancy) were observed. For reference, hydrostatic compression data of Brace et al. [1968] are also included as open diamonds. Elastic closure of microcracks induced permeability to drop to a very low value of 4 x 10^{-21} m². b) Permeability as a function of porosity for Berea sandstone [*Zhu and Wong*, 1996]. The solid circles are for a sample triaxially compressed at an effective pressure of 5 MPa. Permeability decreased monotonically with stress, even though the porosity first decreased with elastic compression and then increased with the onset of dilatancy (marked by *C'*). For reference, hydrostatic compression data are also included as open circles.

3. PORE SPACE STATISTICS AND PERCOLATION NETWORK MODELING OF THE EVOLUTION OF PERMEABILITY

Various analytic relations between permeability k and porosity ϕ have been proposed, such as the "equivalent channel" model [*Paterson*, 1983; *Walsh and Brace*, 1984] that expresses the permeability as

$$k = C \frac{m^2 \phi}{\tau^2}, \qquad (2)$$

where m is the hydraulic radius, and τ is the "tortuosity". The geometric factor C has a value of 1/2 for circular tubes and 1/3 for cracks. This model has provided useful insight into the evolution of permeability with deformation. However, it apparently has the limitation that when applied to compact rocks [*Walsh and Brace*, 1984; *Fischer and Paterson*, 1992; *Zhang et al.*, 1994], the model requires an equivalent crack that has an aperture significantly thinner than microcracks observed under the optical microscope or scanning electron microscope (SEM). In a porous rock, the negative correlation between permeability and porosity can

be explained qualitatively by this model only if stress-induced cracking induces the tortuosity to increase and hydraulic radius to decrease.

A deeper understanding of the complex relation between permeability and porosity would require a more realistic model of the pore space. It should capture the fundamental differences in pore geometry of compact and porous rocks. As shown in Figure 5a, the pore space of a compact rock is dominated by microcracks [*Hadley*, 1976; *Kranz*, 1983]. Dilatancy arises from the proliferation of stress-induced cracking [*Tapponier and Brace*, 1976], which increases the permeability by increasing the crack density and enhancing the connectivity of the flow paths. In contrast, the pore space of a clastic rock is made up of several distinct types of void (Figure 5b). A significant contribution to the overall porosity is from nodal pores located at four-grain vertices. Roughly equidimensional in geometry, these nodes are joined to one another by connective pores, which include microcracks situated along interfaces between two neighboring grains and tubular pores along three-grain edges. Collectively the tubular and sheet-like cavities are sometimes referred to as "throats", with the common attribute that they control the transport properties even though their contribution to the overall porosity is usually less than the nodal pores [*Doyen*, 1988;

Table 1. Permeability change during the development of dilatancy and brittle faulting.

Rock Sample	Initial Porosity (%)	Permeability Change, ξ	Effective Pressure, $P_c - P_p$ (MPa)	Reference
sandstone				
Adamswiller	22.9	-0.62	5	*Zhu and Wong* [1997]
Berea 1	21.0	-0.22	5	*Zhu and Wong* [1997]
		-0.22	10	
Rothbach	19.9	-0.58	5	*Zhu and Wong* [1997]
Berea II	17.0	-0.18	14	*Gatto* [1984]
		-0.16	28	
Darley Dale I	14.5	0.12	5	*Zhu and Wong* [1997]
		0.12	10	
		-0.23	20	
Darley Dale II	11.4	0.37	7	*Mordecai and Morris* [1971]
		0.58	14	
		0.23	28	
		0.08	42	
Tennessee	4.5 - 7.5	3	40	*Keanny et al.* [1998]
		5	100	
Mesaverde	3.8	> 1.5 [b]	25	*Senseny et al.* [1983]
crystalline rock				
Westerly granite	0.9	> 3.0 [c]	39	*Zoback and Byerlee* [1975]
		> 2.0 [c]	14	
Inada granite	0.6	7.9	5	*Kiyama et al.* [1996]
Creighton gabbro	0.5	> 5.5 [d]	25	*Trimmer et al.* [1980]
granular material				
crushed Westerly	21.5 [a]	-0.19	190	*Jones* [1981]
granite	14.0 [a]	-0.35	190	
	11.6 [a]	-0.09	190	

[a] Porosity value attained at peak stress.

[b] Measurement for a sample that was deformed to beyond peak stress and unloaded.

[c] The granite samples were deformed to ~ 75-95% of the peak stress.

[d] The gabbro sample was deformed to ~ 75% of the peak stress.

Bernabé, 1991]. Microstructural observations [*Menéndez et al.*, 1996] show that dilatancy in a porous rock arises from relative grain movement as well as stress-induced cracking (Figure 5b). Whether permeability increases or decreases with dilatancy is controlled by the partitioning of flow between the tubular pores and microcracks.

A theoretical technique that has proved to be very useful is network modeling [e.g. *Seeburger and Nur*, 1984; *David et al.*, 1990; *Zhu et al.*, 1995], because it can readily incorporate pore space statistics that are constrained by quantitative microstructural data to simulate the evolution of permeability with damage. Here we illustrate this approach applied to Westerly granite and Berea sandstone. The statistics of microcrack length, aspect ratio and aperture of Westerly granite have been characterized by *Hadley* [1976], *Tapponier and Brace* [1976], and *Wong et al.* [1989]. Microstructural attributes of Berea sandstone were characterized quantitatively by *Menéndez et al.* [1996]. The

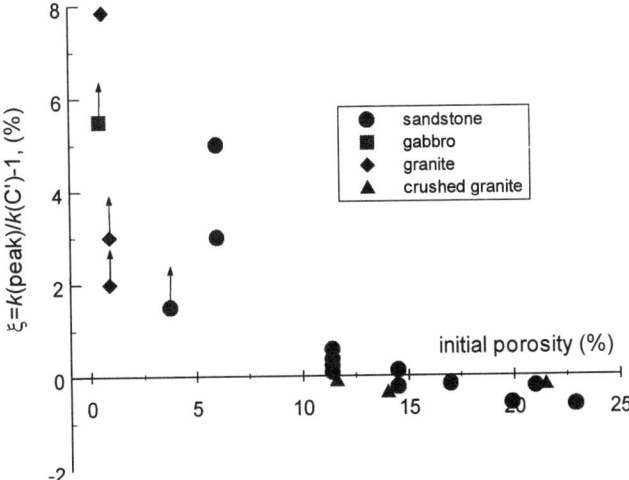

Figure 4. Prefailure permeability change as a function of initial porosity in the brittle faulting regime for lithified rocks (sandstone, gabbro, and granite) and granular material (crushed Westerly granite). The numerical data are also compiled in Table 1. The correlation coefficient ξ (defined in equation 1) represents the relative change in permeability from onset of dilatancy C' to peak stress. If $\xi > 0$ then permeability and porosity changes are positively correlated, and if $\xi < 0$ then permeability decreases in a dilating sample. Initial porosities and porosity values at failure were used for the lithified rocks and granular materials, respectively. A data point marked by an arrow indicates that the permeability was measured before the peak stress had been attained and it probably represents a lower bound. Multiple values for the same porosity value are data from triaxial compression tests at different effective pressures.

aspect ratio spectra of several rock types were also calculated by *Cheng and Toksöz* [1979] by inversion of seismic velocity data.

The network topology and modeling approach are identical to those of *Zhu et al.* [1995], and *Zhu and Wong* [1996]. The pore space is idealized as a simple cubic network of (12 x 12 x 12) conducting bonds. In accordance with percolation theory [*Stauffer and Aharony*, 1992], the percolative and connective properties of such a network undergo two transitions. The first transition occurs when a critical number of bonds have been incorporated such that the network embeds a cluster of bonds that connects two diagonal edges, and thereby undergoes a transition from being unconnected to percolative. As an analogue of the pore space of a rock, the percolation threshold in a network model signifies the onset of hydraulic permeability. The topology of a simple cubic network is such that when 25% of the bonds are occupied, the percolation threshold is attained.

Beyond the percolation threshold, the bonds in the network are divided into two groups. One group is made up of the interconnected cluster that provides a continuous conduit for fluid transport, and the other is made up of isolated clusters of connected bonds that represent unconnected porosity which does not contribute to permeability. As more conduction bonds are incorporated, a second transition occurs when the occupation probability attains the maximum value of 1 and all bonds in the network are connected together. Beyond this "crossover", the network is fully connected and the connectivity cannot be increased further. The pore space is fully or partially connected depending on whether the probability for bond occupation is equal to or less than 1.

In a compact rock, all the conducting bonds are identified with microcracks, idealized to be parallel rectangular plates (with dimensions L and a) separated by the aperture w. The dimension L (presumably fixed at a value comparable to the grain size) is not specified in the network computation because the final result for the permeability is independent of L. In our model, the pore space statistics are characterized completely by the probability distributions of the length a and of the aspect ratio w/a, which are assumed to be independent.

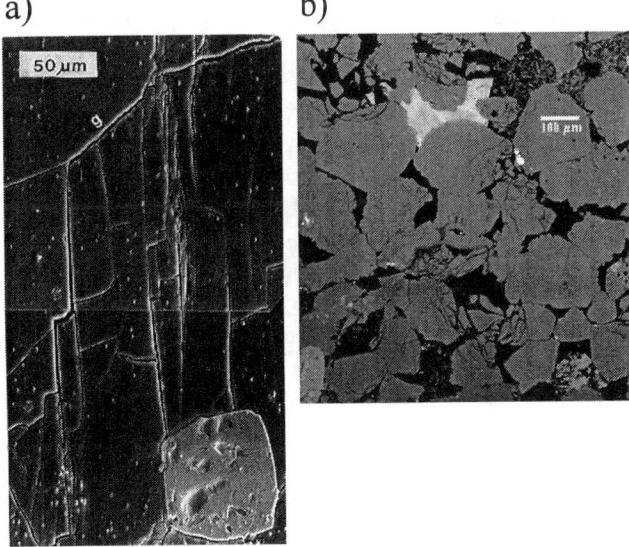

Figure 5. Microstructure observed in samples of a) Westerly granite, and b) Berea sandstone unloaded just before the peak stresses were attained. The maximum principal stress was applied in an axial direction. Note the pore space of the granite is dominated by stress-induced microcracks, whereas that of the sandstone is made up of stress-induced microcracks that are connected to equant and tubular pores.

We assumed that in a compact rock all stress-induced dilation is converted into microcrack porosity and that the pore space statistics remain the same during this dilation process. Similar assumptions were adopted by *Guéguen and Dienes* [1989] and *Peach and Spier* [1996] in their linkage models for uniform-size cracks. The dilation process was simulated by random occupation of bonds in the network, with pore space statistics as prescribed in Figure 6a. Guided by *Hadley's* [1976] measurements of crack length, we adopted an exponential distribution that is skewed towards the shorter cracks. Guided by *Cheng and Toksöz's* [1978] inversion results, the logarithm of aspect ratio was assumed to be uniformly distributed between 10^{-4} and 10^{-2}. The Monte Carlo simulations are compared with laboratory data of Westerly granite in Figure 6b. At each porosity, the flow in 6 different random networks was simulated, and the geometrical mean and range of the calculated permeability values are plotted here.

A critical number of cracks have to be incorporated into the pore space before the percolation threshold is attained. Beyond the percolation threshold, the addition of conductive cracks enhances the connectivity and thereby increases the permeability significantly. As the crack density increases further, the occupation probability attains the maximum value of 1 and the cracks network becomes a fully connected system. Beyond this "crossover", the connectivity cannot be increased further. The permeability and porosity increases in this fully connected regime approximately follow a linear trend.

It can be seen from Figure 6b that the overall agreement between our random network model and experimental data is reasonable. The unstressed Westerly granite (with a porosity of ~0.9% and permeability of ~4 x 10^{-19} m²) has a pore space that seems to be fully connected. On the one hand, permeability data of *Brace et al.* [1968] indicate that the application of confining pressure reduced the connectivity of the pore space somewhat and moved it down to the percolative regime. On the other hand, *Zoback and Byerlee's* [1975] data indicate that the pore space evolved to become fully connected soon after the differential stress was applied. The stress-induced enhancement of permeability of Westerly granite was small relative to Inada granite (Figure 3a), possibly because Westerly granite was more connected in the unstressed state. However, a quantitative analysis of permeability evolution in Inada granite cannot be pursued at this stage because microstructural data are unavailable.

The prescribed pore space statistics for Westerly granite (Figure 6a) implies that 46% of the cracks (with apertures in the range 0.01 - 0.03 μm) would not be resolvable by the SEM. After adjusting for the resolution limitation of the microstructural measurements, *Zhu and Wong*

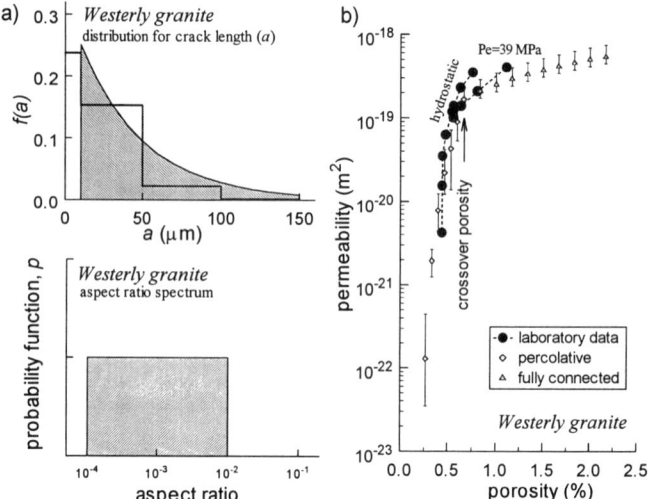

Figure 6. a) Crack length distribution and aspect ratio spectrum of Westerly granite. The histogram represents crack length data of a Westerly granite sample (T5) stressed to near the peak stress at a confining pressure of 50 MPa [*Hadley*, 1976]. b) Simulation results on permeability as a function of porosity for Westerly granite. The solid circles correspond to experimental data (plotted as solid and open diamonds in Figure 3). The open symbols are geometrical means of calculated permeability values of 6 random network simulations, and the range of permeability values at each porosity is given as the error bar. The diamonds represent simulation results in the percolative regime, and the triangles are results in the fully connected regime. The crossover porosity signifies the transition between the two regimes.

[1999] showed that the theoretical distribution of crack aperture and crack density are in very good agreement with SEM measurements of *Wong et al.* [1989] and *Tapponier and Brace* [1976], respectively.

In a porous clastic rock, the pore space is made up of not only microcracks, but also tubular pores. To simulate the interplay of these two different types of void, *Zhu and Wong* [1996] considered a fully connected network made up of "tunnel cracks" with a fixed length L and elliptical cross-sections (with major and minor axes a and b). Tubular pores have relatively large aspect ratios between 0.1 and 1, and microcracks are relatively thin with aspect ratios <0.1. At the onset of dilatancy, all pre-existing microcracks were assumed to be closed by the applied mean stress and therefore the pore space of the reference sample (at C') was assumed to be solely made up of tubular pores. When stressed to beyond C', the pore space of the damaged sample was assumed to include both tubular pores and microcracks. Accumulation of damage is characterized by a parameter ω (defined by the number of bonds with aspect ratio < 0.1 normalized to the total number of bonds in the network).

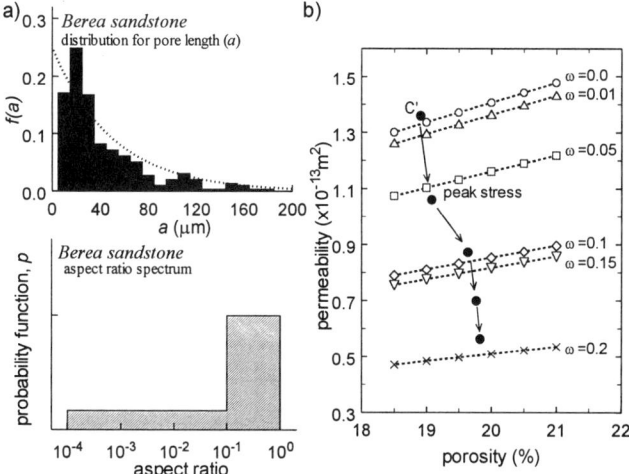

Figure 7. a) Pore size distribution and aspect ratio spectrum of Berea sandstone. The solid bars are chord length distribution for an unstressed sample measured by *Menéndez et al.* [1996]. The dashed curve is the input distribution of pore length in the network modeling. The spectral densities for the cracks (with aspect ratio between 10^{-4} and 10^{-1}) and tubular pores (with aspect ratio between 10^{-1} and 1) were chosen such that the areas under the crack and pore distributions are given by the damage parameter ω and $1-\omega$, respectively. b) Simulation results of permeability evolution in Berea sandstone. Each dashed line indicates permeability change with porosity reduction at a fixed damage parameter ω. The solid circles are experimental data for Berea sandstone deformed from the onset of dilatancy C' to the post-peak stage at effective pressure of 10 MPa. The data can be explained by the network model if the damage progressively increased from $\omega = 0$ to $\omega = 0.2$.

Guided by microstructural measurements of *Menéndez et al.*, [1996], an exponential distribution (given by the dashed curve in Figure 7a) was adopted for the distribution of the major axis a. The aspect ratios of tubular pores and microcracks were each assumed to follow a log-uniform distribution, with a lower cutoff value at 10^{-4}. As indicated in the schematic diagram of (Figure 7a), the spectral densities of the distributions for the cracks and tubular pores were chosen such that the areas under the crack and pore distributions are given by ω and $1-\omega$, respectively. While the damage parameter $\omega = 0$ in the reference sample at C', it increased monotonically as the sample was stressed beyond C' to brittle failure, with corresponding changes in the aspect ratio spectra.

The network simulation results are summarized in Figure 7b. Each dashed line in this plot shows permeability as a function of porosity for a fixed ω. For this subset of networks, the distributions for all the geometric attributes of the conducting bonds are the same, implying that the pore space has the same "tortuosity". As long as ω is fixed, the permeability and porosity changes are related linearly with a positive slope.

For comparison, the Berea sandstone data (for effective pressure of 10 MPa) are shown as dark symbols in Figure 7b. From C' to the peak stress, the pore space dilated and the porosity increased by 0.17%. According to our network model, the permeability would *increase* by ~1% in the absence of stress-induced cracking (with ω fixed at 0). However, the experimental data actually showed a permeability *decrease* of 22%, which would require the damage ω to increase from 0 to ~0.05 (Figure 7b). *Zhu and Wong's* [1996] calculation show that this damage accumulation corresponds to a crack porosity of ~0.04%, which is ~ 1/4 of the measured dilation from C' to the peak stress. This theoretical prediction agrees qualitatively with the microstructural observations of *Menéndez et al.* [1996], that suggested that most of the dilation was from relative movement of the grains, with minor contribution from localized clusters of intragranular cracking that developed near the peak stress. *Zhu and Wong's* [1996] simulations show that the partitioning of flow between paths along tubular pores and microcracks is such that the latter paths are very effective in increasing the "tortuosity" and consequently decreasing the permeability. If the stress-induced cracking effect dominates, then permeability may actually decrease in a dilating rock. The negative correlation between permeability and porosity changes is attributed to the complex interplay of flow processes in a pore space that is made up of tubular pores and microcracks with significantly different geometric and hydraulic attributes.

4. DISCUSSION

4.1. Effect of Stress on the Development of Dilatancy, Brittle Faulting and Permeability

It is commonly assumed that brittle faulting leads to the concomitant development of dilatancy and permeability enhancement. Recent laboratory data on compact crystalline rocks lend support to this assumption. Indeed experimental data show that significant permeability enhancement is generally observed not only in the brittle faulting regime but also in the dilatant cataclastic flow regime [*Zhang et al.*, 1994; *Peach and Spiers*, 1996]. Therefore tectonic deformation in a seismogenic system located in crystalline rock can potentially induce massive fluid discharge along conduits, that may be localized in a high-permeability heterogeneity such as the damage zone [*Barton et al.*, 1995; *Evans and Chester*, 1995; *Caine et al.*, 1996; *Seront et al.*, 1998].

However, recent data also show that permeability evolves along a fundamentally different path in porous rocks. Even though the sample dilates prior to brittle failure, permeability may actually decrease (Figures 3 and 4). *Zhu and Wong's* [1997] data show that permeability in a porous sandstone also decreases with increasing deformation in the post-peak stage. After a through-going shear band has developed, the permeability evolution is expected to be similar to *Teufel's* [1987] observations in fractured Coconino sandstone (of initial porosity 10%), with permeability decreasing with increasing shear displacement in the gouge zone. Unconsolidated materials also show negative values of ξ comparable to those for porous sandstones (Table 1).

Therefore it is unlikely that tectonic faulting in porous rock formations and shear displacement along gouge zones can provide conduits for massive flux of fluid along fault zones. Field observations in sandstone formations [e.g. *Antonellini et al.*, 1994; *Fowles and Burley*, 1994] seem to support this conclusion. Permeability may decrease further in the presence of grain crushing [*Antonellini and Aydin*, 1994; *David et al.*, 1994;], smearing and plastic flow of clay [*Gibson*, 1994], and hydrothermal healing and sealing [*Moore et al.*, 1994; *Zhu et al.*, 1995].

Such complications due to crystal plasticity, granular flow, and diffusive mass transfer have profound influences on the permeability evolution during the brittle-ductile transition. To emphasize the differences between a compact and a porous rock, *Zhu and Wong* [1997] proposed separate deformation-permeability maps in the stress space as conceptual models for the permeability evolution. In a compact crystalline rock (Figure 8a), significant dilatancy and permeability enhancement are observed in both the brittle faulting and cataclastic flow regime when a sample is stressed to beyond C'. However, under elevated effective pressures (and also elevated temperatures for quartzofeldspathic rocks), a transition from dilatant cataclastic flow to fully plastic flow occurs if the differential stress exceeds the yield stress, and the permeability is expected to decrease. The development of plastic flow may eliminate the interconnected porosity and ultimately lead to an impermeable rock, corresponding to the percolation threshold. Therefore the critical stresses for the onset of dilatancy C' and for plastic flow map out the boundary of a domain in which permeability always decreases with increasing stress.

In a porous rock (Figure 8b), the critical stresses for the onset of dilatancy (C') and of shear-enhanced compaction (C^*) map out the boundary of a closed domain in the stress space, within which porosity and permeability decrease with increasing hydrostatic loading, with negligible dependence on the deviatoric stresses. In the brittle faulting regime, permeability may actually decrease even though

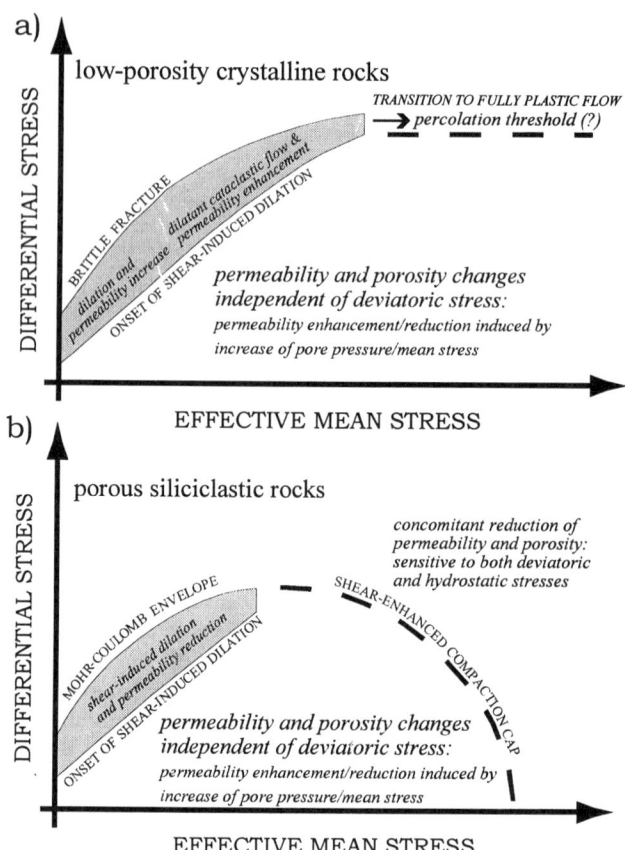

Figure 8. Deformation-permeability maps for a) compact crystalline rocks, and b) porous siliciclastic rocks.

the rock dilates when stressed to beyond C'. After the peak stress (mapped out by the Mohr-Coulomb fracture envelope) has been attained, the rock usually shows accelerating permeability reduction with the development of shear localization. In the cataclastic flow regime, the compactive yield stresses C^* map out an elliptical cap in the stress space which expands with decreasing porosity and grain size [*Wong et al.*, 1997a]. If a porous rock is loaded to beyond this cap, significant permeability reduction results from the development of shear-enhanced compaction and cataclastic flow.

4.2. Permeability Evolution: Interplay of Pore Pressure Excess, Mean Stress and Deviatoric Stresses

It should be emphasized that the triaxial compression experiment is designed to focus on the influence of differential stress on the evolution of hydromechanical properties. In an earthquake cycle, the loading path typically involves variations in the deviatoric stresses, as well as the mean stress and pore pressure [*Muir Wood*, 1994; *Sibson*, 1994]. In the inter-seismic stage, the mean stress progressively increases (or decreases) in a compressional (or extensional) tectonic setting. Significant variations in mean stress also accompany coseismic stress drops, and significant pore pressure excess may also result from numerous mechanisms (such as disequilibrium compaction, dehydration reaction, and influx of fluid from external sources) that have been identified in igneous, metamorphic and sedimentary settings [*Bredehoeft and Hanshaw*, 1968; *Person et al.*, 1996]. Tectonic deformation coupled with healing and sealing processes may reduce the permeability to a sufficiently low level, thus providing a favorable hydromechanical environment for the maintenance of such pore pressure excesses. If the pore pressure builds up to nearly lithostatic, then hydrofracturing may provide conduits for focused and localized fluid transport within fault zones [e.g., *Brown et al.*, 1994].

It is therefore important to also explore the influences of pore pressure and mean stress on permeability. In theoretical models that consider the coupling of mechanical deformation and fluid flow, permeability is commonly assumed to evolve as a function of the porosity (that evolves with pore pressure and mean stress) or the stresses. Motivated by laboratory measurements and the "equivalent channel" model (equation 2), permeability and porosity are often postulated to follow a power law of the form

$$k = k_o (\phi / \phi_o)^\alpha \qquad (3)$$

where k and ϕ are the current permeability and porosity, and k_o and ϕ_o are those at a reference state. Other empirical relations (such as a linear relation between the logarithm of permeability and porosity) have also been adopted [*Bethke*, 1985; *Person et al.*, 1996]. Experimental data pertinent to the evolution of permeability with porosity reduction due to mechanical and chemical compaction were reviewed by *David et al.* [1994], who also compiled values of the porosity sensitivity coefficient α. If the coefficient has a small value (on the order of 1), then fluid conduits can be established at relatively low porosity. In a partially molten system, it was inferred from laboratory measurement that $\alpha \sim 1$, with the implication that melt extraction can be highly effective through high-permeability conduits in the mid-oceanic ridge even thought the melt fraction may be relatively small [*Riley et al.*, 1990]. In contrast, relatively high values of α (up to 7 or more) observed in calcite and quartz aggregates undergoing healing and sealing imply that permeability may drop drastically during compaction, with the

consequence that pore pressure excess may be readily generated and maintained [*Walder and Nur*, 1984; *Zhu et al.*, 1995].

Permeability generally decreases with decreasing pore pressure or increasing mean stress, and the evolution in the poroelastic regime can be expressed as a function of the "effective" mean stress

$$\sigma_{eff} = \sigma_m - \kappa p_p \qquad (4)$$

where σ_m is the mean stress (arithmetic mean of the three principal stresses) and p_p is the pore pressure. For relatively homogeneous materials, it can be shown from poroelasticity theory that the effective stress coefficient for permeability $\kappa \leq 1$ [*Walsh*, 1981], as commonly observed in compact crystalline rock and "clean" sandstone [*Morrow et al.*, 1986; *Bernabé*, 1987; *David and Darot*, 1989]. However, *Berryman* [1992] demonstrated that the effective pressure coefficient for permeability could be significantly greater than unity for a rock containing multiple constituents and he compiled data for clay-bearing porous rocks with κ values up to 7.

In theoretical models [e.g., *Rice*, 1992], permeability is often assumed to decrease exponentially with effective mean stress

$$k = k_o \exp\left(-\gamma(\sigma_{eff} - \sigma_o)\right) \qquad (5)$$

where k_o is the permeability at the reference stress state σ_o. Other empirical relations (such as a power law) have also been postulated [e.g., *Walsh*, 1981; *Shi and Wang*, 1986]. *David et al.* [1994] presented a compilation of the coefficient γ. Since then, new data on the permeability of fault zone materials as a function of stress have become available from recent laboratory studies. As shown in our updated compilation (Table 2), the data can be separated into two groups. Whereas porous sandstone and unconsolidated materials have relatively low values of $\gamma < 0.02$ MPa^{-1}, compact crystalline rock (including samples from fault cores and damage zones in seismogenic systems), tight sandstones and fractures have significantly higher values with $\gamma > 0.02$ MPa^{-1}.

The coefficient γ plays an important role in a model formulated by *Rice* [1992] for the generation and maintenance of pore pressure excess in a seismogenic system. Pore pressure excess is generated by the continuous influx q of fluid from the "ductile root" of the fault zone. To maintain the pore pressure within the fault zone at close to

Table 2. Sensitivity of permeability to changes of effective mean stress, as characterized by the coefficient γ.

Geomaterial Type	γ (10^{-2} MPa^{-1})	Reference
Compact rock		
tight sandstone	3.8 - 6.3	*Yale* [1984]
Westerly granite	3.3	*Brace et al.* [1968]
Chelmsford granite	2.9	*Bernabé* [1986]
Barre granite	2.3	*Bernabé* [1986]
Carrara marble	4.7	*Fischer and Paterson* [1992]
Urach gneiss	4.9	*Huenges and Will* [1989]
gneiss (Kola)	3.2	*Morrow et al.* [1994]
basalt (Kola)	10.2	*Morrow et al.* [1994]
amphibolite (KTB)	5.8 - 11.0	*Morrow et al.* [1994]
granodiorite (Cajun Pass)	12.7	*Morrow and Lockner* [1994]
Fault-related rock		
East Fort thrust faults, Wyoming		*Evans et al.* [1997]
protolith	16.5 - 20.0	
damaged zone	3.5 - 8.2	
fault core	3.8 - 8.2	
Dixie Valley normal fault, Nevada		*Seront et al.* [1998]
protolith	4.0 - 15.2	
damaged zone	5.7	
fault core	4.5 - 8.3	
Porous rock		
sandstone	0.66 - 1.2	*David et al.* [1994]
sandstone	0.14 - 2.0	*Yale* [1984]
Granular material		
Ottawa sand	0.18	*Zoback and Byerlee* [1976a]
Fault gouge		
clay-free gouge	0.45 - 1.4	*Morrow et al.* [1984]
quartz gouge	0.7[a], 2.0[b]	*Zhang and Tullis* [1998]
clay gouge	1.2 - 5.5	*Morrow et al.* [1984]
muscovite gouge	0.3[a], 0.7[b]	*Zhang et al.* [1999]
Fractured rocks		
jointed Barre granite	7.8 - 9.2	*Kranz et al.* [1979]
natural rock joint	10.9	*Raven and Gale* [1985]

[a] Permeability perpendicular to the fault plane at a rotary shear displacement of 100 mm.
[b] Permeability parallel to the fault plane at a rotary shear displacement of 100 mm.

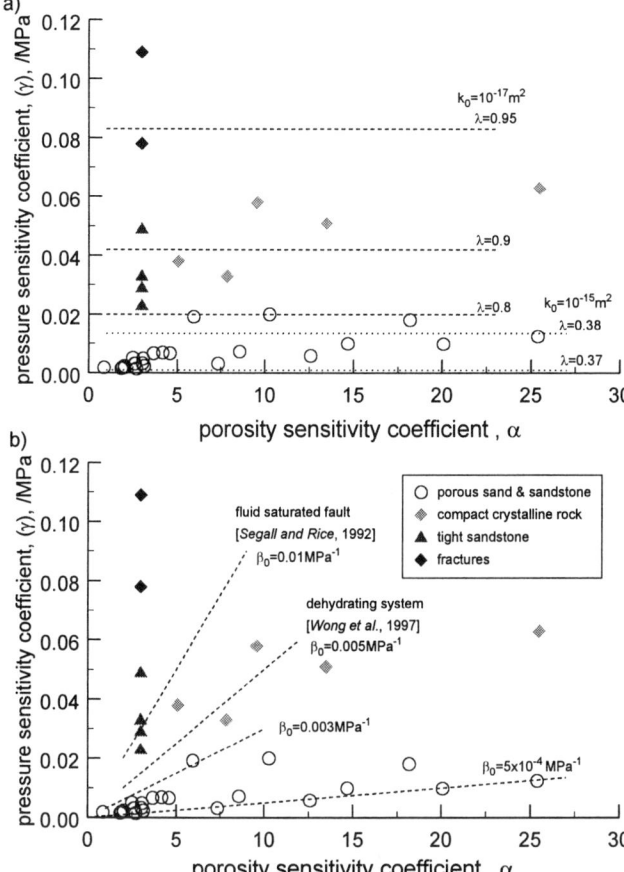

Figure 9. Correlation between the pressure sensitivity coefficient γ and the porosity sensitivity exponent α for sandstone, sand, crystalline rocks and jointed rocks (Table 2). a) Pore pressure excess characterized by the Hubbert-Rubey coefficient λ was calculated in accordance with *Rice's* [1992] model for different values of γ using equation 6. A reference permeability of $k_o = 10^{-17}$ m^2 was used for compact rocks and fractures (dashed horizontal lines). It can be seen that significant pore pressure excess (corresponding to λ = 0.8, 0,9 and 0.95) can readily be maintained in these rocks with relatively high values of γ > 0.02 MPa^{-1}. A reference permeability $k_o = 10^{-15}$ m^2 was used for porous rocks and unconsolidated materials (dotted lines). The pore pressure (represented by the lower horizontal lines with λ = 0.37 and 0.38) only slightly exceeds the hydrostatic value in these materials with relatively low values of γ < 0.02 MPa^{-1}. b) The ratio γ/α represents the pore compressibility $β_\phi$. Data for porous rock and unconsolidated materials can be bracketed by two linear boundaries corresponding to $β_\phi = 5$ x10^{-4} - 3 x 10^{-3} MPa^{-1}. Compact crystalline rocks and tight sandstones have relatively high ratios of γ/α > 3 x 10^{-3} MPa^{-1}. Values of $β_\phi$ recently adopted in theoretical analyses of a fluid saturated fault zone [*Segall and Rice*, 1995] and dehydrating system [*Wong et al.*, 1997b] are also indicated.

lithostatic level, the permeability is assumed to be sensitively dependent on the effective mean stress (with γ relatively small). In such a model, the maximum pore pressure that occurs at the base of the seismogenic zone (of thickness *H*) can be characterized by the Hubbert-Rubey pore pressure coefficient (the ratio between pore pressure and lithostatic stress)

$$\lambda = \frac{\rho_f}{\rho_r} + \frac{1}{\gamma \rho_r gH} \ln[1 + \frac{q\eta}{(\rho_r - \rho_f)gk_0}(e^{\gamma(\rho_r - \rho_f)gH} - 1)] \quad (6)$$

where the rock density, fluid density, and viscosity are denoted by ρ_r, ρ_f and η, respectively [*David et al.*, 1994].

Rice's [1992] model hinges on the influx of fluid from below the fault system. *Kennedy et al.* [1997] recently argued that the lack of correlation between helium isotopes and fluid chemistry or local geology would require that fluids enter the San Andreas fault system from the mantle. Using their estimate of $q \sim 1$ mm yr^{-1} and typical crustal values ($\rho_r = 2700$ kg m^{-3}, $\rho_f = 1000$ kg m^{-3}, η = 0.001 Pa·s, *H* = 15 km), we calculated using equation (6) values of λ for a typical compact rock (with reference permeability $k_o = 10^{-17}$ m^2) and porous rock (with reference $k_o = 10^{-15}$ m^2). The results are plotted in Figure 9a as horizontal lines corresponding to different values of γ. As indicated by the upper three lines, significant pore pressure excess (corresponding to λ = 0.8, 0.9 and 0.95) can readily be maintained in compact rocks and fractures that typically have relatively high values of γ > 0.02 MPa^{-1}. In contrast, the pore pressure in porous rocks and unconsolidated materials (corresponding to the lower lines with λ = 0.37 and 0.38) only slightly exceeds the hydrostatic value since they have relatively low values of γ < 0.02 MPa^{-1}.

The two coefficients α and γ that characterize the sensitivity of permeability to porosity and effective mean stress show a positive correlation. From poroelasticity, *David et al.* [1994] obtained the following relationship between the two coefficients

$$\gamma = \alpha\beta_\phi \quad (7)$$

where $β_\phi$ denotes the pore compressibility. A low γ/α ratio corresponds to a relatively stiff pore space. Experimental data for porous rock and unconsolidated materials can be bracketed by two linear boundaries corresponding to $β_\phi = 5$ x10^{-4} - 3 x 10^{-3} MPa^{-1} (Figure 9b). Compact crystalline rocks and tight sandstones have relatively high ratios of γ/α > 3 x 10^{-3} MPa^{-1}, with the implications that such a compact rock is relatively compliant possibly because the pore space is permeated with microcracks. Pore compressibility values of $β_\phi = 0.01$ MPa^{-1} and 0.005 MPa^{-1} were recently

Table 3. Porosity and permeability changes in response to deviatoric stresses, mean stress and pore pressure: influences of lithology and failure mode.

Type of Geomaterials	Influence of Stress			Mean Stress/ Pore Pressure	
	Brittle Faulting	**Cataclastic Flow**	**Fully Plastic Flow**	γ (MPa^{-1})	γ / α (MPa^{-1})
Compact Rock	*dilatancy +* permeability *enhancement*	*dilatancy +* permeability *enhancement*	progressive *collapse* of porosity $\rightarrow$ *percolation*	< 0.02	> 0.003
Porous Rock/ Granular Material	*dilatancy +* permeability *reduction*	*compaction +* permeability *reduction*	*threshold*	> 0.02	0.0005 - 0.003

adopted in theoretical analyses of a fluid saturated fault zone [*Segall and Rice*, 1995] and dehydrating system [*Wong et al.*, 1997b], respectively.

5. CONCLUSION

We have reviewed recent hydromechanical measurements in crustal rocks and their implications on the development of dilatancy, faulting, and permeability. In relation to the influences of stress and lithology, the laboratory data show that there are fundamental differences between compact and porous rocks in the evolution of permeability with brittle faulting. Significant differences have also been observed in relation to the permeability evolution for other failure modes. These differences are itemized in Table 3. The different paths by which permeability changes in relation to dilatancy and brittle faulting are manifestations of the fundamentally different pore structures in compact and porous rocks. With appropriate constraints from microstructural observations, the complex relations between permeability and porosity can be simulated adequately using random network models.

We have also underscored the interplay of deviatoric stresses, mean stress, and pore pressure in relation to permeability evolution and fluid transport. The influences of the latter two are related to the coefficients α and γ, which characterize the sensitivity of permeability to porosity and effective mean stress, respectively. Typical ranges of values for γ and the ratio γ/α (corresponding to the pore compressibility) are compiled in Table 3. In compact rocks and fractures, the coefficient γ is sufficiently high that near-lithostatic pore pressure may be maintained in a seismogenic system if it receives continuous fluid influx of ~ 1 mm yr^{-1}.

Acknowledgments. We thank Joanne Fredrich and Tomotsu Kiyama who kindly provided us with their deformation and permeability data. We have benefited from critical comments of Steve Martel and an anonymous reviewer. The second author is supported by a post-doctoral fellowship at Woods Hole Oceanographic Institution. The research at Stony Brook is supported by the National Science Foundation under grants EAR9508044 and EAR9805072.

REFERENCES

Antonellini, M., and A. Aydin, Effect of faulting on fluid flow in porous sandstones: petrophysical properties, in *Am. Assoc. Petrol. Geol. Bull.*, pp. 355-377, 1994.

Antonellini, M., A. Aydin, and D.D. Pollard, Microstructure of deformation bands in porous sandstones at Arches National Park, Utah, *J. Struct. Geol.*, *16*, 941-959, 1994.

Barton, C.A., M.D. Zoback, and D. Moos, Fluid flow along potentially active faults in crystalline rock, *Geology*, *23*, 683-686, 1995.

Bernabé, Y., The effective pressure law for permeability in Chelmsford granite and Barre granite, *Int. J. Rock Mech. Min. Sci. & Geomech. Asbtr.*, *24*, 304-315, 1986.

Bernabé, Y., The effective pressure law for permeability during pore pressure and confining pressure cycling of several crystalline rocks, *J. Geophys. Res.*, *92*, 649-657, 1987.

Bernabé, Y., Pore geometry and pressure dependence of the transport properties in sandstones, *Geophysics*, *56* (4), 436-446, 1991.

Berryman, G.J., Effective stress for transport properties of inhomogeneous porous rock, *J. Geophys. Res.*, *97*, 17409-17424, 1992.

Bethke, C.M., A numerical model of compaction-driven groundwater flow and heat transfer and its application to the paleohydrology of intracratonic sedimentary basins, *J. Geophys. Res.*, *90*, 6817-6828, 1985.

Brace, W.F., Volume changes during fracture and frictional sliding: a review, *Pure Appl. Geophys.*, *116*, 603-614, 1978.

Brace, W.F., Permeability of crystalline and argillaceous rocks, *Int. J. Rock Mech. Min. Sci. & Geomech. Asbtr.*, *17*, 241-251, 1980.

Brace, W.F., J.B. Walsh, and W.T. Frangos, Permeability of granite under high pressure, *J. Geophys. Res.*, *73*, 2225-2236, 1968.

Bredehoeft, J.D., and B.B. Hanshaw, On the maintenance of anomalous fluid pressures, I Thick sedimentary sequences, *Geol. Soc. Am. Bull.*, *79*, 1097-1106, 1968.

Bredehoeft, J.D., and D.L. Norton, *The Role of Fluids in Crustal Processes*, 170 pp., National Academy Press, Washington, D. C., 1990.

Brown, K.M., B. Berkins, B. Clennell, D. Dewhurst, and G. Westbrook, Heterogeneous hydrofracture development and accretionary fault dynamics, *Geology*, *21*, 303-306, 1994.

Byerlee, J.D., Model for episodic flow ot high-pressure water in fault zones before earthquakes, *Geology*, *21*, 303-306, 1993.

Caine, J.S., J.P. Evans, and C.B. Forster, Fault zone architecture and permeability structure, *Geology*, *24*, 1025-1028, 1996.

Chen, C.H., and M.N. Toksöz, Inversion of seismic velocities for the pore aspect ratio spectrum of a rock, *J. Geophys. Res.*, *84*, 7533-7543, 1979.

Chester, F.M., J.P. Evans, and R.L. Biegel, Internal structure and weakening mechanisms of the San Andreas Fault, *J. Geophys. Res.*, *98* (B1), 771-786, 1993.

David, C., and M. Darot, *Permeability and Conductivity of sandstones*, 203-209 pp., A. A. Balkema, Rotterdam, 1989.

David, C., Y. Guéguen, and G. Pampoukis, Effective Medium Theory and Network Theory Applied to the Transport Properties of Rocks, *J. Geophys. Res.*, *95* (B5), 6993-7005, 1990.

David, C., T.-f. Wong, W. Zhu, and J. Zhang, Laboratory Measurements of Compaction-Induced Permeability Change in Porous Rocks: Implications for the Generation and Maintenace of Pore Pressure Excess in the Crust, *Pure Appl. Geophys.*, *143* (1/2/3), 425-456, 1994.

Doyen, P.M., Permeability, Conductivity, and Pore Greometry of Sandstones, *J. Geophys. Res.*, *93* (B7), 7729-7740, 1988.

Evans, J.P., and F.M. Chester, Fluid-rock interaction and weakening of faults of the San Andreas system: Inferences from Sam Gabriel fault-rock geochemistry and microstructure, *J. Geophys. Res.*, *100*, 13007-13020, 1995.

Evans, J.P., C.B. Forster, and J.V. Goddard, Permeability of fault related rocks, and implications for hydraulic structure of fault zones, *J. Struct. Geol.*, *19*, 1393-1404, 1997.

Fischer, G.J., The Determination of Permeability and Storage Capacity: Pore Pressure Oscillation Method, in *Fault Mechanics and Transport Properties of Rocks*, edited by B. Evans, and T.-f. Wong, pp. 187-211, Academic Press Ltd., 1992.

Fowles, J., and S. Burley, Textural and permeability characteristics of faulted, high porosity sandstones, *Mar. Petrol. Geol.*, *11*, 608-623, 1994.

Fredrich, J.T., B. Evans, and T.-f. Wong, Micromechanics of the brittle to plastic transition in Carrara marble, *J. Geophys. Res.*, *94*, 4129-4145, 1989.

Freeze, R.A., and J.A. Cherry, *Groundwater*, 604 pp., Prentice Hall, Englewood Cliffs, NJ, 1979.

Gatto, H.G., The Effect of Various States of Stress on the Permeability of Berea Sandstone, M. S. Thesis thesis, Texas A&M University, 1984.

Gibson, R.G., Fault zone seals in siliciclastic strata of the Columbus Basin, offshore Trinidad, *Am. Assoc. Petrol. Geol. Bull.*, *78*, 1372-1385, 1994.

Guéguen, Y., and J. Dienes, Transport Properties of Rocks from Statistics and Percolation, *Mat. Geol.*, *21* (1), 1-13, 1989.

Hadley, K., Comparison of calculated and observed crack densities and seismic velocities in Westerly granite, *J. Geophys. Res.*, *81*, 3484-3494, 1976.

Hanson, R.B., The hydrodynamics of contact metamorphism, *Geol. Soc. Am. Bull.*, *107*, 595-611, 1995.

Hickman, S., R.H. Sibson, and R. Bruhn, Introduction to special issue: mechanical involvement of fluids in faulting, *J. Geophys. Res.*, *100*, 12,831-12,840, 1995.

Huenges, E., and G. Will, Permeability, bulk modulus and complex resistivity in crystalline rocks, in *Fluid Movements, Element Transport and the Composition of the Deep Crust*, edited by N. ASI, pp. 361-375, Dordrecht, Kluwer, 1989.

Jamtveit, B., and B. Yardley, *Fluid Flow and Transport in Rock, Mechanisms and Effects*, 319 pp., Chapman & Hall, London, 1997.

Jones, L.M., Field and Laboratory Studies of the Mechanics of Faulting, Ph. D. Thesis thesis, Massachusetts Institute of Technology, 1981.

Keaney, G.M.J., P.G. Meredith, and S.A.F. Murrell, Laboratory study of permeability evolution in a "tight" sandstone under non-hydrostatic stress conditions, in *Society of Petroluem Engineers/ International Society of Rock Mechanics Europe '98*, pp. SPE paper 47265, Trondheim, Norway, 1998.

Kennedy, B.M., Y.K. Kharada, W.C. Evans, A. Ellwood, D.J. DePaolo, J. Thordsen, G. Ambats, and R.H. Mariner, Mantle fluids in the San Andreas Fault system, California, *Science*, *27*, 1278-1281, 1997.

Kiyama, T., H. Kita, Y. Ishijima, T. Yanagidani, K. Aoki, and T. Sato, Permeability in anisotropic granite under hydrostatic compression and triaxial compression including post-failure region, in *Proc. N. Am. Rock Mech. Symp.*, pp. 1161-1168, 1996.

Kranz, R.L., Microcracks in rocks: A review, *Tectonophysics*, *100*, 449-480, 1983.

Kranz, R.L., A.D. Frankel, J.T. Engelder, and C.H. Scholz, The permeability of whole and jointed Barre granite, *Int. J. Rock Mech. Min. Sci. & Geomech. Abstr.*, *16*, 225-235, 1979.

Kranz, R.L., J.S. Saltzman, and J.D. Blacic, Hydraulic diffusivity measurements on laboratory rock samples using an oscillating pore pressure method, *Int. J. Rock Mech. Min. Sci. & Geomech. Abstr.*, *27*, 345-352, 1990.

Menéndez, B., W. Zhu, and T.-f. Wong, Micromechanics of brittle faulting and cataclastic flow in Berea sandstone, *J. Struct. Geol.*, *18*, 1-16, 1996.

Moore, D.E., D.A. Lockner, and J.D. Byerlee, Reduction of permeability in granite at elevated temperatures, *Science*, *265*, 1558-1561, 1994.

Mordecai, M., and L.H. Morris, An Investigation into the changes of permeability occuring in a sandstone when failed under triaxial conditions, in *Proc. U. S. Rock Mech. Symp.*, pp. 221-239, 1971.

Morrow, C.A., and D.A. Lockner, Permeability differences between surface-derived and deep drillhole core samples, *Geophys. Res. Lett., 21*, 2151-2154, 1994.

Morrow, C.A., D.A. Lockner, S. Hickman, M. Rusanov, and T. Röckel, Effects of lithology and depth on the permeability of core samples from the Kola and KTB drillholes, *J. Geophys. Res., 99*, 7263-7274, 1994.

Morrow, C.A., L.Q. Shi, and J.D. Byerlee, Permeability of fault gouge under confining pressure and shear stress, *J. Geophys. Res., 99*, 3193-3200, 1984.

Morrow, C.A., B.-C. Zhang, and J.D. Byerlee, Effective pressure law for permeability of Westerly granite under cyclic loading, *J. Geophys. Res., 91*, 3870-3876, 1986.

Muir Wood, R., Earthquakes, strain-cycling and the mobilization of fluids, in *Geofluids: Origin, Migration and Evolution of Fluids in Sedimentary Basins*, edited by J. Parnell, pp. 85-98, The Geological Society, London, 1994.

Neuzil, C.E., How permeable are clays and shales?, *Water Resour. Res., 30*, 145-150, 1994.

Parnell, J., *Geofluids: Origin, Migration and Evolution of Fluids in Sedimentary Basins*, 372 pp., The Geological Society, London, 1994.

Paterson, M.S., *Experimental Rock Deformation-The Brittle Field*, 254 pp., Spinger-Verlag, New York, 1978.

Paterson, M.S., The equivalent channel model for permeability and resistivity in fluid-saturated rocks: A re-appraisal, *Mech. Mat., 2*, 345-352, 1983.

Peach, C.J., and C.J. Spiers, Influence of crystal plastic deformation on dilatancy and permeability development in synthetic salt rock, *Tectonophysics, 256*, 101-128, 1996.

Person, M., J.P. Raffensperger, S. Ge, and G. Garven, Basin-scale hydrogeologic modeling, *Rev. Geophys., 34*, 61-87, 1996.

Pittman, E.D., Effect of fault-related granulation on porosity and permeability of quartz sandstones, Simpson Group (Ordovician), Oklahoma, *Am. Assoc. Petrol. Geol. Bull., 65*, 2381-2387, 1981.

Raven, K.G., and J.E. Gale, Water flow in a natual rock fracture as a function of stress and sample size, *Int. J. Rock Mech. Min. Sci. & Geomech. Abstr., 22*, 251-261, 1985.

Rice, J.R., Fault stress states, pore pressure distribution, and the weakness of the San Andreas Fault, in *Fault Mechanics and Transport Properties of Rocks*, edited by B. Evans, and T.-f. Wong, pp. 475-504, Academic Press, 1992.

Riley, G.N.J., D.L. Kohlstedt, and R.M. Richter, Melt migration in a silicate liquid-olivine system: An experimental test of compaction theory, *Geophys. Res. Lett., 17*, 2101-2104, 1990.

Roberts, S.J., J.A. Nunn, L. Cathles, and F.-D. Cipriani, Expulsion of abnormally pressured fluids along faults, *J. Geophys. Res., 101*, 28,231-28,252, 1996.

Seeburger, D.A., and A. Nur, A pore space model for rock permeability and bulk modulus, *J. Geophys. Res., 89*, 527-536, 1984.

Segall, P., and J.R. Rice, Dilatancy, compaction, and slip instability of a fluid-infiltrated fault, *J. Geophys. Res., 100*, 22,155-22,171, 1995.

Senseny, P.E., P.J. Cain, and G.D. Callahan, Influence of deformation history on permeability and specific storage of Mesavere sandstone, in *Proc. U. S. Rock Mech. Symp.*, pp. 525-531, 1983.

Seront, B., T.-f. Wong, J.S. Caine, C.B. Forster, R. Bruhn, and J.T. Fredrich, Laboratory charaterization of hydrothermal properties of a seismogenic normal fault system, *J. Struct. Geol., 20*, 865-881, 1998.

Shi, Y., and C.-Y. Wang, Pore pressure generation in sedimentary basins: overloading versus aquathermal, *J. Geophys. Res., 91*, 2153-2162, 1986.

Sibson, R.H., Faulting and fluid flow, in *Geofluids: Origins, Migration and Evolution of Fluids in Sedimentary Basins*, edited by J. Parnell, pp. 69-84, The Geological Society, London, 1994.

Stauffer, D., and A. Aharony, *Introduction to Percolation Theory*, 181 pp., Taylor & Francis, London, 1992.

Tapponier, P., and W.F. Brace, Development of stress-induced microcracks in Westerly granite, *Int. J. Rock Mech. Min. Sci. & Geomech. Abstr., 13*, 103-112, 1976.

Teufel, L.W., Permeability changes during shearing deformation of fractured rock, in *Proc. U. S. Rock Mech. Symp.*, pp. 473-480, 1987.

Torgersen, T., Crustal-scale fluid trnsport: magnitude and mechanisms, *Geophys. Res. Lett., 18*, 917-918, 1991.

Trimmer, D., B. Bonner, H.C. Heard, and A. Duba, Effect of pressure and stress in water transport in intact and fractured gabbro and granite, *J. Geophys. Res., 85*, 7059-7071, 1980.

Walder, J., and A. Nur, Porosity reduction and crustal pore pressure development, *J. Geophys. Res., 89*, 11,539-11,548, 1984.

Walsh, J.B., Effect of pore pressure and confining pressure on fracture permeability, *Int. J. Rock Mech. Min. Sci. & Geomech. Abstr., 18*, 429-435, 1981.

Walsh, J.B., and W.F. Brace, The effect of pressure on porosity and the transport properties of rock, *J. Geophys. Res., 89*, 9425-9431, 1984.

Wong, T.-f., C. David, and W. Zhu, The transition from brittle faulting to cataclastic flow in porous sandstones: Mechanical deformation, *J. Geophys. Res., 102*, 3009-3025, 1997a.

Wong, T.-f., J.T. Fredrich, and G.D. Gwanmesia, Crack aperture statistics and pore space fractal geometry of Westerly granite and Rutland quartzite: implications for an elastic contact model of rock compressibility, *J. Geophys. Res., 94*, 10,267-10,278, 1989.

Wong, T.-f., S.-C. Ko, and D.L. Olgaard, Generation and maitenance of pore pressure excess in a dehydrating system, 2 Theoretical analysis, *J. Geophys. Res., 101*, 841-852, 1997b.

Yale, D.P., Network Modeling of Flow, Storage and Deformation in Porous Rocks, Ph. D. Thesis thesis, Stanford University, 1984.

Zhang, S., S.F. Cox, and M.S. Paterson, The influence of room temperature deformation on porosity and permeability in calcite aggregates, *J. Geophys. Res., 99*, 15,761-15,775, 1994.

Zhang, S., and T.E. Tullis, The effect of fault slip on permeability and permeability anisotropy in quartz gouge, *Tectonophysics*, 295, 41-52, 1998.

Zhang, S., T.E. Tullis, and V.J. Scruggs, Permeability anisotropy and pressure dependency of permeability in experimentally sheared gouge materials, *J. Struct. Geol.*, *in press*, 1999.

Zhu, W., C. David, and T.-f. Wong, Network modeling of permeability evolution during cementation and hot isostatic pressing, *J. Geophys. Res.*, *100*, 15,451-15,464, 1995.

Zhu, W., and T.-f. Wong, Permeability reduction in a dilatant rock: Network modeling of damage and tortuosity, *Geophys. Res. Lett.*, *23*, 3099-3102, 1996.

Zhu, W., and T.-f. Wong, The transition from brittle faulting to cataclastic flow: Permeability evolution, *J. Geophys. Res.*, *102*, 3027-3041, 1997.

Zhu, W., and T.-f. Wong, Network modeling of the evolution of permeability and dilatancy in compact rock, *J. Geophys. Res.*, *104*, 2963-2971, 1999.

Zoback, M.D., and J.D. Byerlee, The effect of microcrack dilatancy on the permeability of Westerly granite, *J. Geophys. Res.*, *80*, 752-755, 1975.

Zoback, M.D., and J.D. Byerlee, Effect of high-pressure deformation on permeability of Ottawa sand, *Am. Assoc. Petrol. Geol. Bull.*, *60*, 1531-1542, 1976a.

Zoback, M.D., and J.D. Byerlee, A note on the deformational behaviour and permeability of crushed granite, *Int. J. Rock Mech. Min. Sci. & Geomech. Abstr.*, *13*, 291-294, 1976b.

Teng-fong Wong, Department of Geosciences, State University of New York, Stony Brook, New York, NY 11794-2100. (e-mail: wong@seism1.ess.sunysb.edu)

Wenlu Zhu, Department of Geology and Geophysics, Woods Hole Oceanographic Institution, Woods Hole, MA 02543. (e-mail: wzhu@whoi.edu)

Fault Zone Architecture and Fluid Flow: Insights From Field Data and Numerical Modeling

Jonathan Saul Caine and Craig B. Forster

Department of Geology and Geophysics, University of Utah,
Salt Lake City, Utah

Fault zones in the upper crust are typically composed of complex fracture networks and discrete zones of comminuted and geochemically altered fault rocks. Determining the patterns and rates of fluid flow in these distinct structural discontinuities is a three-dimensional problem. A series of numerical simulations of fluid flow in a set of three-dimensional discrete fracture network models aids in identifying the primary controlling parameters of fault-related fluid flow, and their interactions, throughout episodic deformation. Four idealized, but geologically realistic, fault zone architectural models are based on fracture data collected along exposures of the Stillwater Fault Zone in Dixie Valley, Nevada and geometric data from a series of normal fault zones in east Greenland. The models are also constrained by an Andersonian model for mechanically compatible fracture networks associated with normal faulting. Fluid flow in individual fault zone components, such as a fault core and damage zone, and full outcrop scale model domains are simulated using a finite element routine. Permeability contrasts between components and permeability anisotropy within components are identified as the major controlling factors in fault-related fluid flow. Additionally, the structural and hydraulic variations in these components are also major controls of flow at the scale of the full model domains. The four models can also be viewed as a set of snapshots in the mechanical evolution of a single fault zone. Changes in the hydraulic parameters within the models mimic the evolution of the permeability structure of each model through a single deformation cycle. The model results demonstrate that small changes in the architecture and hydraulic parameters of individual fault zone components can have very large impacts, up to five orders of magnitude, on the permeability structure of the full model domains. Closure of fracture apertures in each fault zone magnifies the magnitude and orientation of permeability anisotropy in ways that are closely linked to the implicitly modeled deformation. Changes in fault zone architecture can cause major changes in permeability structure that, in turn, significantly impact the

Faults and Subsurface Fluid Flow in the Shallow Crust
Geophysical Monograph 113
Copyright 1999 by the American Geophysical Union

magnitude and patterns of fluid flux and solute transport both within and near the fault zone. Inferences derived from the model results are discussed in the context of the mechanical strength of an evolving fault zone, fault zone sealing mechanisms which control the conduit-barrier systematics of a fault zone as a flow system, and how these processes are related to fluid flow in natural fault zones.

INTRODUCTION

Improving our understanding of how brittle fault zones form and, once formed, how they influence fluid flow in the upper crust is facilitated by field-based modeling of fault zone architectural style, fracture mechanics, and fluid flow processes through time. This process-based approach to field work and modeling will help to develop predictive solutions to problems involving seismic hazards, hydrocarbon migration and trapping, and the utilization of groundwater, mineral, and geothermal energy resources. Although a number of modeling studies have addressed the possible impacts that fault zones have on fluid flow [e.g., *Mase and Smith*, 1985; *Forster and Evans*, 1991; *Ge and Garven*, 1994; *Haneberg*, 1995; *Lopez and Smith*, 1995 and 1996; *Roberts et al.*, 1996; *Zhang and Sanderson*, 1996; *Flemming et al.*, 1998; *Matthäi et al.*, 1998; *Ferrill et al.*, 1999], the permeability structures assigned in the simulated fault zones are either highly simplified or are modeled in two-dimensions. In most cases a homogeneous and isotropic permeability structure is assumed for the entire fault zone. *Lopez and Smith* [1996] included complex fault zone permeability structures in their three-dimensional fluid flow simulations, but, the theoretical permeability structures they created are not based on an analysis of field data. Modeling results presented in this paper help to illustrate a step towards creating more realistic fault zone architectural styles and associated permeability structures needed as input to numerical models of fluid flow in fault zones.

A primary goal of this paper is to outline how three-dimensional, geologically plausible fault zone permeability structures, developed using data collected from outcrop, influence fluid flow in and near fault zones. We have not attempted to model the impact that fault zones have on regional flow systems. The work presented is based on the fault zone architectural models and permeability structures summarized by *Caine et al.* [1996] and shown in Figure 1. Fault zones are commonly composed of a complex set of components, a fault core and a damage zone, which are distinct mechanical and permeability heterogeneities in the upper crust [*Sibson*, 1977; *Chester and Logan*, 1986; *Davison and Kozak*, 1988; *Forster and Evans*, 1991; *Byerlee*, 1993; *Scholz and Anders*, 1994; *Caine et al.*, 1996]. The combined effect of different permeabilities associated with variations in the distribution of fault zone components leads to fault zones that may act as conduits, barriers, or combined conduit-barrier systems [*Randolph and Johnson*, 1989; *Smith et al.*, 1990; *Scholz*, 1990;

Antonellini and Aydin, 1994; *Bruhn, et al.*, 1994; *Newman and Mitra*, 1994; *Goddard and Evans*, 1995; *Caine* et al., 1996; *Sibson*, 1996; *Evans et al.*, 1997; *Forster et al.*, 1997; *Jones et al.*, 1998; *Rowley*, 1998].

In this paper we focus on permeability structures that correspond to complex fracture networks associated with brittle faulting in low permeability rocks. Fault-related permeability structures are derived by constructing outcrop-scale, three-dimensional stochastic models of discrete fracture networks for each architectural style shown in Figure 1. Each architectural style modeled in this study (Fig. 1) is tied to reality using field data and inferences derived from the Stillwater Normal Fault in Dixie Valley, Nevada [*Caine and Forster*, 1997] and from a series of normal fault zones in east Greenland [*Caine*, 1996].

Numerical fluid flow simulations provide estimates of the permeability structure of each individual fault zone component and each full model domain where components are combined within a protolith. Because the rock matrix is assumed to be impermeable, fluid flow is simulated only in the discrete fracture networks. Neglecting matrix permeability provides an opportunity to evaluate how macroscopic

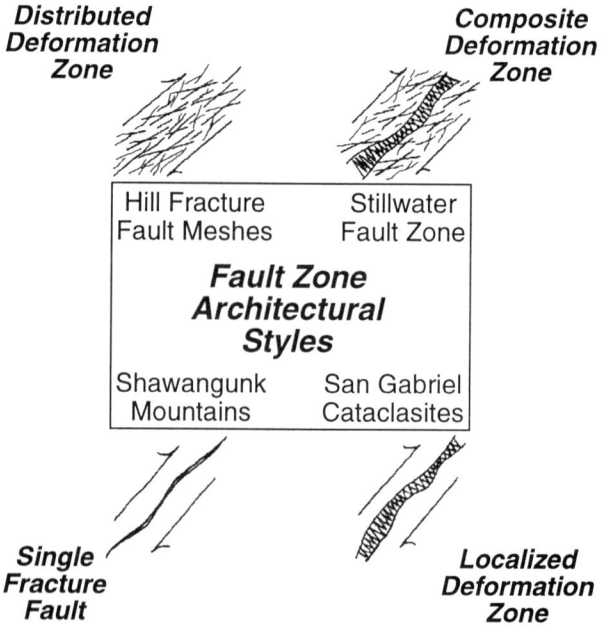

Figure 1. Conceptual model for idealized fault zone architectural styles with field examples [after *Caine et al.*, 1996].

fracture networks contribute to fault zone permeability structure. The flow experiments yield values of bulk permeability, computed in three orthogonal directions relative to the orientation of each modeled fault zone. Additionally, an evolving fault zone is simulated by stepwise closure of fracture apertures restricted to each fault core. The modeling results are then discussed with reference to the mechanics, permeability structure, and evolution of natural fault zones.

FAULT ZONE ARCHITECTURAL STYLES AND IDEALIZED PERMEABILITY STRUCTURES

The core and damage zone components of a fault zone are surrounded by a protolith, or country rock, where fault-related permeability structures are generally absent [*Chester and Logan*, 1986; *Caine et al.*, 1996; *Evans et al.*, 1997]. The geometry and intensity of fracturing associated with each component yield heterogeneity and anisotropy in bulk component permeabilities. Consequently, the bulk permeability anisotropy of a fault zone reflects the combined effect of anisotropy developed within each component and anisotropy caused by permeability contrasts between components. Field-based observations and permeability data obtained from fault rocks [*Morrow et al.*, 1981; *Forster and Evans*, 1991; *Antonellini and Aydin*, 1994; *Bruhn et al.*, 1994; *Caine et al.*, 1996; *Evans et al.*, 1997; *Seront et al.*, 1998] suggest that the distinctive internal structure, external geometry, and composition of each component play an important role in controlling the patterns and rates of fluid flow in and around fault zones.

A fault core is the component of a fault zone where comminution, fluid flow, geochemical reaction, and other fault-related processes alter the original lithology. For example, progressive grain-size reduction, dissolution, reaction, and mineral precipitation during fault zone evolution typically cause the core to have reduced permeability, relative to that of the adjacent damage zone and protolith [e.g., *Chester and Logan*, 1986; *Antonellini and Aydin*, 1994; *Bruhn et al.*, 1994; *Goddard and Evans*, 1995; *Caine et al.*, 1996; *Evans et al.*, 1997; *Seront et al.*, 1998; *Fisher and Knipe*, 1998]. Protolithology has a strong influence on the structure and composition of fault cores and their resulting permeability structure [*Caine et al.*, 1996; *Knipe et al.*, 1998]. The detailed relationships, however, between the formation of fault cores and the varying processes that alter permeability structure are incompletely understood for different protolithologies. In mature fault zones hosted in granitic rocks the presence of feldspars may result in fault core lithologies that become rich in clay minerals, primarily due to reactions between fluids and comminuted minerals [*Goddard and Evans*, 1995; *Evans et al*, 1997; *Seront et al.*, 1998]. This may significantly lower fault core permeability relative to what might be expected for less feldspar-rich protolithologies. In small

displacement fault zones hosted in quartz-rich sandstones low permeability fault cores often develop by grain size reduction and the formation of deformation bands [*Antonellini and Aydin*, 1994]. In protoliths that are heterogeneous, interbedded sandstones and shales or granitic rocks juxtaposed with sandstones, the combined effects of mechanical and geochemical processes can lead to the formation of fault cores with heterogeneously distributed high and low permeabilities [*Chester and Logan*, 1986; *Caine*, 1996; *Foxford et al.*, 1998]. At present, little work has been done to understand the formation of fault cores and their resulting permeability structures in carbonates, mafic igneous and metamorphic rocks, and pelitic metamorphic rocks that have undergone brittle deformation.

Damage zones are the network of subsidiary structures including small faults, veins, fractures, cleavage, pressure solution seams, and folds that surround the fault core. Damage zone structures result from the growth and linkage of fracture networks that accompany episodic deformation distributed in a fault zone [*Chester and Logan*, 1986; *Evans*, 1990; *Bruhn et al.*, 1994; *McGrath and Davison*, 1995; *Caine et al*, 1996; *Cowie and Shipton*, 1998; *Knipe et al.*, 1998]. Composed of both open and filled fractures, damage zones yield a heterogeneous and anisotropic permeability structure [*Bruhn et al.*, 1994]. Damage zone permeability structures are dominated by macroscopic fractures and are typically enhanced relative to both the fault core and the undeformed protolith [*Chester and Logan*, 1986; *Smith et al.*, 1990; *Andersson et al.*, 1991; *Scholz and Anders*, 1994; *Goddard and Evans*, 1995; *Caine et al.*, 1996; *Evans et al.* 1997; *Seront et al.*, 1998].

Figure 1 relates fault zone architectures with deformational style. The Distributed Deformation Zone (DDZ; not to be confused with *Mitra's* [1978] ductile deformation zones) and Localized Deformation Zone (LDZ) represent two idealized, architectural end members. The Single Fracture Fault (SFF) is a special case of the LDZ where deformation is accommodated along a single fault. The Composite Deformation Zone (CDZ) represents a hybrid between the Distributed and Localized Deformation Zones. Each idealized fault zone architecture represents only one moment in time and space [*Caine et al.*, 1996]. Because fault zone architectures evolve through time it should be noted that a range of styles varying between those shown in Figure 1 might be found along any section of a single fault zone. Additionally, individual fault zone strands with varying architectural styles are often combined as several sets of strands in a complex fault zone. *Caine et al.* [1996] associates each of the four architectural styles shown in Figure 1 with permeability structures in inactive fault zones as discussed by a variety of workers [*Chester and Logan*, 1986; *Bruhn et al.*, 1990 and 1994; *Forster and Evans*, 1991; *Moore and Vrolijk*, 1992; *Newman and Mitra*, 1994]. Idealized permeability structures in non-deforming fault zones include localized conduits (SFF model), distributed conduits (DDZ model), localized barriers (LDZ model), and combined conduit-barriers (CDZ model). *Caine et al.* [1996] discuss these conceptual models in greater detail and outline a set of fault zone architectural indices that link fault zone architecture with permeability structure.

The modeling presented here is based on the assumption that the magnitude and geometry of permeability anisotropy within each fault zone component, combined with permeability contrasts between the components, control the evolution of the hydraulic properties of a fault zone. For example, macroscopic fracture intensity in the fault core at the prefailure and postfailure stages of deformation is usually much less than that of the damage zone [*Andersson et al.*, 1991; *Chester et al.*, 1993; *Caine et al.*, 1996]. Thus, before and after failure the reduced permeability of the fault core is dominated by the intergranular porosity and microscopic fractures of the fault rocks, whereas the permeability of the damage zone is enhanced by the hydraulic properties of macroscopic fracture networks. The resulting permeability structure leads to a focusing of fluid flow within the damage zone during the prefailure and postfailure stages of deformation. During brittle deformation at failure, however, microscopic to macroscopic fractures that open and close within the fault core may allow this component to play a more significant, but transient, role in transmitting fluid through and across the fault zone.

When not actively deforming, fault cores commonly act as localized barriers that restrict fluid flow across the fault zone because of their reduced permeability [*Antonellini and Aydin*, 1994; *Caine*, 1996; *Evans et al.*, 1997]. Thus, during periods of inactivity Localized Deformation Zones may act as localized barriers (Fig. 1). Distributed Deformation Zones contain networks of both open and closed macroscopic fractures throughout their evolution. These distributed fracture networks allow DDZs to act as distributed conduits that enhance fluid flow along the fault zone at any time during the cycle of deformation [e.g., *Sibson*, 1996; *Caine et al.*, 1996]. When a damage zone surrounds a well developed, low permeability fault core a Composite Deformation Zone may have a combined conduit-barrier permeability structure. The specific characteristics of each permeability structure will depend on *in situ* stress state, fault rock heterogeneity, fracture interconnectivity, and the extent of fracture filling by mineral precipitation.

THREE-DIMENSIONAL DISCRETE FRACTURE MODELS

We explore the way that fault zone architectures influence fluid flow by modeling idealized fault zones where fractures provide the principal pathways for flow. A primary goal is to illustrate how the geometry, variability, and intensity of fracturing associated with faulting processes might lead to permeability heterogeneity and anisotropy within individual fault zone components and the surrounding rocks. To do this we simulate fluid flow through a set of field-based, three-dimensional, discrete fracture network

models that represent the four idealized architectural models shown in Figure 1. The models are not representations of fault zones from actual field sites. Rather, we have used the field data to generate realistic and generalized fracture model parameters in order to model each of the idealized architectural styles. Specifically, we used fracture density and trace length data from the Stillwater Fault Zone. Representative fault zone component widths are generalized from field data collected along the fault zones in east Greenland where each of the four architectural styles are observed. Fracture types and orientation data are generalized primarily from the Stillwater Fault Zone but are similar to what is observed in east Greenland. Other fracture parameters, such as aperture and transmissivity, are not taken from field data because they are poorly constrained. The choice of these parameters and how they are derived are discussed in detail below. The fracture data used in constructing the models are summarized in Table 1.

The discrete fracture models are constructed in three steps. First, fracture orientation and morphology data from field work are organized and plotted on equal area stereonets to segregate genetically-related fracture sets. Second, statistical fracture network parameters are delineated for each type of field data used to constrain the models. Finally, the stochastic three-dimensional fracture network models are created using the fracture network parameter data and the code FracMan™ [*Dershowitz et al.*, 1996].

Several key features of natural fault zones are represented in the fracture models and subsequent fluid flow modeling experiments. These features include distinct sets of fault-related macroscopic fractures whose densities and orientations are mechanically realistic and based on field data. We also simulated component specific fracture networks that obliterate and replace earlier structures. We investigate the mechanical evolution of a single fault zone where each idealized model can be viewed as a single step in that evolution. Finally, we explore the evolution of permeability structure in each of the idealized models through a single deformation event.

FIELD DATA AND SEGREGATION OF FRACTURE TYPES

Fracture data collected along exposures of the footwall of the Stillwater Fault Zone provide the primary field-based dataset used in parameterizing the models. The Stillwater Fault Zone is an active, crustal-scale, seismogenic normal fault that lies along the eastern margin of the Stillwater Mountains [*Parry and Bruhn*, 1990; *Power and Tullis*, 1989; *Bruhn et al.*, 1994; *Caskey et al.*, 1996; *Caine and Forster*, 1997; *Seront et al.*, 1998; *Caine*, 1999]. The architectural style of the Stillwater Fault Zone observed at several localities is that of a Composite Deformation Zone.

Detailed fracture mapping was done along scanlines located on sets of outcrop faces oriented roughly parallel and perpendicular to the nominal plane of the fault zone [*Caine*

Table 1. Fault zone architectural styles and fracture model parameters.

Architectural model and component	Fault zone component widths[a]	Type of fracture set[b]	Mean orientation of each set (trend/plunge)[c]	Mean trace length (m), distribution, and standard deviation[d]	Density (m²/m³) and termination (%)[d]	Total number of fractures and total density (D)
Single Fracture Fault						
Protolith	10m on each side of fault core	Extension	090 / 30	2, Log Normal, 0.5	0.2, 0	408 Fractures
		Cross A	344 / 26	2, Log Normal, 0.5	0.2, 1	
		Cross B	196 / 26	2, Log Normal, 0.5	0.2, 1	D = 0.65 m²/m³
Fault	1mm	Shear	090 / 00	20, Constant	0.05, 0	
Distributed Deformation Zone						
Protolith	7m on each side of fault zone	Extension	090 / 30	2, Log Normal, 0.5	0.2, 0	708 Fractures
		Cross A	344 / 26	2, Log Normal, 0.5	0.2, 1	
		Cross B	196 / 26	2, Log Normal, 0.5	0.2, 1	D = 2.70 m²/ m³
Fault Zone	5m	Shear A	270 / 00	6, Constant	4.0, 0	
		Shear B	090 / 30	4, Constant	2.0, 1	
		Step	270 / 80	4, Constant	1.0, 1	
		Shear	090 / 00	20, Constant	0.25, 0	5 fractures with slip
Localized Deformation Zone						
Protolith	19m on each side of fault core	Extension	090 / 30	2, Log Normal, 0.5	0.2, 0	4990 Fractures
		Cross A	344 / 26	2, Log Normal, 0.5	0.2, 1	
		Cross B	196 / 26	2, Log Normal, 0.5	0.2, 1	D = 1.02 m²/m³
Fault Zone	1m	Shear A	270 / 30	0.6, Constant	4.0, 0	
		Shear B	090 / 60	0.4, Constant	2.0, 0	
Composite Deformation Zone						
Protolith	6m on each side of DZ	Extension	090 / 30	2, Log Normal, 0.5	0.2, 0	5436 Fractures
		Cross A	344 / 26	2, Log Normal, 0.5	0.2, 1	
		Cross B	196 / 26	2, Log Normal, 0.5	0.2, 1	D = 3.12 m²/m³
Damage Zone	3m on each side of fault core	Shear A	270 / 00	6, Constant	4.0, 0	
		Shear B	090 / 30	4, Constant	2.0, 1	
		Step	270 / 80	4, Constant	1.0, 1	
Core	1m	Shear A	270 / 30	0.6, Constant	4.0, 0	
		Shear B	090 / 60	0.4, Constant	2.0, 0	

Note: For each model: Fracture model domain size = 20 m by 20 m by 20 m; Flow simulation region = 18 m by 18 m by 18 m; No matrix permeability; DZ = Damage Zone.

[a] Component widths are representative of fault zones from east Greenland where each architectural style is represented.

[b] From field observations along the Stillwater and east Greenland fault zones of fracture network sets that are mechanically compatible with an Andersonian mechanical model generalized into the model domains.

[c] Fracture set orientations are from the types of fractures associated with a generalized Andersonian normal fault zone.

[d] From field observations and data collection along the Stillwater Fault Zone.

and Forster, 1997]. Note that the actual geometry of the fault zone is curviplanar to lenticular. Two orthogonal scanlines were used to minimize orientation bias. Data collected along each scanline include fracture position, orientation, type, trace length, apparent aperture, geometry (planar, curviplanar, irregular, etc.), roughness, mineral fillings, slickenline orientations on slip surfaces, truncation and termination style, and age relationships. The fracture orientation data are plotted on equal area nets as poles to fracture planes (Fig. 2). The data are then contoured and individual clusters of data are segregated into distinct fracture sets (Fig. 2). The mean vector, or average orientation, for each fracture set is computed, and the corresponding Fisher

dispersions, a measure of the spread of each set, are calculated [e.g., *Marshak and Mitra*, 1988].

The fracture sets are assigned to a genetic mode of formation based on an assumed mechanical compatibility with an Andersonian model of a normal fault [*Anderson*, 1951; *Davis*, 1984; *Sibson*, 1994], as well as field observations of each fracture type. Figure 2b shows a fault-related fracture network composed of shear, extension, and step fractures that are primarily found in the damage zone data, but not in the protolith data shown in Figure 2a. The protolith primarily contains cross fractures while shear fractures are notably absent (Fig. 2a). Additionally, the presence of quartz-kaolinite mineralization found in the fault

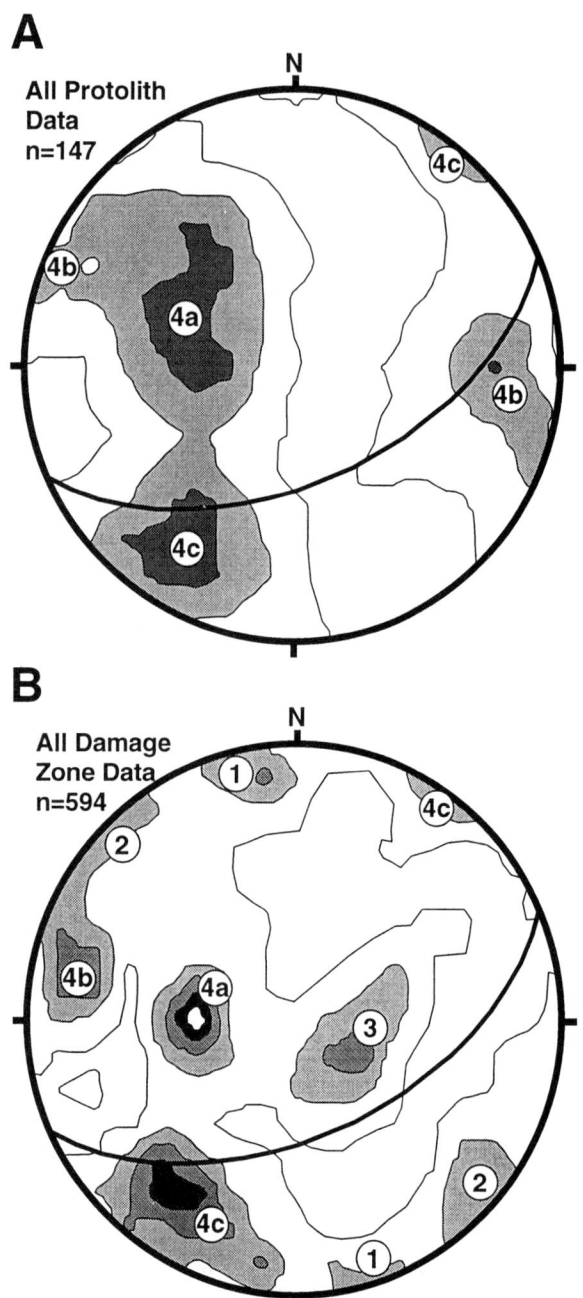

Figure 2. Equal area, Kamb contoured poles to all fracture data from the Mirrors Locality, Stillwater Fault Zone, Dixie Valley, Nevada (C.I. = 2.0 sigma). A: Data from protolith only. B: Data from damage zone only. Fractures broken into sets from raw data are assigned to a mode of formation based on mechanical compatibility with an Andersonian model of a normal fault. Relative to the master fault zone the following sets are defined: 1 = Extension; 2 = Shear; 3 = Step; 4a, 4b, and 4c = Cross fractures. Great circles indicate average orientation of master fault zone (066/55). In general, shear and step fractures are found within the damage zone while extension and cross fractures are found in both the protolith and damage zone.

core and damage zone fracture networks suggest syntectonic, fault-related fluid flow [*Power and Tullis*, 1989; *Caine and Forster*, 1997; *Seront et al.*, 1998; *Caine*, 1999]. The cross fractures labeled set 4 in Figure 2 are of uncertain origin. The cross fractures are both unfilled and filled with fault-related mineral assemblages and may be related to an early stage of ENE / WSW regional extension [*Parry et al.*, 1991]. The fault core is dominated by highly silicified and comminuted fault rocks where poorly developed macroscopic fracturing occurs in generally random orientations.

The idealized geometric and spatial relationships between an Andersonian master fault zone and the subsidiary fracture types accounted for in the fracture models are illustrated in Figure 3. The fracture networks observed along the Stillwater Fault Zone may have formed in non-Andersonian stress fields, particularly localized stress fields near the fault zone. The Andersonian model, however, allows us to link the formation of distinct fracture sets in our generalized models to a mechanical model for fault-related fracturing. Extension fractures are curviplanar to planar, generally rough, opening mode fractures that are typically mineral filled in the damage zone and unfilled in the protolith. Shear fractures are often curviplanar, smooth walled, mineral filled, and typically show polished and striated slip surfaces. These are interpreted to be Mode 2 or hybrid Mode 2 and Mode 3 fractures. Step fractures are of uncertain origin but they are generally curviplanar, filled and unfilled, and form at a high angle to the average orientation of the master fault zone and are typically subhorizontal. The step fractures may form because of bending stresses related to shear during faulting. Alternatively, the step fractures may be similar to the linking fractures and small faults found in compound fault zones of the Sierra Nevada of central California [*Martel*, 1990]. Cross fractures are typically steeply dipping, cut across the fault zone at a high angle, are often smooth walled, and are typically filled when in the damage zone while unfilled when found in the protolith. Because they are generally cut by fault-related fractures, the cross fractures are assumed to have formed prior to faulting but were hydraulically active during faulting.

DETERMINATION OF FRACTURE NETWORK PARAMETERS

The field-based stochastic fracture models are created using FracMan™. Our first step in model construction is to determine fracture density for each fracture set. The total number of fractures, from the fracture position field data, for each pair of scanlines are input into a trial model. Simulated trace planes, with the same orientations as the outcrop faces, are cut from the model volume. The modeled patterns of fracture traces are visually compared to photographs of each outcrop face and qualitatively evaluated to determine how closely the model represents the photograph. The fracture models are iteratively regenerated and compared to the field data until there is a satisfactory match between the model, the photographs, and the statistics of the field data. When a satisfactory result is obtained the fracture density is calculated by FracMan™ and used in subsequent model construction.

An Enhanced Baecher model [*Dershowitz et al.*, 1996] is found to yield fracture models that best match the field observations. The Enhanced Baecher model locates fracture centers in a model domain using a Poisson distribution and allows for fracture terminations at intersections with preexisting fractures [*Dershowitz et al.*, 1996]. The Enhanced Baecher model produces fracture sets with relatively uniform spatial distributions and minimal clustering, as observed in the field. All fractures in this study are modeled as hexagonal plates with lengths matched to the observed variability in fracture traces. Fracture lengths are modeled in the protolith and damage zones using a log normal distribution. A constant fracture length is used in the fault cores.

Fracture density and orientation are most closely honored in the models because these data are most reliably measured in the field. Because it is uncommon to see both tips of a single fracture exposed along a scanline, fracture length is moderately well honored in the models. Length estimates are obtained, however, by measuring individual traces associated with a specific fracture type wherever they are found with both tips exposed. The mean length computed for each fracture type is used in generating the fracture models (Table 1).

Fracture aperture and roughness are elusive parameters that we do not attempt to constrain with field data. Because it is nearly impossible to measure a meaningful value for fracture aperture in the field, uniform fracture apertures of 100 μm are initially assigned throughout the fracture models. The range of aperture values used in our models are reasonable for fractures found in the shallow subsurface and are based on measurements [*Snow*, 1968] and rock mechanical experiments [*Krantz et al.*, 1979; *Witherspoon et al.*, 1980]. We use this approach to illustrate how better constrained parameters, such as fracture geometry and density, might influence fault zone permeability structure. We also examine the effect of reducing fracture apertures within the fault zone as a way of modeling the consequences of fault zone sealing. This illustrates how changes in component-specific permeabilities might influence the bulk permeability structure of a fault zone and its evolution though a deformation cycle.

THREE-DIMENSIONAL FRACTURE NETWORK MODEL CONSTRUCTION

We created idealized, three-dimensional fracture network models for each fault zone model shown in Figure 4 using the statistics of the field data and the FracMan™ code [*Dershowitz et al.*, 1996]. Building each model requires a series of input parameters that include the size of the model region; mean orientation and dispersion for each fracture set;

mean fracture density and spacing model for each set; the mean, standard deviation, and probability distribution function for lengths in each set; termination percent for each set; and the aperture, transmissivity, and storativity for each set. For each fault zone component, within each model, we use the statistics generated from the iterative process discussed above to construct the models. For example, each fracture set in each model protolith is constructed using the same mean and standard deviation for a log normal length distribution, a mean density, and a mean orientation randomly chosen within the dispersion range. Thus, each final model is an outcrop scale representation (cubes that are 20 m on a side) of the field-based data; constructed by a component-wise piecing together of each part of the model domain (Fig. 4). Each model is populated with a mechanically compatible network of fractures appropriate for each

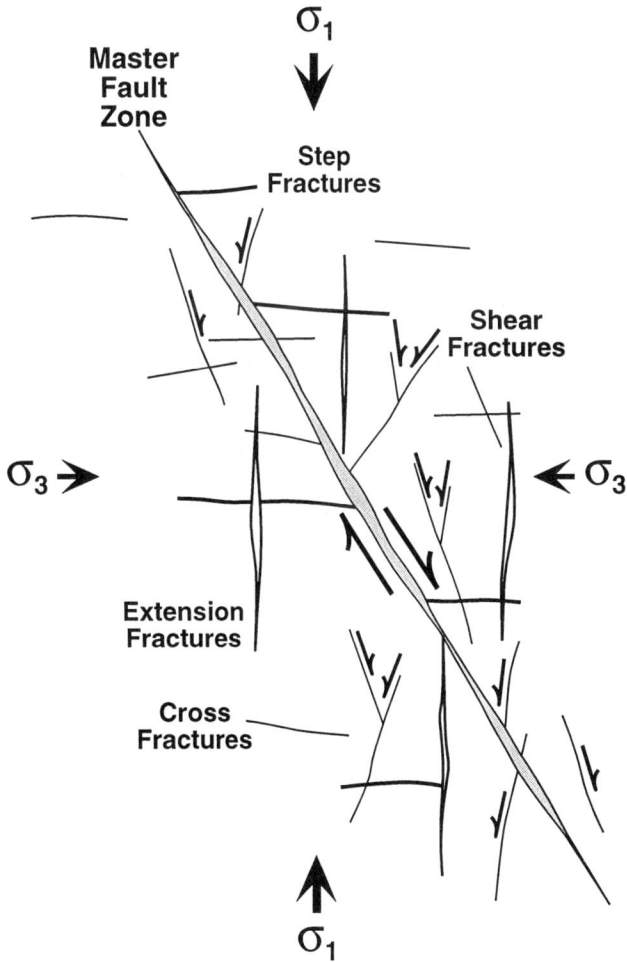

Figure 3. Schematic, field-based cross section of an idealized network of mechanically compatible Andersonian fracture traces. Fracture orientations with respect to the master normal fault zone and extensional stress field are shown.

Single Fracture Fault

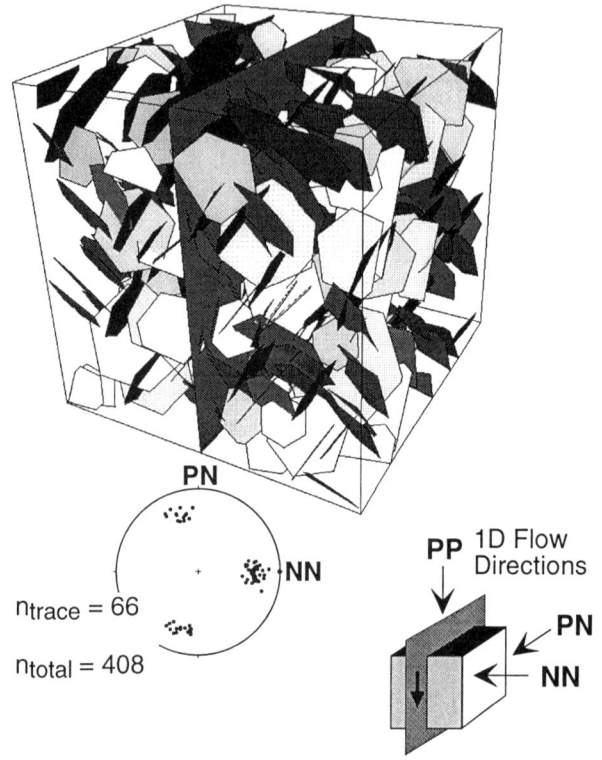

$n_{trace} = 66$

$n_{total} = 408$

PP 1D Flow Directions

PN

NN

Distributed Deformation Zone

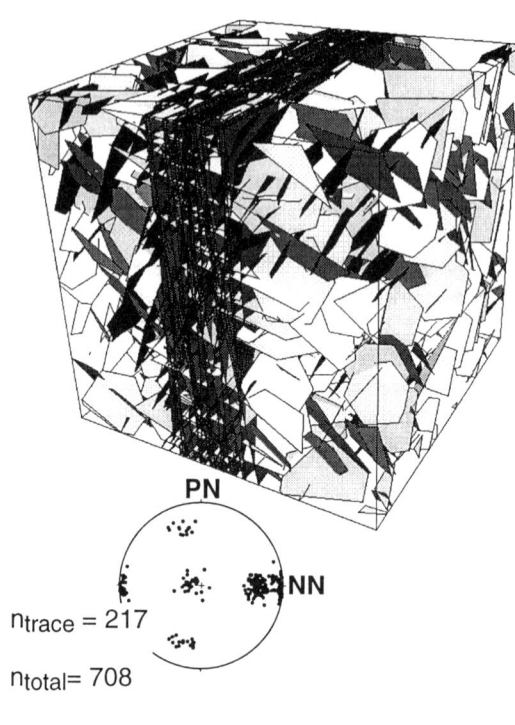

$n_{trace} = 217$

$n_{total} = 708$

Localized Deformation Zone

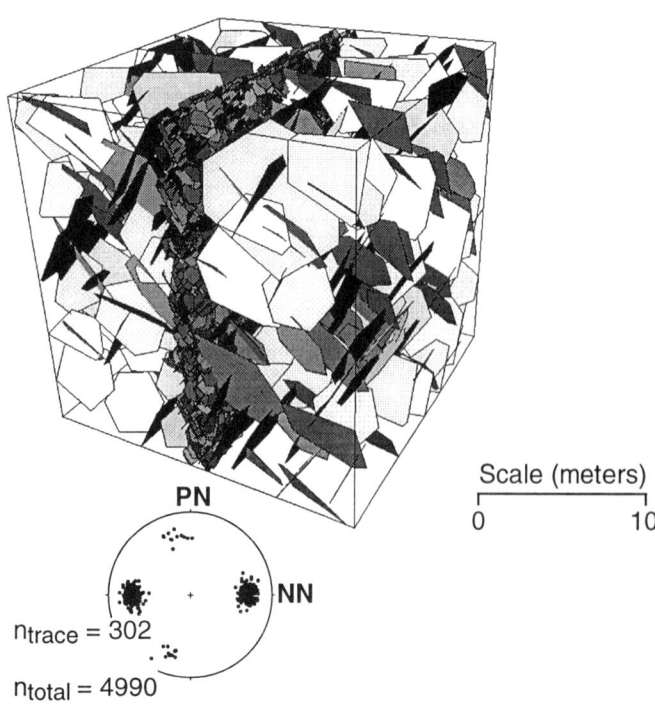

$n_{trace} = 302$

$n_{total} = 4990$

Scale (meters)

0   10

Composite Deformation Zone

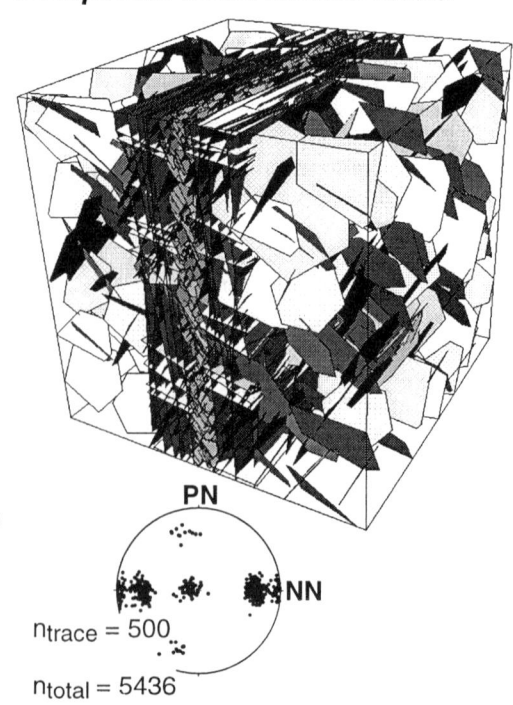

$n_{trace} = 500$

$n_{total} = 5436$

fault zone component based on the field data and the Andersonian mechanical model.

FracMan™ enables us to construct three-dimensional, fully interconnected fracture networks within rectangular subregions that represent different fault zone components within the cubic model domains. The subregions can be truncated by, or interconnected with, adjacent subregions. The width of each fault zone component (Table 1) is based on field data from *Caine et al.* [1996] and *Caine* [1996]. Because the topography of curviplanar boundaries found between components commonly exhibit large wavelength to amplitude ratios, at the model scale it is reasonable to define planar component boundaries. Transitions in fracture intensity and geometry observed at component boundaries are approximated by generating fractures within one component that project into the adjacent component (Fig. 4).

Fractured Protolith Model

Fracture data collected from the protolith of the Stillwater Fault Zone lead us to include two sets of cross fractures and one set of extension fractures in the protolith components of each model (Figs. 3 and 4; Table 1). Each fault zone model also simulates varying degrees of displacement within the protolith blocks. In the models with relatively small displacement, the protolith fracture networks are generated continuously throughout the full model domain (displacements of 0.1 m to 10 m in the SFF and DDZ models in Fig. 4). The fault zone and associated deformation-related fracture networks are superimposed on the protolith by offsetting the adjacent regions and adding appropriate fracture sets. In the models with larger displacements the protolith blocks are generated separately in a pair of rectangular subregions (displacements of 10 m to more than 1 km in the LDZ and CDZ models in Fig. 4). These are separated by a central region which is subsequently constructed as a fault core. The protolith on either side of the fault core is distinct, thus simulating the juxtaposition of unique fracture networks. Because the same fracture parameters are used to construct all protolith regions we obtain protolith subregions with similar fracture networks.

Single Fracture Fault Model

The SFF model comprises a single protolith cube cut by a single fracture fault (Fig. 4). Pure dip slip along the fault drops the right-hand side of the protolith 0.5 m relative to the left-hand side (Fig. 4). The offset is accomplished using a 'faulting' option within the FracMan™ code and results in protolith fractures being cut at their intersections with the fault. The fault has a displacement trace-length ratio of 0.025 which is within the range of observed trace length-displacement ratios found along natural SFFs [e.g., *Dawers et al.*, 1993; *Schlische et al.*, 1996].

The SFF model represents a fault with a highly localized, single slip surface where deformation has been accommodated in a network of unlinked individual faults, or perhaps along a preexisting discontinuity. The SFF is meant to represent a host of small displacement, individual faults that occur in a range of different tectonic settings and rock types. Field examples of SFFs include unlinked individual normal fault networks in the Bishop Tuff in Owens Valley, California [*Dawers et al.*, 1993]; individual strike slip faults in the Northern Shawangunk Mountains of south eastern New York State [*Caine et al.*, 1991; *Vermilye and Scholz*, 1994]; and individual strike slip faults that formed along preexisting joints in the Mount Abbot area of the Sierra Nevada in central California [*Martel*, 1990].

Distributed Deformation Zone Model

The fault zone in the DDZ model comprises the 5 m wide central portion of the model. It contains a dense network of step fractures, extension fractures, and large shear fractures (Fig. 4). A total of 10 m of displacement is modeled in this zone. The model is constructed within a single protolith block that originally occupies the entire model volume. The first step in constructing the distributed deformation zone is to superimpose a set of step fractures over a central subregion of the protolith fracture network (Fig. 4 and Table 1). 'Deformation' is accommodated along 5 evenly spaced shear fractures, with 20 m trace lengths, each of which accommodates 2 m of constant dip slip (Fig. 4). The protolith to the right of the fault zone is dropped down, in a stepwise fashion, relative to the left-hand side of the fault zone. After the protolith regions are displaced, the fault zone subregion is populated with two sets of shear fractures, along which no displacement is modeled. These shear fracture sets represent the accompanying subsidiary deformation commonly found in DDZ type fault zones.

While it may be more representative of natural fault zones to model smaller amounts of displacement on all of the shear fractures in the two superimposed shear fracture

Figure 4. Three-dimensional views of the four fault zone architectural styles modeled as discrete fracture networks. Equal area nets show the poles to a set of fractures that fall on a *y, z* trace plane that passes through the origin of each model. This is the same plane for which many of the following figures display model results. The number of fractures in each trace plane (n_{trace}) and the total number of fractures (n_{total}) in each model are shown on the bottom of each net. Note the NN, PN, PP global coordinate system depiction with the origin (*x*=0, *y*=0, and *z*=0) at the center of each cube. Each fracture model domain is 20 m by 20 m by 20 m. Shading of the fractures is exclusively related to arbitrary 'lighting' chosen in the visualization code (GeomView).

sets, it is not possible to accomplish this with efficiency using the FracMan™ code. A series of sensitivity studies, not presented here, shows that incremental fault displacements ranging from 10 cm through 100 m has little effect on the resulting flow simulation results. Thus, the compromises made in modeling the style of displacement found in these types of natural fault zones appear to yield a satisfactory result.

Field observations of natural fault zones indicate a gradational contact between damage zone (the entire fault zone in this case) and protolith with fracture intensities that decrease with increasing distance from the center of the fault zone [Caine et al., 1996; Caine and Forster, 1997]. This transition zone is simulated in the models by placing some fracture centers close to the protolith / fault zone boundary such that the fractures associated with each component project past the boundary into the adjacent component.

The DDZ model represents a fault zone where strain is accommodated by a distributed fracture network composed primarily of shear fractures. Natural examples of fault zones of this type are found in normal displacement Hill fault / fracture meshes [Hill, 1977; Sibson, 1996; Caine et al., 1996]; some fault zones dominated by swarms of deformation bands such as those found in southern Utah [Antonellini and Aydin, 1994]; and strike slip compound fault zones in the Sierra Nevada [Martel, 1990]. Other types of DDZs that accommodate much larger displacements include ancient and modern accretionary prism décollements found along the Barbados margin [Moore and Vrolijk, 1992; Screaton et al., 1990] and in the Newfoundland Appalachians [Caine, 1989] respectively.

Localized Deformation Zone Model

The fault zone in the LDZ model contains a 1 m wide fault core region, simulated as a very dense network of small shear fractures. The fault zone is sandwiched between two separately generated, but similar, protolith subregions as described above (Fig. 4 and Table 1). This model represents a highly localized fault zone that has undergone a history of episodic displacement and associated cataclastic deformation. In this case the original protolith rock fracture fabric has been obliterated through a large amount of deformation (from 100 m to greater than 1 km of displacement).

While the fractures used to represent the fault zone are much larger (mean fracture length = 0.5 m, see Table 1) than the microscopic fractures and pore space that we believe controls flow in the cores of inactive fault zones, the computational capacity required to model fluid flow through such a detailed fracture network is immense. Although it would be preferable to represent the fault core as an equivalent porous medium, the fluid flow simulator we use cannot easily model the dual porosity of a fractured porous media.

The sharp contact often observed [Caine et al., 1996; Caine, 1996] between fault core and damage zone (or protolith) is preserved in the models while also preserving fracture continuity between the components. This is accomplished by allowing the fault core fractures to extend 0.25 m into the adjacent components (Fig. 4). The high density, constant length fractures modeled in the fault core yield a distinct boundary because fracture centers located near the boundary project a similar distance into the adjacent component.

In the LDZ model, and the CDZ model discussed below, we simulate displacements that are much greater than the size of the model domain. Thus, we generate separate protolith regions as described above. Field examples of the LDZ type of fault zone are found along several segments of the very large displacement San Gabriel strike slip fault zone in southern California [Anderson et al., 1980; Chester et al., 1993; Caine et al., 1996] and along isolated segments of two large displacement normal faults in east Greenland [Caine, 1996].

Composite Deformation Zone Model

The CDZ model is composed of two different protolith subregions separated by damage zones superimposed on the protolith. A 1 m wide fault core containing a dense shear fracture network similar to the core created for the LDZ model is placed between the two 2.5 m wide damage zones (Fig. 4 and Table 1). This model represents a fault zone where deformation is initially accommodated by distributed deformation. Later phases of faulting are then accommodated within a narrow zone of localized deformation [Caine et al., 1996]. Field examples include the Stillwater Fault Zone and normal faults found on Traill Island in east Greenland [Caine et al., 1996; Caine, 1996]. The protolith and damage zone fracture networks in the CDZ model are equivalent to those used in constructing the DDZ model.

SIMULATING FLUID FLOW IN FAULT ZONE
MODELS

Numerical simulations of fluid flow through the discrete fracture network models are performed using the finite element simulator Mafic™ [Miller et al., 1995]. The simulation results illustrate how fracture networks influence the relative magnitudes and anisotropies of the bulk equivalent permeabilities associated with each fault zone component and each full fault zone model. In the finite element model all fractures are assumed to act as parallel, smooth walled conduits with rectangular cross sections. This assumption is commonly made when modeling fluid flow through discrete fracture networks [Snow, 1968; Witherspoon et al., 1980; Long et al., 1982]. Thus, each element in the mesh is assigned a fracture transmissivity, T, that can be directly related to fracture aperture a:

$$T = \frac{a^3}{12} \frac{\rho g}{\mu} \qquad (1)$$

where T is fracture transmissivity [L^2/T], a is aperture [L], ρ is the fluid density [M/L^3], g is the acceleration due to gravity [L/T^2], and μ is the dynamic fluid viscosity [M/LT] (M=mass, L=length, T=time). Assuming the fluid is aqueous, at standard temperature and pressure, equation (1) yields the following approximate relationship between transmissivity and aperture:

$$T \approx 1 \times 10^6 \left(a^3\right) \qquad (2)$$

where aperture is expressed in meters and transmissivity is expressed in meters squared per second. Single values for transmissivity and aperture are assigned to each individual fracture in each model. Transmissivities used in this study range from 1×10^{-12} m²/s to 1×10^{-3} m²/s (corresponding to apertures from 1 μm to 1000 μm) and are varied uniformly within each fault zone to illustrate how variations in fracture aperture influence the hydraulic properties of the idealized models.

Mafic™ is used to compute the steady state distribution of hydraulic head at each node in the model domains. In addition, volumetric fluid fluxes are computed along external boundaries and fluid velocities are computed within each fracture element. The Galerkin finite element method is used to solve the governing equation for two-dimensional, steady state fluid flow subject to the boundary conditions imposed on each model domain. MeshMaker™ is used to create two-dimensional triangular elements constructed within each of the fracture planes that comprise the fully three-dimensional fracture network models. The MeshMaker™ and Mafic™ codes are described by *Miller et al.* [1995] and *Dershowitz et al.* [1996].

A series of numerical, one-dimensional flow experiments are used to calculate directional, bulk fault zone permeabilities. Boundary conditions are applied to a cube, 18 m on a side, cut from the 20 m fracture model domains (Fig. 5). This slight reduction in model size ensures that each flow simulation boundary will have a sufficient number of intersecting fractures to allow interconnection and flow. The size of the model domains were chosen to represent outcrop scale permeability structures as well as to allow for computational efficiency. The calculated permeabilities are intended to provide a measure of the combined effect of fluxes and gradients within the discrete fracture models. Equivalent permeabilities are not computed for use in continuum-based flow simulators, but to provide a useful parameter for comparing the model results obtained for each architectural style. Thus, the results are strictly valid only for the length and volume scales modeled.

Three mutually-perpendicular flow directions, relative to the orientation of the fault zone, slip vector, and fault-related fracture fabric are simulated for each case. These directions are referred to as the NN, PN, and PP directions (where N=normal and P=parallel). The first symbol in this notation refers to the direction of flow relative to the orientation of the nominal plane of the fault zone and the second refers to the direction of flow relative to the slip vector. For example, NN symbolizes one-dimensional flow that is normal to the strike of the fault zone and normal to the orientation of the slip vector (Fig. 5).

In each simulation, a uniform hydraulic gradient of 0.06 is applied across a pair of opposing model faces for each flow direction. This gradient is consistent with regional

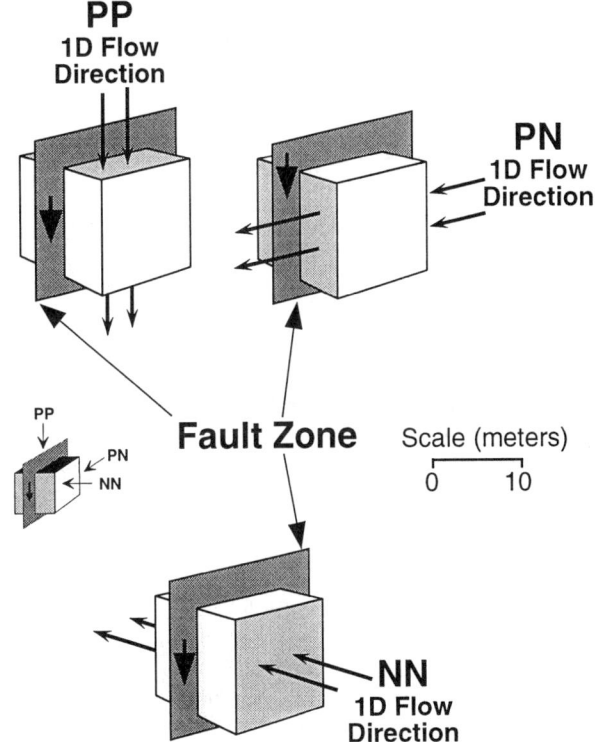

Figure 5. Schematic block diagrams depicting the direction of flow in each numerical flow experiment relative to the fault zone, slip vector, and boundary conditions. In each diagram the darkest gray panel represents the orientation of the 'plane' of the fault zone and the orientation of the slip vector (short arrow). The intermediate gray panels represent a constant head boundary across which there is a hydraulic gradient of 1 and through which each one-dimensional flow experiment was simulated. The lightest gray panels represent constant flux boundaries where $q = 0$. NN represents flow normal to the fault zone and normal to the slip vector; PN represents flow parallel to the fault zone and normal to the slip vector; and PP represents flow parallel to the fault zone and parallel to the slip vector. Note the NN, PN, PP global coordinate system depiction and scale bar.

scale flow system gradients. Uniform values of hydraulic head are set on a pair of opposing model faces. Zero flux conditions are set on the remaining four faces. The total volumetric fluid flux computed between the two opposing faces is used to compute the equivalent bulk permeability, k, of the specified fault zone component, or full model domain, in each mutually-perpendicular (NN, PN, PP) direction (Fig. 5). Equivalent bulk permeabilities are calculated in each of the three directions using:

$$k = \frac{\mu}{\rho g} \frac{Q}{IA} \qquad (3)$$

where Q is the simulated volumetric flow rate [L^3/T], I [dimensionless] is the specified hydraulic gradient, A [L^2] is the specified cross sectional area across which the discharge, Q, flows, k is the calculated permeability [L^2], ρ is the fluid density [M/L^3], g is the acceleration due to gravity [L/T^2], and μ is the dynamic fluid viscosity [M/LT].

FLUID FLOW SIMULATION RESULTS

The primary goal of the fluid flow simulations is to examine the way that changes in fracture orientation, density, geometry, and aperture influence the relative magnitude of permeability heterogeneity and anisotropy within individual fault zone components and within the full model domains. Results obtained for each model are also used as static analogs to illustrate snapshots in the stepwise evolution of the permeability structure of a fault zone during a deformation cycle. The model results highlight the interplay of structural and hydraulic properties at the scale of individual fault zone components and at the scale of a full fault zone in its protolith.

Permeability Structure of the Fault Zone Components

An important goal of this modeling study is to examine the ways that different fault zone architectural styles might lead to the permeability structures observed *in situ*. The results of *in situ* permeability tests performed in natural fault zones formed in crystalline rocks [e.g., *Davison and Kozak*, 1988 and *Andersson et al.*, 1991] indicate that fault zone permeabilities may vary over at least six orders of magnitude within distances as small as 1 m or less. The significant role that permeability heterogeneity within fault zones might play in controlling patterns and rates of fluid flow, within and near a fault, is also suggested in the results of regional-scale, two-dimensional [*Forster and Evans*, 1991] and three-dimensional fluid flow simulations [*Lopez and Smith*, 1996]. Realistic estimates of fault zone architecture and permeability structure, however, were unavailable at the time the earlier studies were performed. The work presented here provides a geological basis for estimating the bulk permeability of individual fault zone

components (using Equation 3) as a first step in evaluating the heterogeneous permeability structure of different fault zone types.

Equivalent bulk permeabilities obtained for the protolith subregions, and each fault zone component, are shown in Figure 6 as sets of three log (k) values, in m^2, for each flow direction (NN, PN, PP). These results are summarized in Table 2. The calculated geometric mean permeabilities are shown to provide insight into the bulk permeability of each fault zone component, averaged over the three flow directions. Permeability ratios (Table 2) provide insight into the permeability contrasts that yield a bulk permeability anisotropy at the fault zone scale. Permeability ratios are computed by comparing permeabilities calculated in the PP (k_{PP}) and PN (k_{PN}) directions to that of the corresponding permeability in the NN direction (k_{NN}).

In order to determine the degree of potential variability in computed permeabilities a series of 20 protolith blocks were constructed, as described above, using the identical parameters and a different random seed for each model. Flow was simulated in each of the three directions in each block and the corresponding permeabilities were computed. Table 3 shows the results of these experiments which indicate that the spread of permeability values are small (approximately factors of 2 to 4 from one model to another with maximum anisotropy of less than one order of magnitude). The variability in computed permeabilities for the damage zone and fault core components is likely smaller because their fracture densities are much higher.

Two categories of computed results are presented in Figure 6. First, equivalent bulk permeabilities are computed for the protolith and each fault zone component using a single fracture aperture of 100 μm. Using constant aperture illustrates how fault zone architecture alone (e.g., changes in fracture density, fracture orientation, and component geometries) affects fluid flow. Second, fracture apertures in the fault zone only are reduced from 100 μm to 10 μm. Note that the fault zone in the CDZ model comprises both a damage zone and a fault core (Fig. 4).

Figure 6 reveals the computed variations of anisotropy and mean permeability in the protolith of each model. In the protoliths modeled here the similar fracture networks yield no consistent orientation for k_{max} or k_{min}, as different results might be expected for protoliths of different lithology or that have undergone different geologic histories. Permeability ratios (k_{NN} : k_{PN} : k_{PP}) in the protolith vary from 1:0.2:0.8 to as high as 1:4.7:0.4 with the orientation of k_{max} varying between parallel or perpendicular to the fault zone (Fig. 6 and Table 2).

In contrast to the protolith, the fault zone fracture networks result in permeability anisotropy where k_{min} is always oriented perpendicular to the fault zone, in the NN direction, except in the SFF model (Fig. 6). Permeability anisotropy in the non-SFF fault zones yields ratios that range from 1:3:1 to 1:6:5 (Table 2). Thus, anisotropy within individual fault zone components is relatively small

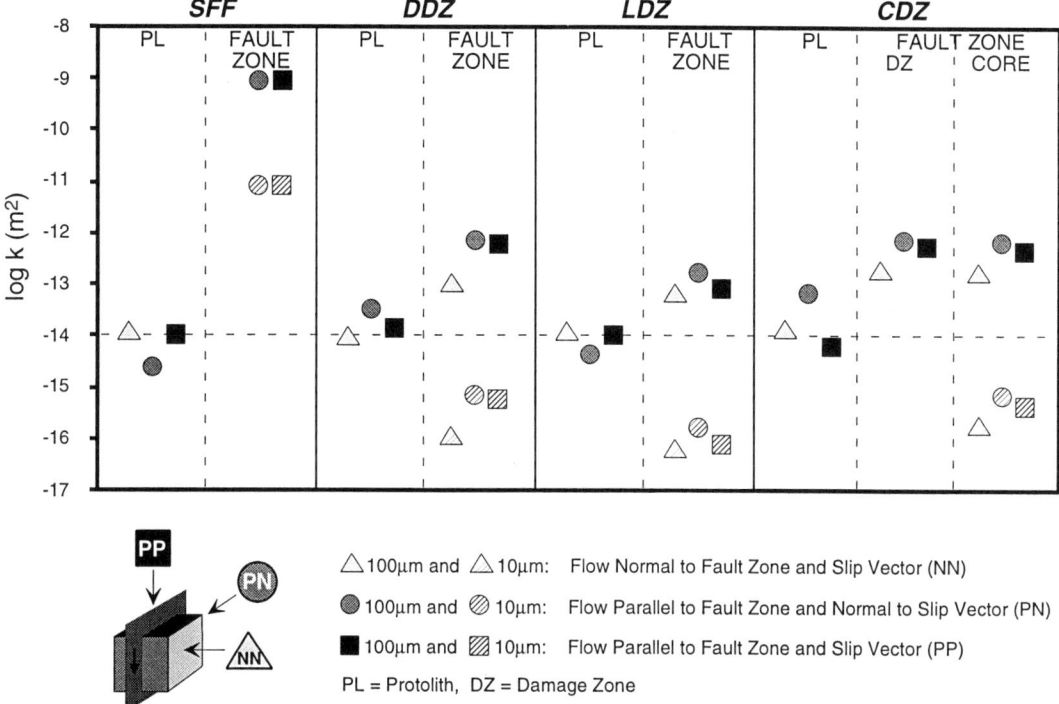

Figure 6. Bulk log permeabilities of each protolith and each fault zone calculated from model results for each of the three flow directions (NN = triangles, PN = circles, and PP = squares). SFF is the single fracture fault model, DDZ is the distributed deformation zone model, LDZ is the localized deformation zone model, and CDZ is the composite deformation zone model (note that the fault zone in the CDZ model includes a damage zone and a fault core). The filled set of points show the permeability structure where fracture apertures are uniformly 100 μm in each fault zone component. The hatched points show the permeability structure where the protolith and damage zone fracture apertures are uniformly 100 μm and fault zone fracture apertures are 10 μm. The horizontal dashed line is shown for reference to an average protolith value for each model.

and similar to that of the protolith. It is, however, noteworthy that the permeability ratio is the largest in the DDZ model (Fig. 6 and Table 2). This is because the fault zone is composed of a network of very long (~5 m), high density (7 m²/m³), shear fractures as compared with the very short (~0.5 m), high density (6 m²/m³), shear fracture network in the LDZ and CDZ models (Tables 1 and 2). This suggests that fracture length and orientation is more important than fracture density in controlling bulk anisotropy at the fault zone component scale.

Bulk permeability anisotropy of a fault zone is influenced by the anisotropy of individual fault zone components combined with the integrated effect of permeability contrasts between components. Permeability contrasts between components is estimated by comparing the geometric mean permeabilities (Table 2) computed for each component. The uniform, 100 μm cases yield permeability contrasts between the fault zones and protoliths that range between one and two orders of magnitude for the DDZ, LDZ, and CDZ models. These relatively large contrasts can cause significant changes in the patterns and rates of fluid flow in porous or fractured media that would not otherwise occur in

homogeneous, isotropic permeability structures [*Freeze and Cherry*, 1979]. The enhanced permeability of the fault zones in these models occurs because they contain higher density fracture networks that include both long shear fractures and small step fractures (Table 1).

The SFF model is a special case where k_{NN} is undefined because the permeability across the open space in the single fracture is infinite. However, if the single fracture is filled with fine-grained comminuted or precipitated minerals, it will have an internal permeability with magnitude and anisotropy dictated by the pore structure and/or microscopic to macroscopic fractures. The permeability contrast between the protolith and the fault in the SFF model is up to five orders of magnitude because the permeability of a single relatively open fracture is very high compared with that of even a dense network of fractures (Fig. 6).

The possible impact of fault zone sealing by progressive comminution and mineral precipitation is modeled in each architectural style by varying fracture aperture within each fault zone from 1 to 1000 μm (Table 2). A subset of these results is shown in Figure 6. A large reduction in equivalent permeability occurs when fracture aperture is changed

Table 2. Fault zone architectural modeling results for bulk component permeabilities.

Component aperture (μm)	Architectural component	Geometric mean permeability of NN, PN, PP directions (m²)	Permeability in NN direction (m²)	Permeability in NN, PN, PP directions normalized to the NN direction			$k_{max} : k_{min}$ directions
				NN	PN	PP	
Single Fracture Fault							
100	Protolith	7.4E-15	1.3E-14	1	0.2	0.8	NN : PN
1000	Fault	9.8E-08	9.8E-08	1	1.0	1.0	ISOTROPIC
100	Fault	9.8E-10	9.8E-10	1	1.0	1.0	ISOTROPIC
10	Fault	9.8E-12	9.8E-12	1	1.0	1.0	ISOTROPIC
1	Fault	1.0E-13	1.0E-13	1	1.0	1.0	ISOTROPIC
Distributed Deformation Zone							
100	Protolith	1.9E-14	1.1E-14	1	3.3	1.5	PP : PN
1000	Fault Zone	4.1E-10	1.3E-10	1	6.3	5.4	PN : NN
100	Fault Zone	4.1E-13	1.3E-13	1	6.3	5.4	PN : NN
10	Fault Zone	4.1E-16	1.3E-16	1	6.3	5.4	PN : NN
1	Fault Zone	4.1E-19	1.3E-19	1	6.3	5.4	PN : NN
Localized Deformation Zone							
100	Protolith	9.2E-15	1.4E-14	1	0.4	0.8	NN : PN
1000	Fault Zone	1.1E-10	7.4E-11	1	2.6	1.3	PN : NN
100	Fault Zone	1.1E-13	7.4E-14	1	2.6	1.3	PN : NN
10	Fault Zone	1.1E-16	7.4E-17	1	2.6	1.3	PN : NN
1	Fault Zone	1.1E-19	7.4E-20	1	2.6	1.3	PN : NN
Composite Deformation Zone							
100	Protolith	2.0E-14	1.6E-14	1	4.7	0.4	PN : PP
1000	Damage Zone	4.7E-10	2.1E-10	1	3.7	2.9	PN : NN
100	Damage Zone	4.7E-13	2.1E-13	1	3.7	2.9	PN : NN
10	Damage Zone	4.7E-16	2.1E-16	1	3.7	2.9	PN : NN
1	Damage Zone	4.7E-19	2.1E-19	1	3.7	2.9	PN : NN
1000	Core	4.3E-10	2.1E-10	1	3.7	2.4	PN : NN
100	Core	4.3E-13	2.1E-13	1	3.7	2.4	PN : NN
10	Core	4.3E-16	2.1E-16	1	3.7	2.4	PN : NN
1	Core	4.3E-19	2.1E-19	1	3.7	2.4	PN : NN

Note: For each fault zone component model run: No matrix permeability; Initial conditions @ time=0, head=0, and flux=0; Hydraulic gradient = 1; Steady state flow; k_{max} and k_{min} refer to the directions of maximum and minimum permeability.

because total volumetric flux through a fracture is related to the cube of fracture aperture:

$$Q = \frac{a^3}{12} \frac{\rho g}{\mu} IW \qquad (4)$$

where W is the width [L] of the fracture measured normal to the direction of flow within the fracture plane. Thus, a 10-fold change in fracture aperture yields a 1,000-fold change in the value of Q that, in turn, yields a 1,000-fold change in bulk permeability (Equation 3).

Figure 6 and Table 2 illustrate how closure of the fault zone fractures from 100 to 10 μm in the DDZ and LDZ models reduces the bulk permeability of the fault zones by three orders of magnitude. This produces a fault zone with significantly lower permeability than the surrounding

protolith. As a consequence, hydraulic gradients oriented subparallel to the fault zone (e.g., PP and PN directions) would cause most fluid to flow through the adjacent protolith while flow across the fault zone would be restricted by its reduced permeability. Note that the anisotropy of the fault zones remain unchanged because all fracture apertures are uniformly reduced.

Reducing fracture apertures in the core of the CDZ model reduces the bulk permeability of the core while leaving the damage zone as a primary pathway for fluid flow. The resulting permeability structure produces a combined conduit-barrier flow system similar to those described by *Forster and Evans* [1991], *Caine et al.* [1996], and *Evans et al.* [1997]. In this type of system, fluid flow across the fault zone is restricted and redirected in directions parallel to the nominal orientation of the plane of the fault zone [*Caine et al.*, 1996].

Table 3. Fault zone architectural modeling results: variation in bulk permeability of protolith multiple realizations.

Flow direction	Permeability range of 20 realizations (m²)	Arithmetic mean permeability of 20 realizations (m²)	Standard deviation of permeability (m²)
NN	4.7E-15 to 1.2E-14	7.6E-15	2.2E-15
PN	2.2E-15 to 8.8E-15	5.0E-15	1.6E-15
PP	5.1E-15 to 1.3E-14	9.7E-15	1.8E-15

Note: For each protolith realization: Fracture model domain size = 20 m by 20 m by 20 m; Flow simulation region = 18 m by 18 m by 18 m; No matrix permeability; Initial conditions @ time=0, head=0, and flux=0; Hydraulic gradient = 1; Steady state flow; Apertures are uniformly 100μm with transmissivities of 10^{-6} m²/s. All parameters are the same for each of the 20 realizations modeled.

Permeability Structure of the Full Fault Zone Models

The previous section outlined how fluid flow simulations help to understand the permeability magnitude and anisotropy of different fault zone components. The following section describes the results of a similar approach used to assess how combining different component types influences the bulk permeability magnitude and anisotropy of each full model domain (Table 4 and Fig. 4). In this section, bulk model permeabilities are computed at each of four stages in an idealized fault zone evolution. This provides insight into the time-dependent evolution of fault zone permeability structure and is an important first step in defining geologically plausible and heterogeneous permeability structures.

Equivalent bulk permeability values obtained for each full model domain (Fig. 4) are shown in Figure 7 as sets of three log (k) values, in m², for each flow direction. These results are also summarized in Table 4. Again, calculated geometric mean permeabilities and permeability ratios (k_{NN} : k_{PN} : k_{PP}) provide insight into the averaged bulk permeability and anisotropy of each model (Table 4).

The impact of fracture opening and closure during fault zone evolution is approximated by uniformly varying fracture apertures in each fault zone from 1 to 1000 μm (Table 4 and Fig. 7). The uniform fracture aperture cases of 100 μm are assumed to represent an intermediate stage in the evolution of a fault zone that undergoes an idealized mechanical deformation cycle from prefailure, to failure, to postfailure. Stepwise fracture closure from 100 μm to 10 μm and then to 1 μm represents a series of postfailure to prefailure stages with progressively more extensive fracture sealing. Uniformly increasing fracture apertures to 1000 μm in the fault zones represents the increase in permeability that likely occurs during failure.

Flow simulation results obtained for the uniform aperture case suggest that the SFF and LDZ full models are effectively isotropic with bulk permeability values similar to that of the protolith (~10^{-14} m²). This result is obtained because fluid flow both parallel (PP, PN directions) and normal (NN direction) to the fault zone is effectively controlled by the protolith. With an aperture of 100 μm, the fault in the SFF model transmits insufficient fluid to

influence the bulk flow through the model domain. The higher density network of shear fractures in the fault zone of the LDZ model provides only a small (factor of two) increase in bulk permeability parallel to the fault zone (SFF fault zone fracture density = 0.05 m²/m³ versus LDZ fault zone fracture density = 6 m²/m³, Table 1).

In contrast, incorporating high density, long shear and step fractures in the fault zones of the DDZ (7 m²/m³) and CDZ (13 m²/m³) models yields a large (one order of magnitude) increase in bulk permeability parallel to the fault zone at the full model scale (PP and PN directions). This suggests that fracture length and orientation has a greater impact on anisotropy than does fracture density at the full model scale, as it did at the component scale. As a consequence, the fault zone can act as a conduit for fluid flow in directions parallel to the fault zone at this intermediate stage in fault evolution.

During failure a net increase in fracture aperture is expected within the fault zone, even though some fractures open while others close (e.g., dilation versus shearing). This effect is simulated by uniformly increasing fracture apertures in each fault zone from 100 to 1000 μm. In the DDZ, LDZ, and CDZ models the bulk permeability normal to the fault zone is slightly enhanced by a factor of two to three (Fig. 7 and Table 4). Increasing the aperture in the SFF model, however, has little impact on bulk permeability normal to the fault because the protolith fracture network controls flow normal to the fault. In all cases the 10-fold increase in fracture aperture yields a large increase in permeability parallel to the fault zone (two to three orders of magnitude: Fig. 7 and Table 4). Permeability anisotropy ratios of two to three orders of magnitude demonstrate that fluid flow will be focused into the fault zone whenever the hydraulic gradient is not exactly perpendicular to the fault. Thus, at failure great potential exists, from a fault zone architecture and permeability structure standpoint, for significant fault-related fluid flow as demonstrated from many field studies [e.g., *Sibson*, 1994; *Newman and Mitra*, 1994; *Goddard and Evans*, 1995; *Caine*, 1996].

In summary, mineral comminution, stress relaxation, and mineral precipitation localized in the fault core progressively seal fault zones during the postfailure stage of

Table 4. Fault zone architectural modeling results: full model permeabilities.

Fracture apertures in fault zone (μm)	Architectural model	Geometric mean permeability of NN, PN, PP directions (m^2)	Permeability in NN direction (m^2)	Permeability in NN, PN, PP directions normalized to the NN direction			$k_{max} : k_{min}$ directions
				NN	PN	PP	
1000	SFF	4.3E-13	1.3E-14	1	1.9E+2	1.9E+2	PN : NN
	DDZ	1.1E-11	3.1E-14	1	7.5E+3	6.5E+3	PN : NN
	LDZ	1.8E-12	3.0E-14	1	6.2E+2	3.4E+2	PN : NN
	CDZ	3.5E-12	4.6E-14	1	8.8E+2	5.0E+2	PN : NN
100	SFF	1.7E-14	1.3E-14	1	1.0E+0	2.0E+0	PP : NN
	DDZ	1.1E-13	1.9E-14	1	1.4E+1	1.2E+1	PN : NN
	LDZ	2.1E-14	1.4E-14	1	2.0E+0	2.0E+0	PN : NN
	CDZ	1.1E-13	2.4E-14	1	1.2E+1	9.0E+0	PN : NN
10	SFF	1.4E-14	1.2E-14	1	1.0E+0	1.0E+0	ISOTROPIC
	DDZ	6.4E-15	4.8E-16	1	5.8E+1	4.0E+1	PN : NN
	LDZ	4.4E-15	3.8E-16	1	7.2E+1	2.0E+1	PN : NN
	CDZ	3.8E-14	9.8E-16	1	2.9E+2	2.0E+2	PN : NN
1	SFF	1.3E-14	1.2E-14	1	1.0E+0	1.0E+0	ISOTROPIC
	DDZ	6.6E-16	5.2E-19	1	5.3E+4	3.7E+4	PN : NN
	LDZ	4.2E-16	3.4E-19	1	8.1E+4	2.5E+4	PN : NN
	CDZ	3.8E-15	1.0E-18	1	2.8E+5	1.9E+5	PN : NN

Note: For each full model run: Fracture model domain size = 20 m by 20 m by 20 m; Flow simulation region = 18 m by 18 m by 18 m; No matrix permeability; Initial conditions @ time=0, head=0, and flux=0; Hydraulic gradient = 1; Steady state flow; k_{max} and k_{min} refer to the directions of maximum and minimum permeability.

deformation [e.g., *Power and Tullis*, 1989]. Reducing fracture apertures in each fault zone from 100 to 10 μm, then to 1 μm (Fig. 7 and Table 4) mimics the way that reduced fault zone permeability might cause significant reductions in bulk permeability normal to the fault zone (one to two orders of magnitude for the 10 μm case and four to five orders of magnitude for the 1 μm case). Permeability parallel to the fault zone, however, is little affected by sealing because fluid directed toward the fault zone is primarily transmitted through the adjacent protolith in the SFF, LDZ, and DDZ models, or in the damage zone of the CDZ model (Fig. 7 and Table 4). The net effect of this significant anisotropy resulting from sealing causes redirection of fluid flow. For example, progressive sealing during the postfailure stage of deformation leads to a permeability structure that directs flow along the outside of a low-permeability fault zone when hydraulic gradients are not exactly perpendicular to the fault zone. Under the same hydraulic conditions, significant permeability enhancement in fault zone-parallel directions can occur during failure to cause fluid flow to be focused in the fault zone. As the deformation cycle continues, sealing in the fault zone causes progressively less fluid to flow through it as flow is redirected through the adjacent protolith. Additionally, when fracture aperture between components is not uniform; fracture length, orientation, and density become less important than aperture in controlling anisotropy at the full model scale.

Patterns of Fault-Related Fluid Flux

The patterns and rates of fluid flux within and near a fault zone are thought to play an important role in the processes of fault zone sealing, mineral deposition, hydrocarbon migration, solute transport, geothermal fluid migration, and evolution of fault zone strength due to reequilibration of pore fluid pressures before, during and after failure events. For example, fault zone sealing can, in part, be attributed to mechanical processes such as grain-size reduction [*Chester and Logan*, 1986; *Hippler*, 1993]. However, aqueous geochemical reactions are also an integral part of sealing and other processes [*Knipe*, 1993; *Fisher and Knipe*, 1998; *Caine*, 1999]. Graphic illustrations of the magnitudes and directions of fluid flux within each fault zone architectural model help to conceptualize the spatial variations of fluid flux and how those variations might influence fault zone evolution (Figs. 8 and 9).

Flux magnitude plots (Fig. 8) are created using the three-dimensional fluid flux vectors computed at the centroid of each triangular finite element. The plots show the flux results located within a 1 m wide, vertical slab parallel to the PP and NN flow directions centered within each model domain. The values of fluid fluxes computed at each element centroid are first projected onto a single plane and then contoured. Using a thin slab from the full model domain allows a limited amount of the three-dimensional

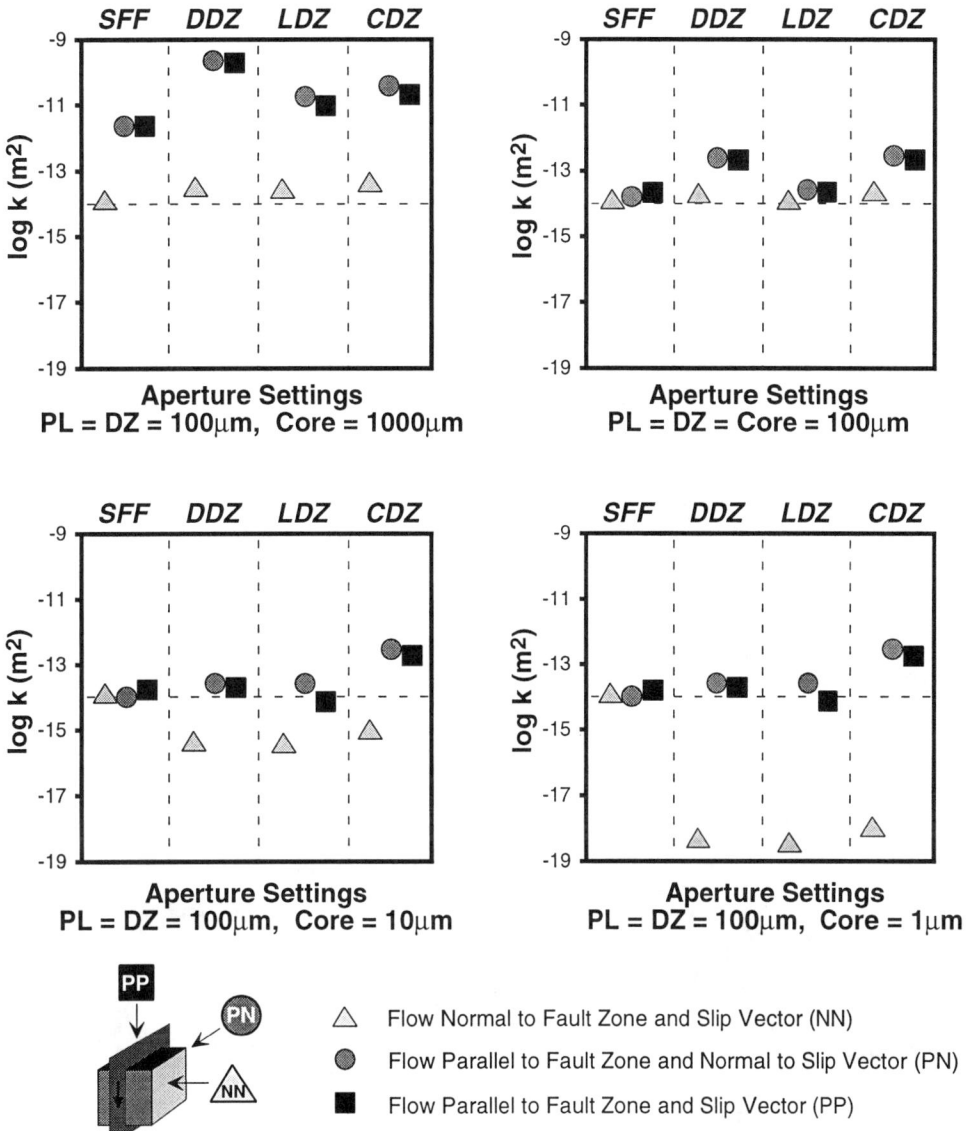

Figure 7. Bulk log permeabilities for each full model domain calculated from model results for each of the three flow directions (NN = triangles, PN = circles, and PP = squares). Results for each model (SFF, DDZ, LDZ, and CDZ) show the progressive closure of fault zone fracture apertures uniformly from 1 μm to 1000 μm. The protolith and damage zone fracture apertures are uniformly 100 μm in each model.

flux structure to come into view without obscuring the details.

The two-dimensional flux vector plots of Figure 9 show the directions of fluid flux within a 1 cm thick, vertical slab located within the 1 m wide slab used to create the flux magnitude maps of Figure 8. Only the y and z components of fluid flux are used because it is difficult to present the three-dimensional vectors that fully capture all aspects of

the fluid flow patterns. The reader must note that the actual three-dimensional flow directions are different. With this cautionary note in mind, however, insight can be gained from the flux vector plots.

The detailed variability of fluid flux shown in Figures 8 and 9 can be directly linked to the variations in fracture density, orientation, trace length, interconnectivity, and aperture associated with the protolith and each fault zone.

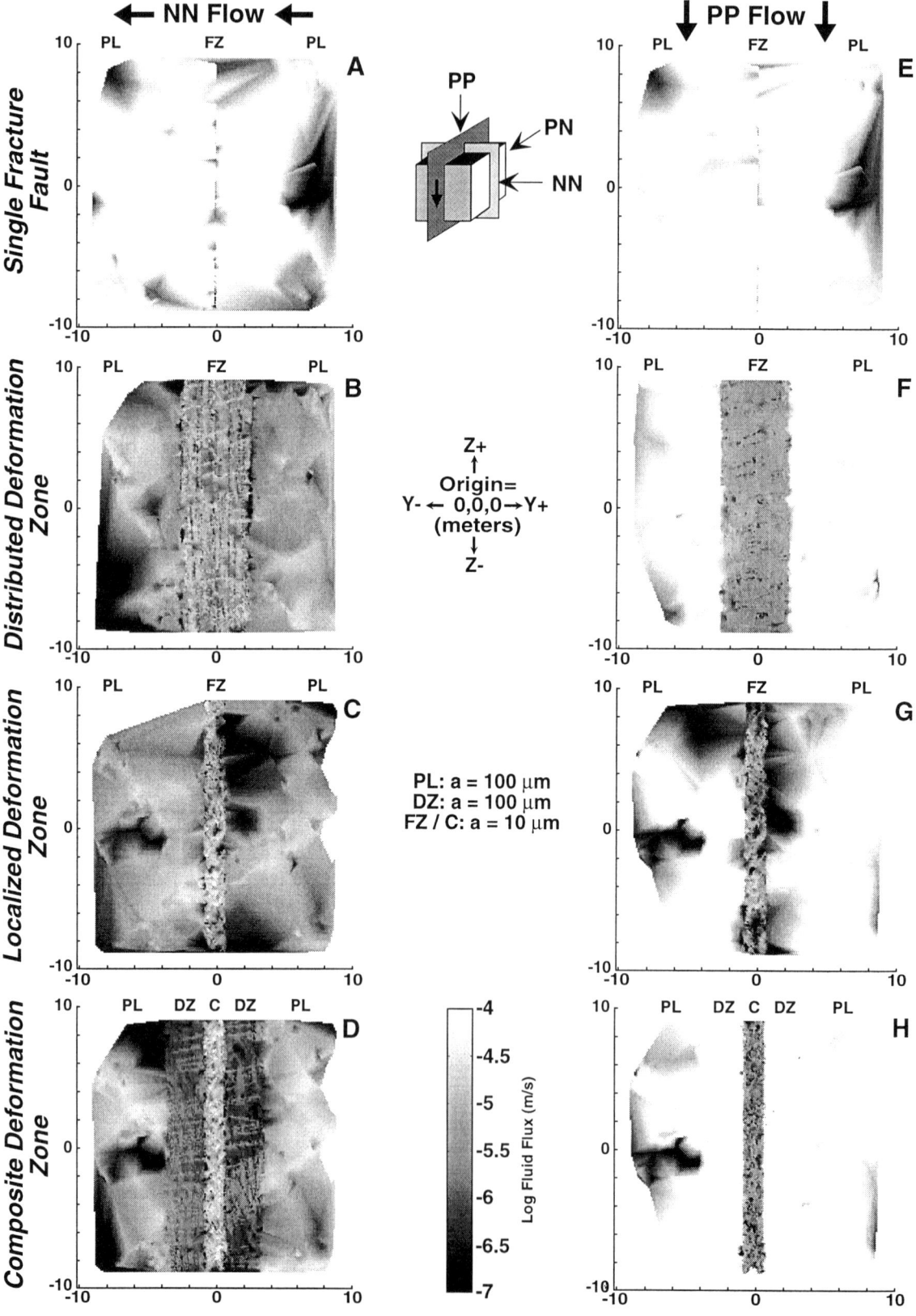

Note that Figures 8 and 9 are created only for prefailure and postfailure stages of deformation where fracture apertures in each fault zone are assigned a value of 10 μm while all other fractures are 100 μm. Large gradients in fluid flux (several orders of magnitude) found both within and between fault zone components highlight the location of important transitions between regions of low and high fluid flux in each model (Fig. 8).

Consider the results presented in Figures 8 and 9 for the SFF model. Fluid fluxes (Figs. 8a and 8e) are uniformly high within the protolith regardless of gradient orientation while fluxes in the 10 μm, single fracture fault are several orders of magnitude lower. The corresponding fluid flow vectors of Figures 9a and 9b clearly illustrate how the overall direction and the detailed patterns of flow differ between the two flow directions (NN versus PP flow). Detailed variations within the protolith of the SFF model reflect the internal variation of fracture interconnectivity.

Results obtained for the DDZ model emphasize the observations made in the previous sections regarding the way that the reduced fracture apertures (10 versus 100 μm) in the fault zone cause the bulk of the fluid moving parallel to the fault to be transmitted at higher flux rates through the protolith (Figs. 8f and 9f). When the hydraulic gradient is normal to the fault zone flow is restricted by smaller fracture apertures in the fault zone. This causes fluxes in the protolith of the DDZ model (Figs. 8b and 9b) to be reduced by several orders of magnitude below that of the SFF model (Figs. 8a and 9a). In the case of flow in the NN direction, flux variations are much reduced within the model domain (Fig. 8b) when compared to the variations computed for flow in the PP direction (Fig. 8f). Similar results are obtained for the thinner fault zone in the LDZ model (Figs. 8c, 8g, 9c, and 9g).

Fluid flux magnitudes and flux vector patterns obtained for the CDZ model reveal an interesting aspect of fault-related fluid flow. When the hydraulic gradient is normal to the fault zone, fluid fluxes in the fault core are greater than those found in both the damage zone and the protolith. This occurs because the lower permeability protolith defines the overall rate of fluid delivered to both the damage zone and the fault core (Table 2). The large aperture (100 μm), long length (~5 m), high density (7 m^2/m^3) fracture network in the damage zone, on the right-hand side of the fault core, efficiently transmits the fluid from the protolith to the core at reduced fluxes. Yet, the small aperture (10 μm), short length (~0.5 m), high density (6 m^2/m^3) fracture network in

the fault core must transmit the fluid at much higher fluxes to the damage zone on the left-hand side of the fault core (Figs. 8d and 9d). Note that the fracture density in the fault core and damage zone are one order of magnitude greater than that of the protolith (0.6 m^2/m^3, see Table 1). This suggests that, at a one order of magnitude difference between component apertures, fracture aperture has a much larger effect on flux than does fracture orientation, density, and length at the full model scale. Although a similar effect occurs for flow normal to the fault zone in all models, the two-component fault zone of the CDZ model amplifies the effect (Figs. 8a, 8b, 8c, and 8d). The fault core of the CDZ model (Figs. 8d and 9d) forms a localized zone of high fluxes when flow is directed normal to the fault, similar to what is hypothesized for natural fault zones [e.g., *Sibson*, 1996].

DISCUSSION AND IMPLICATIONS FOR NATURAL FAULT ZONES

The fault zone models shown in Figures 1 and 4 can be viewed as sequential steps in the mechanical initiation and growth of a brittle fault zone. For example, when a single fault 'plane' with a very small initial displacement (i.e., the SFF model) experiences multiple deformation events it may grow into a DDZ type of fault zone. As the fault zone continues to grow it may take a path toward either the LDZ type of fault zone, with highly localized strain, or the CDZ type of fault zone where the strain is partitioned between the core and the damage zone (Fig. 4). Increasing mineralogical 'maturity' in the fault core of LDZ and CDZ type faults (through mechanical and geochemical assimilation of protolith) likely leads to important changes in the permeability structure of a fault zone [e.g., *Knipe*, 1993; *Caine*, 1996].

Progression through the evolution of a fault zone is accompanied by episodic cycles of deformation [*Byerlee*, 1993; *Sibson*, 1994; *Caine et al.*, 1996]. In the previous sections we use the model results to illustrate possible relationships between the mechanics of the fault zone architectural styles and the evolution of fault zone permeability structure in the context of stress cycling and associated deformation [e.g., *Sibson*, 1994 and 1996]. By closing fracture apertures in the fault zone we mimic the evolution of fault zone permeability structure through various stages of deformation (e.g., Fig. 7). This evolution can be thought of as reflecting the mechanical deformation

Figure 8. Maps of the two-dimensional magnitude of Darcy fluid flux for each model. A through D shows results for flow normal to the fault zone and normal to the slip vector (NN) or flow from right to left. E through H shows results for flow parallel to the fault zone and parallel to the slip vector (PP) or flow from top to bottom. Each map represents the contoured data from a 1 m thick slab (in the x direction) projected on to the y, z plane (Note: the coordinate system depiction has the orientation of the mapped slabs as a gray plane). In each case the 'plane of the fault zone' is perpendicular to the page. The fault zone components are listed as PL = Protolith, FZ = Fault Zone, DZ = Damage Zone, C = Core. The uniform fracture aperture settings are 100 μm in each protolith and 10 μm in each fault zone.

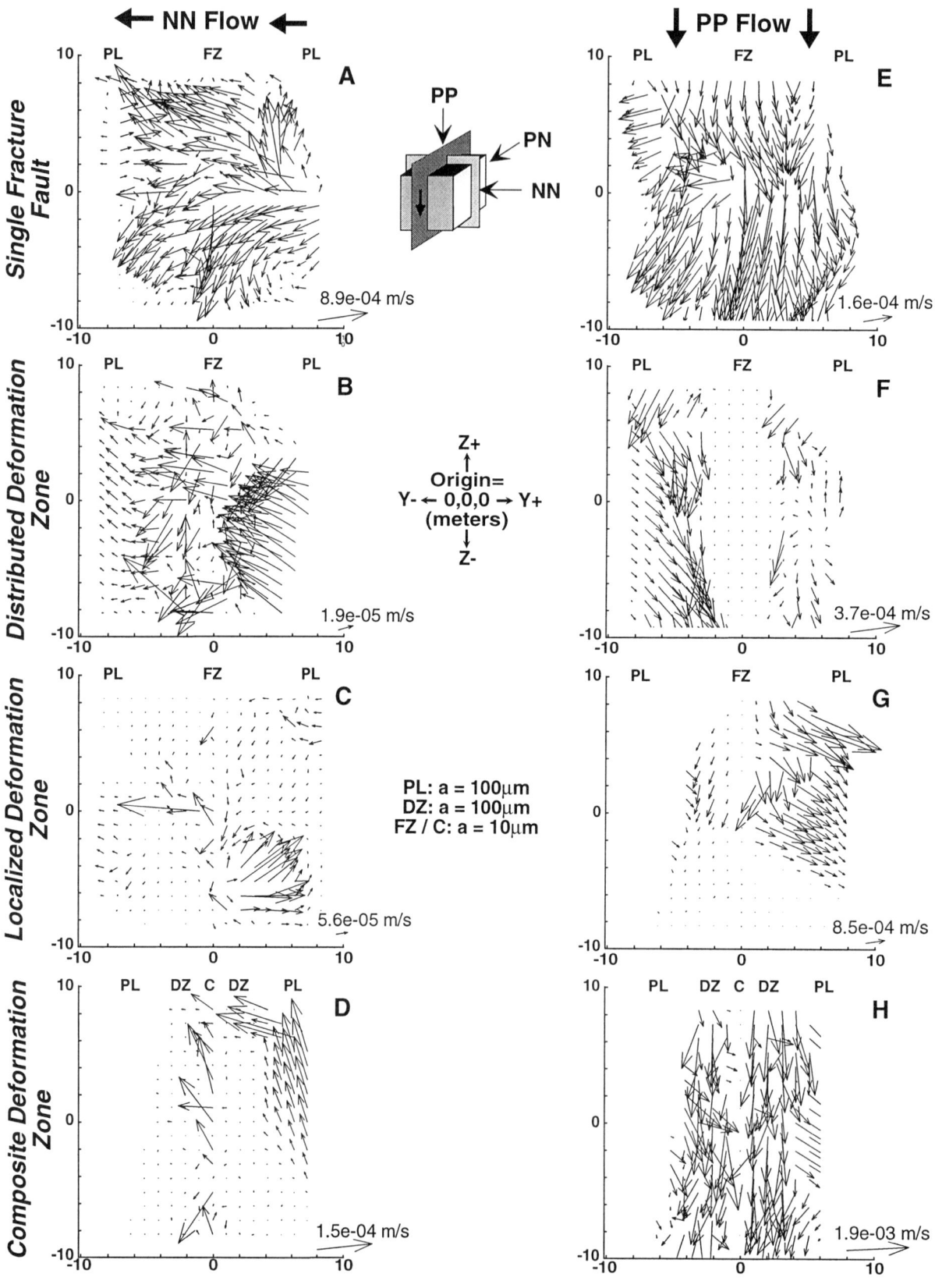

cycle from prefailure, to failure, to postfailure for a brittle fault zone. In the prefailure stage microscale to macroscale fractures will begin to open and close depending on their orientations, the stress field, and their tendency to shear and/or dilate [e.g., *Brown and Bruhn*, 1998; *Caine*, 1999; *Ferrill et al.*, 1999]. The migration of pore fluids into incipient fault zone fractures may initiate and enhance brittle failure where localized increases of pore fluid pressure in the fault zones occur [e.g., *Byerlee*, 1993; *Sibson*, 1994 and 1996]. During the initial phase of failure, fault zone permeability structure might be exemplified by the models with fault core apertures of 1000 μm (Fig. 7). At brittle failure new fractures develop, and preexisting fractures may grow, to cause enhanced permeability within the fault zone [e.g., *Sibson*, 1994]. Both the model results and outcrop observations suggest that fluid flow is enhanced and localized in the fault zone during failure. In the failure to postfailure stages comminution and fluid flow may facilitate geochemical reaction, thus localizing precipitation of minerals trapped in the fault core to cause fault zone sealing and reduced permeability [e.g., *Knipe*, 1993; *Sibson*, 1994; *Caine*, 1996; *Fisher and Knipe*, 1998; *Caine*, 1999]. If we assume a flux limited, fault-related flow system [e.g., *Bunds et al.*, 1997] the greatest accumulation of mineral precipitates will form where fluid fluxes are high. Additionally, fluid pressure redistribution may also be associated with dilatancy hardening and subsequent localized rupture arrest [*Parry and Bruhn*, 1990; *Seront et al.*, 1998; *Caine*, 1999].

Lithologic heterogeneity of fault-rocks leads to inherent permeability heterogeneity within a fault zone that, in turn, produces a complex distribution of sealed and unsealed zones that may lead to spatial variations in fault zone strength [e.g., *Cowie and Shipton*, 1998] and pore pressures [e.g., *Byerlee*, 1993]. *Nur and Booker* [1972] and *Miller et al.* [1996] suggest that the spatial and temporal distribution of earthquake aftershocks may be linked to the process of heterogeneous fault zone sealing as deformation ceases and pore pressure readjusts within the fault zone. The relatively simple variations in permeability structure addressed in this paper provide preliminary insight regarding the more complex processes that operate in natural fault zones (Figs. 8 and 9). For example, large permeability contrasts may cause large, heterogeneously distributed pressure gradients due to transient, deformation-related fluid flow pulses. As deformation and fluid flow are accommodated, discrete zones of rupture and arrest may cause longitudinal and lateral

growth of the fault zone as well as further alteration of its permeability structure.

A natural fault zone may contain any or all of the architectural styles shown in Figures 1 and 4 within a single fault strand. Additionally, many natural fault zones are not the simple, single fault zone features studied here. Rather, they are sometimes composed of many strands of deformed zones also containing any or all of the architectural styles. Although many variations of cyclic fault zone evolution might be proposed, the one discussed here forms a useful backdrop for interpreting the modeling results that, in turn, helps in thinking through the complex interplay of coupled mechanical and fluid-flow processes that operate in natural fault zones.

In reviewing the modeling results, it is important to recall that the impact of ambient regional stress and localized deformation-related stress found in the subsurface are not considered when assigning values for fracture apertures in the models. Stress will cause significant variations in fracture aperture as a function of orientation that, in turn, must influence permeability anisotropy of both individual fault components and the entire fault zone [*Brown and Bruhn*, 1998]. Fracture surface roughness, particularly when coupled with stress and deformation, is also an additional source of potential anisotropy that may make the use of the cubic law to model real fracture networks under stress invalid [*Brown*, 1995; *Brown and Bruhn*, 1998]. Aperture may also vary as a function of host rock lithology, degree of mineralization, and fracture type [*Neuzil and Tracy*, 1981; *Mozley and Goodwin*, 1995; *Brown and Bruhn*, 1996 and 1998; *Caine*, 1999]. This is particularly true for shear fractures where variations in the surface topography of slickensided surfaces may cause channelized flow parallel to the slip vector [*Brown and Bruhn*, 1996]. An alternative to using the cubic law for aperture is presented in *Brown and Bruhn* [1998] where approximations for fracture surface roughness using fractal surface geometries are coupled with stress and deformation.

In spite of the additional sources of anisotropy, the simple parallel plate models used to represent fractures in this modeling study do help in discerning the overall impact of the different fault zone components. While the use of stressed fracture models and their associated aperture variations would be ideal, the models presented are meant to be a first look at the impacts of fault zone architecture and component heterogeneity on fluid flow without being

Figure 9. Maps of the two-dimensional, *y*, *z* components, of the Darcy fluid fluxes for each fault zone model corresponding to the same maps in Figure 8. A through D show results for flow normal to the fault zone and normal to the slip vector (NN) or flow from right to left. E through H shows results for flow parallel to the fault zone and parallel to the slip vector (PP) or flow from top to bottom. Each map represents the vector data from a 1 cm thick slab (in the *x* direction) projected on to the *y*, *z* plane (Note: the coordinate system depiction has the orientation of the mapped slabs as a gray plane). In each case the 'plane of the fault zone' is perpendicular to the page. The fault zone components are listed as PL = Protolith, FZ = Fault Zone, DZ = Damage Zone, C = Core. Note that the three-dimensional flow directions are different and these are shown to illustrate the complexity of each flow system. The empty space in some of the maps is due to lack of fracture connection with a flow boundary.

obscured by variations in aperture. One consequence, however, of neglecting stress-related aperture anisotropy may be exhibited by the result that the maximum permeability computed within the fault zone, in the PP and PN flow directions, is found normal to the slip vector in the PN flow direction (Figs. 6 and 7). This is in agreement with that postulated by *Sibson* [1996], but differs from the suggestion by *Brown and Bruhn* [1998] and *Caine* [1999] that the direction of maximum permeability within the fault zone is subparallel to the slip vector or PP flow direction.

One goal of this study has been to explore permeability anisotropy in directions related to the major fabric elements of the fault zone models. Results obtained from these numerical experiments are used to compare the apparent anisotropy in each component and each architectural style. Although it is of great practical interest to define equivalent continuum permeability tensors at the fault zone component or 20 m full model scale, this may not be possible [*National Research Council Committee on Fracture Characterization and Fluid Flow*, 1996]. For example, a variety of workers [e.g., *Snow*, 1969; *Long et al.*, 1982; *Long and Witherspoon*, 1985] have tried unsuccessfully to define when fracture networks may be properly represented by equivalent continuum permeability tensors. Even if plausible tensors could be defined for the fracture networks included in this study, the flow directions we use do not necessarily coincide with the principal directions of permeability. Despite our inability to define permeability tensors (scale-independent or otherwise), our results do yield insight into apparent permeabilities in directions related to the geometric character of the fracture networks in the fault zones. The orientations of the one-dimensional flow experiments are generally parallel to the orientations of major fabric elements (e.g., mean orientation of the planes of fracture sets and their intersections). Thus, the results provide useful first approximations to the permeability anisotropy of the fault-related fracture network fabric.

Inferences made using the model results resemble those made by mapping natural fault zones exposed in outcrop. Field observations made along outcrops of the Stillwater Fault Zone show that the fracture sets most consistently filled with the latest fault-related mineral assemblage are found in fracture networks subparallel to the fault zone. Because the fault core and damage zone fractures of this CDZ type of fault zone (Fig. 4) are filled with the same mineral assemblage, we can infer syntectonic fluid flow extended into the damage zone. Fault rock textures from the core of the Stillwater Fault Zone also show a preferred orientation subparallel to the fault zone that are related to the coupled processes of deformation and fluid flow [*Caine*, 1999]. Similar observations are reported by *Caine* [1996] from the sequence of clastic rocks that have undergone normal faulting in east Greenland. Because strike slip and thrust fault zones commonly contain the same architectural elements found in the normal fault zones considered in this

paper, the model results may be applicable to any style of faulting.

The results of this modeling study can also be compared to *in situ* fluid flow measurements made in the Dixie Valley geothermal field that is hosted in the Stillwater Fault Zone [*Hickman et al.*, 1997; *Barton et al.*, 1998]. Although the boreholes are several kilometers from the outcrop exposures where the data used in building the models was collected, the downhole studies were performed in similar rock types subjected to similar deformation histories. Our models suggest that the permeability anisotropy in the fault zones are controlled primarily by long shear fractures that may also contribute to the bulk permeability anisotropy inferred by *Hickman et al.* [1997] and *Barton et al.* [1998] in the geothermal reservoir. For example, *Hickman et al.* [1997] and *Barton et al.* [1998] found that the hydraulically conductive fractures in the boreholes are optimally oriented and critically stressed for frictional failure in the present stress field. The orientation data for the conductive fractures that *Hickman et al.* [1997] and *Barton et al.* [1998] found in the boreholes are similar to the shear fracture orientations as mapped in outcrop (Fig. 2). Both sets of data are mechanically compatible with an Andersonian model for shear fractures associated with a normal fault. Thus, the model results presented here support the field-based inference that the relatively large aperture, long trace length, high density networks of shear fractures associated with faulting can exert an significant impact on bulk permeability anisotropy within a natural fault zone. In addition, *Hickman et al.* [1997] and *Barton et al.* [1998] show that the *in situ* stress data at the Dixie Valley Geothermal Reservoir indicates a present day stress regime that is consistent with an Andersonian fault model and associated fault-related fracture networks. Although an Andersonian model is not the only model that could be used to explain the Dixie Valley stress and fracture data, we have used it in our modeling because of its simplicity and general applicability to many different fault zones.

Computed permeabilities based on outcrop data from the Stillwater Fault Zone can be used to infer the cause of enhanced fault zone permeabilities found during *in situ* testing at the Dixie Valley geothermal field [*Barton et al.*, 1996; *Rose et al.*, 1997; *Hickman et al.*, 1997]. For example, the various *in situ* tests at depths of 2.5 km suggest a fault zone permeability of 10^{-12} to 10^{-11} m^2. *Hickman et al.* [1997] suggest that the fractures causing the zone of enhanced permeability are preferentially open and accessible for flow because they are optimally oriented with respect to the local stress regime. Extrapolating our modeling results suggests that increasing fracture apertures in the fault zone components of the non-SFF fault types by only a factor of 5 (from 100 to 500 μm) would yield fault zone permeabilities similar to those measured *in situ*.

This modeling study addresses only outcrop scale volumes (20 m by 20 m by 20 m) of faulted rock that influence the local patterns and rates of fluid flow within and near a fault zone. The flow of fluid to the fault, however, is ultimately controlled by regional-scale flow systems [*Forster and Evans*, 1991] driven by a variety of mechanisms (e.g., topography, thermal gradients, or tectonic activity). Thus, the next step in examining the impact of fault zone permeability structure on fluid flow requires incorporating the effective permeability magnitudes and anisotropies estimated in this study into larger, stressed and unstressed, regional-scale fluid flow simulations. This approach would not only remove the need to apply arbitrary orientations of hydraulic gradients to the fault model blocks, but would also aid in evaluating how fluids moving through the regional-scale flow systems can carry solutes from sources either close to, or far from, the fault zone. For example, *Power and Tullis* [1989] note that, once delivered to the fault zone, solutes carried by the migrating fluids can participate in the processes that contribute to fault zone sealing, mineral comminution, stress relaxation and mineral precipitation.

SUMMARY OF MAJOR RESULTS AND CONCLUSIONS

We have created four stochastic, three-dimensional discrete fracture network models (single fracture fault, SFF; distributed deformation zone, DDZ; localized deformation zone, LDZ; and composite deformation zone, CDZ) that exemplify the field-based fault zone architectural styles summarized by *Caine et al.* [1996]. The models represent outcrop scale (cubes that are 20 m on a side) idealized fault zones developed in low permeability rocks. Each model type contains one or both fault zone components, a damage zone or fault core, surrounded by a protolith. Fracture densities, trace lengths, and orientations included in the models were based on data obtained from detailed outcrop mapping along the Stillwater Fault Zone in Dixie Valley, Nevada and from a series of normal fault zones in east Greenland which exhibit each of the four architectural styles. Fault zone component widths were based on data from east Greenland. Insights regarding fault mechanics and kinematics are also used in constructing the models. Thus, fluid flow simulations performed on the discrete fracture network models provide direct insight into the permeability structure of the Stillwater Fault Zone in particular, and normal faults in general. Furthermore, because strike slip and thrust fault zones contain similar architectural elements, the model results may apply when considering any style of fault zone.

Simulating fluid flow through the discrete fracture network models allowed us to quantify inferences made by *Caine et al.* [1996] and other workers regarding the relationships between fault zone architecture and permeability structure. Fracture networks associated with each component type impart bulk permeabilities that, when combined, lead to bulk permeability anisotropy that can exert a significant impact on patterns and rates of fluid flow. For example, if we exclude the SFF (single fracture fault) then simulations performed with a uniform 100 μm fracture aperture yield permeabilities for the protolith ($\sim 10^{-14}$ m^2), damage zone ($\sim 10^{-13}$ m^2), and fault core ($\sim 10^{-13}$ m^2).

Two distinct types of anisotropy were identified as major controlling factors in fault-related fluid flow and permeability structure. The first is the internal bulk anisotropy within individual fault zone components contributed by the character of the fracture network fabrics. The second source of anisotropy is derived from the permeability contrasts that result from juxtaposing different fault zone components. Thus, component-scale permeability heterogeneity leads to bulk fault zone scale anisotropy. Both types of anisotropy reflect the protolithology, deformation style, and the stress, temperature, pressure, and geochemical conditions encountered throughout the evolution of a fault zone. The resulting fault zone architectures and permeability structures for any given set of conditions and histories, at any given point in time, can cause changes in the magnitude and direction of local fluid fluxes that vary by orders of magnitude.

Ratios of permeability anisotropy were determined in three orthogonal flow directions with respect to the average orientation of each fault zone ($k_{NN} : k_{PN} : k_{PP}$). Anisotropy values obtained within each individual component in each non-SFF model where all fracture apertures are the same (100 μm), were small and range from $\approx 1:0.2:0.8$ in a protolith to $\approx 1:6:5$ in a fault core for NN, PN, and PP flow directions respectively. The permeability contrasts between components that contribute to bulk anisotropy in fault zone permeability are much larger. For example, permeability contrasts range from approximately one to five orders of magnitude between a fault core and a protolith ($\approx 10^{-19}$ m^2 and $\approx 10^{-14}$ m^2 respectively). Thus, the architecture of multiple fault zone components has more impact than internal fault zone component architecture alone.

With uniform apertures between and within fault zone components, the model results suggest that fracture length and orientation are more important in controlling anisotropy than fracture density. This was found to be the case at both the fault zone component scale and the full model scale. When fault zone fracture apertures are uniformly opened or closed relative to the protolith or damage zone, aperture becomes the controlling parameter of anisotropy relative to fracture length, orientation, and density. This is due to the cubic law correlation between aperture and transmissivity used in the models. Variations of fracture aperture in natural fault zones, due to different types of fracture sets, stress, and degree of mineral filling may change these sensitivities, but were not investigated here. Thus, the modeling results suggest that the orientation of fracture sets and particularly fracture length, which is often very difficult to measure, are the most important field data to collect for later model input.

Geologically plausible fracture networks constructed for each non-SFF fault zone component yielded permeabilities within the fault zone and parallel to slip that are consistently smaller than those computed perpendicular to the slip vector. Additional sources of anisotropy not included in the fracture models (e.g., orientation-dependent aperture variability caused by: differential stress, variations in fracture roughness, and anisotropy in fracture surface topography) could be used to explain why the maximum permeability in the fault zone is not found to be parallel to the slip vector in the simulations. The model results also demonstrate that regardless of the regional flow direction, fault zone architectural style, and magnitude of displacement fluid flow is enhanced parallel to the fault zone and impeded normal to the fault zone when fracture apertures are reduced relative to an 'unfaulted' protolith reference case.

Increasing or decreasing fracture apertures by only one order of magnitude within the fault zone of each non-SFF model yields one to five orders of magnitude change in component permeability that, in turn, yields similarly large permeability anisotropy for all non-SFF type fault zone models. Values of permeability anisotropy in excess of one order of magnitude can significantly modify the patterns and rates of fluid flow within and near fault zones, depending upon the orientation of hydraulic gradients acting at the location of interest. The modeled variations in fracture aperture mimic the changes that might occur during fault zone evolution. Thus, increased bulk anisotropy caused by enhanced permeabilities in the fault zone can lead to a focusing of fluid flow into the fault zone during failure. On the other hand, similarly large anisotropy caused by sealing of fractures in the fault zone can cause fluid flow to be distributed throughout the adjacent protolith, or damage zone.

When the model results are considered in conjunction with contemporaneous deformation they highlight how a fault zone may be destined to seal itself and arrest deformation because initial opening of the fault-related permeability structure causes mass transport in the core of the fault zone. Positive feedback between mineral precipitation, permeability reduction, and pore pressure relaxation in the fault core may cause localized, but heterogeneously distributed, mechanical strength. The modeling results and conjectured processes also highlight the means by which, and the controls on how, fault zone architecture, permeability structure and growth might be intimately involved with the distribution of earthquakes and the deposition of economic mineral deposits. Furthermore, the models illustrate how fault zone architectural styles and the hydraulic parameters identified may lead to conduit, barrier, or combined conduit-barrier permeability structures.

This study also forms a first step in outlining a methodology for constraining fault-related permeability values assigned in continuum (porous media) fluid flow simulators. For example, the model results yield permeability estimates that can be used in parameterizing fluid flow simulations for the geothermal reservoir hosted in the Stillwater Fault Zone. If sufficient computational capacity is available, simulator grid blocks can be sized to map the heterogeneous permeability structure defined by the spatial distribution of permeability associated with each fault zone component type. If blocks no smaller than those used in this study can be used, then bulk permeability anisotropies defined for the full width of the fault zone should be assigned in an effort to preserve the detailed heterogeneity contributed by the individual fault zone components.

Acknowledgments. The authors would like to thank the University of Utah, Department of Geology and Geophysics for financial and computer facility support of the first author during his graduate studies there. We most gratefully thank Glori Lee, William Dershowitz, Paul LaPointe, and Trenton Cladouhos of Golder Associates, Inc., for support in using FracMan™, MeshMaker™, EdMesh™, and Mafic™. We thank Stephen Snelgrove and Steven Schulz for their input and efforts in completing the model runs and associated output. Constructive and thoughtful reviews from Simon Cox, an anonymous reviewer, and Bill Haneberg were helpful in clarifying several issues in the manuscript. Comments from Ron Bruhn and Jim Evans were also much appreciated. Fracture network visualizations were created using GeomView by the Geometry Center, University of Minnesota.

REFERENCES

Anderson, E. M., *The dynamics of faulting*, Oliver and Boyd, Edinburgh, 1951.

Anderson, L. J., Osborne, R. H., Palmer, D. F., Petrogenesis of cataclastic rocks within the San Andreas fault zone of southern California, USA, *Tectonophysics*, 67, 221-249, 1980.

Andersson, J. E., Ekman, L., Nordqvist, R., and Winberg, A., Hydraulic testing and modeling of a low-angle fracture zone at Finssjon, Sweden, *J. Hydrol.*, 126, 45-77, 1991.

Antonellini, M., and Aydin, A., Effect of faulting on fluid flow in porous sandstones, petrophysical properties, *AAPG Bull.*, 78, 355-377, 1994.

Barton, C. A., Hickman, S., Morin, R., Zoback, M. D., and Benoit, D., Reservoir scale fracture permeability in the Dixie Valley, Nevada, geothermal field, *Proc. 23rd Workshop on Geothermal Reservoir Eng.*, Stanford Univ., California, 1998.

Brown, S. R., Fluid flow through rock joints: The effect of surface roughness, *J. Geophys. Res.*, 92, 1337-1347, 1987.

Brown, S. R., and Bruhn, R. L., Formation of voids and veins during faulting, *J. Struct. Geol.*, 18, 657-671, 1996.

Brown, S. R., and Bruhn, R. L., Fluid permeability of deformable fracture networks, *J. Geophys. Res.*, 103, 2489-2500, 1998.

Bruhn, R. L., Yonkee, W. E., and Parry, W. T., Structural and fluid-chemical properties of seismogenic normal faults, *Tectonophysics*, 175, 139-157, 1990.

Bruhn, R. L., Parry, W. T., Yonkee, W. A., and Thompson, T.,

Fracturing and hydrothermal alteration in normal fault zones, *Pure Appl. Geophys.*, 142, 609-644, 1994.

Bunds, M. P., Bruhn, R. L., and Parry, W. T., Comparing nature and experiment at the top of the seismogenic zone: The Castle Mountain fault, Alaska (abstract), *Eos, Trans. AGU*, 77, 718, 1996.

Byerlee, J., Model for episodic flow of high-pressure water in fault zones before earthquakes, *Geology*, 21, 303-306, 1993.

Caine, J. S., Regional comparison of melange fabrics in the Taconic Orogen, southeastern New York state and western Newfoundland, Canada (abstract), *2nd Ann. Cen. Can. Geol. Conf. Prog.*, Toronto, 2, 24, 1989.

Caine, J. S., Coates, D. R., Timoffeef N. P., Davis, W. D., Hydrogeology of the northern Shawangunk mountains, *NY State Geol. Surv. open file report 1g806*, 72, 1991.

Caine, J. S., Deformation, fault zone architecture, permeability structure and evidence for fluid flow in a brittle normal fault, Traill Island, east Greenland, in *Faulting, fault sealing and fluid flow in hydrocarbon reservoirs* (abstract), edited by G. Jones, Q. Fisher, and R. Knipe, Univ. of Leeds, U.K., 81-82, 1996.

Caine, J. S., Evans, J. P., and Forster, C. B., Fault zone architecture and permeability structure, *Geology*, 24, 1025-1028, 1996.

Caine, J. S. and Forster, C. B., Architecture and permeability structure of the Stillwater normal fault, Dixie Valley, Nevada (abstract), *Ann. Meeting Geol. Soc. Am.*, Salt Lake City, Utah, 29, A-226, 1997.

Caine, J. S., The architecture and permeability structure of brittle fault zones, Ph.D. dissertation, Univ. of Utah, Salt Lake City, UT, 1999.

Caskey, S. J., Wesnousky, S. G., Zhang-Peizhen, Slemmons, D. B., Surface faulting of the 1954 Fairview Peak (Ms 7.2) and Dixie Valley (Ms 6.8) earthquakes, central Nevada, *Bull. Seism. Soc. Am.*, 86, 761-787, 1996.

Chester, F. M. and Logan, J. M., Composite planar fabric of gouge from the Punchbowl Fault, California, *J. Struct. Geol.*, 9, 621-634, 1986.

Chester, F. M., Evans, J. P., and Biegel, R. L., Internal structure and weakening mechanisms of the San Andreas fault, *J. Geophys. Res.*, 98, 771-786, 1993.

Cowie, P. A., and Shipton, Z. K., Fault tip displacement gradients and process zone dimensions, *J. Struct. Geol*, 20, 983-997, 1998.

Davis, G. H., *Structural geology of rocks and regions*, John Wiley and Sons, New York, 1984.

Davison, C. C. and Kozak, E. T., Hydrogeologic characteristics of major fracture zones in a large granite batholith of the Canadian shield, *Proc. 4th Can. Am. Conf. on Hydrogeo., Banff*, 1988.

Dawers, N. H., Anders, M. H., Scholz, C. H., Growth of normal faults: Displacement-length scaling, *Geology*, 21, 1107-1110, 1993.

Dershowitz, W., Lee, G., Geier, J., Foxford, T., LaPointe, P., Thomas, A., FracMan™: Interactive discrete feature data analysis, geometric modeling, and exploration simulation, User Documentation, version 2.5, *Golder Associates, Inc.*, Redman, Washington, 1996.

Evans, J. P., Forster, C. B., Goddard, J. V., Permeability of fault-related rocks, and implications for hydraulic structure of fault zones: *J. Struct. Geol.*, 19, 1393-1404, 1997.

Evans, J. P., Textures and deformation mechanisms and the role of fluids in cataclastically deformed granitic rocks, in *Deformation mechanisms, rheology, and tectonics*, edited by R. J. , Knipe and E. Rutter, *Geol. Soc., Lond., Spec. Pub.* 54, 29-39, 1990.

Ferrill, D. A., Winterle, J., Wittmeyer, G., Sims, D., Colton, S., Armstrong, A., Stressed rock strains groundwater at Yucca Mountain, Nevada, *GSA Today*, 9, 1-8, 1999.

Fisher, Q. J. and Knipe, R. J., Fault sealing processes in siliciclastic sediments, in *Faulting, Fault Sealing, and Fluid Flow in Hydrocarbon Reservoirs*, edited by G. Jones, Q. J., Fisher, and R. J. Knipe, *Geol. Soc., Lond., Spec. Pub.* 147, 117-134, 1998.

Fleming, C. G., Couples, G. D., and Haszeldine, R. S., Thermal effects of fluid flow in steep fault zones, in *Faulting, Fault Sealing, and Fluid Flow in Hydrocarbon Reservoirs*, edited by G. Jones, Q. J., Fisher, and R. J. Knipe, *Geol. Soc., Lond., Spec. Pub.* 147, 217-229, 1998.

Forster, C. B. and Evans, J. P., Hydrogeology of thrust faults and crystalline thrust sheets: results of combines field and modeling studies, *Geophys. Res. Let.*, 18, 979-982, 1991.

Forster, C. B., Caine, J. S., Schulz, S., and Nielson, D. L., Fault zone architecture and fluid flow: an example from Dixie Valley, Nevada, *Proc. 22nd Workshop on Geothermal Reservoir Eng.*, Stanford Univ., California, 1997.

Freeze, R. A. and Cherry, J. A., *Groundwater*, Prentice Hall, New Jersey, 1979.

Foxford, K. A., Walsh, J. J., Watterson, J., Garden, I. R., Guscott, S. C., and Burley, S. D., Structure and content of the Moab Fault Zone, Utah, USA, and its implications for fault seal prediction, in *Faulting, Fault Sealing, and Fluid Flow in Hydrocarbon Reservoirs*, edited by G. Jones, Q. J., Fisher, and R. J. Knipe, *Geol. Soc., Lond., Spec. Pub.* 147, 87-103, 1998.

Ge, S., and Garven, G., A theoretical model for thrust-induced deep groundwater expulsion with application to the Canadian Rocky Mountains, *J. Geophys Res.*, 99, 13,851-13,868, 1994.

Goddard, J. V. and Evans, J. P., Chemical changes and fluid-rock interaction in faults of crystalline thrust sheets, northwestern Wyoming, U.S.A., *J. Struct. Geol.*, 17, 533-547, 1995.

Haneberg, W. C., Steady state groundwater flow across idealized faults, *Water Resour. Res.*, 31, 1815-1820, 1995.

Hickman, S., Barton, C. A., Zoback, M. D., Morin, R., Sass, J., and Benoit, R., *In situ* stress and fracture permeability along the Stillwater fault zone, Dixie Valley, Nevada, *Int. J. Rock Mech. Min. Sci.*, Paper 126, 34, 1997.

Hill, D. P., A model for earthquake swarms, *J. Geophys. Res.*, 82, 347-352, 1977.

Hippler, S. J., Deformation microstructures and diagenesis in sandstone adjacent to an extensional fault: Implications for the flow and entrapment of hydrocarbons, *AAPG Bull.*, 77, 625-637, 1993.

Jones, G., Fisher, Q. J., and Knipe, R. J., editors, Faulting, Fault Sealing, and Fluid Flow in Hydrocarbon Reservoirs, *Geol. Soc., Lond., Spec. Pub.* 147, 1998.

Knipe, R. J., The influence of fault zone processes and diagenesis on fluid flow, in *Diagenesis and basin development*, edited by A. D. Horbury and A. G. Robinson, *AAPG Studies in Geol. 36*, 135-151, 1993.

Knipe, R. J., Jones, G., and Fisher, Q. J., Faulting, fault

sealing and fluid flow in hydrocarbon reservoirs: an introduction, in *Faulting, Fault Sealing, and Fluid Flow in
Hydrocarbon Reservoirs*, edited by G. Jones, Q. J., Fisher,
and R. J. Knipe, *Geol. Soc., Lond., Spec. Pub.* 147, vii-xxi,
1998.

Krantz, R. L., Frankel, A. D., Engelder, T., and Scholz, C. H.,
The permeability of whole and jointed Barre granite
(abstract), *Int. J. Rock Mech. Min. Sci. Geomech.*, 16,
225-234, 1979.

Long, J. C. S., Remer, J. S., Wilson, C. R., Witherspoon, P.
A., Porous media equivalents for networks of discontinuous
fractures, *Water Resour. Res.*, 18, 645-658, 1982.

Long, J. C. S. and Witherspoon, P. A., The relationship of the
degree of interconnection to permeability in fracture networks, *J. Geophys. Res.*, 90, 3087-3098, 1985.

Lopez, D. L. and Smith, L., Fluid flow in fault zones: Analysis
of the interplay of convective circulation and topographi-
cally-driven groundwater flow, *Water Resour. Res.*, 31,
1489-1503, 1995.

Lopez, D. L. and Smith, L., Fluid flow in fault zones: Influence
of hydraulic anisotropy and heterogeneity on the fluid flow
and heat transfer regime, *Water Resour. Res.*, 32, 3227-
3235, 1996.

Marshak, S and Mitra, G., *Basic methods of structural geology*,
Prentice Hall, New Jersey, 1988.

Martel, S. J., Formation of compound strike-slip fault zones,
Mount Abbot quadrangle, California, *J. Struct. Geol.*, 12,
869-882, 1990.

Mase, C. W. and Smith, L., Pore-fluid pressure and frictional
heating on a fault surface, *Pure Appl. Geophys.*, 122, 583-
607, 1985.

Matthäi, S. K., Aydin, A., Pollard, D. D., Roberts, S. G.,
Numerical simulation of departures from radial drawdown in
a faulted sandstone reservoir with joints and deformation
bands, in *Faulting, Fault Sealing, and Fluid Flow in Hydrocarbon Reservoirs*, edited by G. Jones, Q. J., Fisher, and R.
J. Knipe, *Geol. Soc., Lond., Spec. Pub.* 147, 157-191,
1998.

McGrath, A. G., Davison, I., Damage zone geometry around
fault tips, *J. Struct. Geol.*, 17, 1011-1024, 1995.

Miller, I., Lee, G., Dershowitz, W., and Sharp, G., Mafic™:
Matrix / fracture interaction code with solute transport: User
Documentation, Version β1.5, *Golder Associates, Inc.*,
Redman, Washington, 1995.

Miller, S. A., Nur, A., Olgaard, D. L., Earthquakes as a coupled
shear stress-high pore pressure dynamical system, *Geophys. Res. Let.*, 23, 197-200, 1996.

Mitra, G., Ductile deformation zones and mylonites: The
mechanical processes involved in the deformation of crystalline basement rocks, *Am. J. Sci.*, 278, 1057-1084,
1978.

Moore, J. C. and Vrolijk, P., Fluids in accretionary prisms,
Rev. Geophys., 30, 113-135, 1992.

Morrow, C. A., Shi, L. Q., and Byerlee, J. D., Permeability and
strength of San Andreas fault gouge under high pressure,
Geophys. Res. Lett., 8, 325-328, 1981.

Mozley, P. S. and Goodwin, L B., Patterns of cementation
along a Cenozoic normal fault: A record of paleoflow orientations, *Geology*, 23, 539-542, 1995.

National Research Council Committee on Fracture Characterization and Fluid Flow, Rock fractures and fluid flow, National
Academy Press, Washington, D.C., 1996.

Neuzil, C. E. and Tracy, J. V., Flow through fractures, *Water
Resour. Res.*, 17, 191-199, 1981.

Newman, J. and Mitra, G., Fluid-influenced deformation and
recrystallization of dolomite at low temperatures along a
natural fault zone, Mountain City window, Tennessee, *Geol.
Soc. Am. Bull.*, 106, 1267-1280, 1994.

Nur, A., and Booker, J. R., Aftershocks caused by pore fluid
flow?, *Science*, 175, 885-886, 1972.

Parry, W. T., Hedderly-Smith, D., and Brhun, R. L., Fluid
inclusions and hydrothermal alteration on the Dixie Valley
Fault, Nevada, *J. Geophys. Res.*, 96, 19,733-19,748,
1991.

Parry, W. T. and Bruhn, R. L., Fluid Pressure transients on
seismogenic normal faults, *Tectonophysics*, 179, 335-344,
1990.

Power, W. L. and Tullis, T. E., The relationship between
slickenside surfaces in fine-grained quartz and the seismic
cycle, *J. Struct. Geol.*, 11, 879-893, 1989.

Randolph, L. and Johnson, B., Influence of faults of moderate
displacement on groundwater flow in the Hickory sandstone
aquifer in central Texas (abstract), *23rd Ann. Meeting,
Geol. Soc. Am., south-central Sec.*, 21, 242, 1989.

Roberts, S. J., Nunn, J. A., Cathles, L. M. and Cipriani, F. D.,
Expulsion of abnormally pressured fluids along faults, *J.
Geophys. Res.*, 101, 28,231-28,252, 1996.

Rose, P. E., Apperson, K. D., Johnson, S., and Adams, M. C.,
Numerical simulation of a tracer test at Dixie Valley, Ne-
vada, *Proc. 22nd Workshop on Geothermal Reservoir Eng.*,
Stanford Univ., California, 1997.

Rowley, P. D., Cenozoic transverse zones and igneous belts in
the Great Basin, western United States: Their tectonic and
economic implications, in *Accommodation zones and
transfer zones: The regional segmentation of the Basin and
Range province*, edited by J. E. Faulds and J. H. Stewart,
Geol. Soc. Am. Spec. Paper 323, 195-228, 1998.

Schlische, R. W., Young, S. S., Ackermann, R. V., and Gupta,
A., Geometry and scaling relations of a population of very
small rift-related normal faults, *Geology*, 24, 683-686,
1996.

Scholz, C. H., *The mechanics of earthquakes and faulting*,
Cambridge University Press, Cambridge, 1990.

Scholz, C. H., and Anders, M. H., The permeability of faults, in
The mechanical involvement of fluids in faulting, edited by
S. Hickman, R. Sibson, R. Bruhn, *U.S. Geol. Surv. Red-
Book Conf. Proc.*, open-file report 94-228, 247-253,
1994.

Screaton, E. J., Wuthrich, D. R., and Dreiss, S. J., Permeabilities, fluid pressures, and flow rates in the Barbados ridge
complex, *J. Geophys. Res.*, 95, 8997-9007, 1990.

Seront, B., Wong, T-f., Caine, J. S., Forster, C. B., Bruhn, R.
L., and Fredrich, J. T., Laboratory characterization of hydromechanical properties of a seismogenic normal fault
system, *J. Struct. Geol.*, 20, 865-881, 1998.

Sibson, R. H., Fault rocks and fault mechanisms, *J. Geol. Soc.,
Lond.*, 133, 191-231, 1977.

Sibson, R. H., Crustal stress, faulting, and fluid flow: *in
Parnell, J., Geofluids: Origin, migration, and evolution of*

fluids in sedimentary basins, *Geol. Soc., Lond., Spec. Pub.* 78, 69-84, 1994.

Sibson, R. H., Structural permeability of fluid-driven fault-fracture meshes, *J. Struct. Geol.*, 18, 1031-1042, 1996.

Smith, L., Forster, C. B., and Evans, J. P., Interaction between fault zones, fluid flow, and heat transfer at the basin scale, in *Int. Assoc. Hydrol. Sci., Selected Papers in Hydrology*, edited by S. Newman and I. Neretnieks, 2, 41-67, 1990.

Snow, D. T., Rock fracture spacings, openings, and porosities, *J. Soil Mech. Found. Div., Am. Soc. Civ. Eng.*, 94, 73-91, 1968.

Snow, D. T., Anisotropic permeability of fractured media, *Water Resour. Res.*, 5, 1273-1289, 1969.

Vermilye, J. M. and Scholz, C. H., The process zone; an analysis of naturally formed fault zones (abstract), *Ann. Meeting, Geol. Soc. Am.*, Seattle, 26, 267, 1994.

Witherspoon, P. A., Wang, J. S. Y., Iwai, K., Gale, J. E., Validity of cubic law for fluid flow in a deformable rock fracture, *Water Resour Res.*, 16, 1016-1024, 1980.

Zhang, X., Sanderson, D. J., Numerical modeling of the effects of fault slip on fluid flow around extensional faults, *J. Struct. Geol.*, 18, 109-119, 1996.

J. S. Caine and C. B. Forster, Department of Geology and Geophysics, University of Utah, 135 South 1460 East, WBB 719, Salt Lake City, UT, 84112-0111
(e-mail: jscaine@mines.utah.edu and cforster@mines.utah.edu)

Geochemistry and Hydromechanical Interactions of Fluids Associated With the San Andreas Fault System, California

Yousif K. Kharaka, James J. Thordsen, and William C. Evans

U. S. Geological Survey, Water Resources Division, Menlo Park, California

B. Mack Kennedy

Lawrence Berkeley National Laboratory, Univ. of California, Berkeley, California

Field determinations of heat flow and in situ stress orientations indicate that the San Andreas fault is anomalously weak. The weakness of this major transform fault has been attributed to several mechanisms, including the presence of serpentines, clays or other minerals with low coefficients of friction. The most plausible explanation for this weakness, however, relates to the distribution of fluid pressures within the fault zone and in the adjoining crust. In order to investigate the geochemistry and the role of fluids in the dynamics of this fault, we have carried out detailed chemical and isotopic analyses of water and associated gases from 41 thermal and saline springs and a few wells, together with about 20 cold springs, located along or near the San Andreas fault system between San Francisco and Los Angeles. Results show that the main trace of the San Andreas fault is conspicuously devoid of high-discharge springs, and the δD and $\delta^{18}O$ values establish that waters are predominantly of meteoric origin. The chemical compositions of water and gases are controlled mainly by the ambient rock types, and chemical geothermometry gives reservoir temperatures of 80-150°C indicating shallow to moderate circulation depths of up to 6 km. However, compositions and isotope abundances of noble gases and $\delta^{13}C$ values of HCO_3 indicate a significant (up to 50%) mantle component for the volatiles. The relatively high fluxes of CO_2 ($C/^3He \approx 10^{10}$) and other volatiles of mantle origin support a deep continuous flow model, especially at depths >6 km. Numerical simulations indicate that these high fluxes of CO_2 of mantle and deep crustal origin are sufficient to generate lithostatic fluid pressures, and thus a weakened fault, in time scales comparable to those of earthquake cycles.

INTRODUCTION

It is generally accepted that faults and shear zones provide conduits which focus fluid flow in the upper crust and that water and gases play a critical role in a variety of faulting processes, including earthquakes [*Berry*, 1973; *Irwin and Barnes*, 1975; 1980; *Barnes et al.*, 1978; *O'Neil and Hanks*, 1980; *Sibson*, 1981, 1992; *Kerrich et al.*, 1984; *Kharaka et al.*, 1988a; *Hickman et al.*, 1995; *Fournier*, 1996; *Kennedy et al.*, 1997]. Fault hosted gold-quartz and other metalliferous mineral veins together with associated hydrothermal rock alteration, for example, provide good evidence for large fluxes of deeply sourced fluids along

Faults and Subsurface Fluid Flow in the Shallow Crust
Geophysical Monograph 113
Copyright 1999 by the American Geophysical Union

fault systems and shear zones [*Fyfe et al.*, 1978; *Kerrich et al.*, 1984; *Sibson*, 1988; *Robert et al.*, 1995]. Detailed site investigations have indicated that fluid flow in these faults, which have experienced repeated failure, is episodic and driven by transient lithostatic fluid pressure, consistent with *Sibson's* [1992] 'fault valve' model during the seismic cycle [e.g., *Hacker*, 1997].

The role of fluid pressure in controlling the strength of faults has been recognized since the seminal paper by *Hubbert and Ruby* [1959], which applied the physics of effective stress to faulting. The complete formulation of fault failure is generally represented [e.g., *Fyfe et al.*, 1978] by a criterion of Coulomb form:

$$\tau_s = C + \mu \sigma_n' = C + \mu(\sigma_n - P_f) \qquad (1)$$

where τ_s is the frictional shear strength, C is the cohesive strength of fault gouge, μ is the static coefficient of friction (with values of 0.6-0.9), σ_n' and σ_n are the effective and normal stresses on the fault, respectively, and P_f is the fluid pressure. Equation (1) indicates that fault strengths would be reduced drastically, if fluid pressures would increase from hydrostatic ($\lambda \simeq 0.4$) to lithostatic values ($\lambda \simeq 1.0$), where:

$$\lambda = P_f/\sigma_n. \qquad (2)$$

Several theoretical and laboratory studies indicate that the chemical effects of fluids on fault zone rheology are probably as important as their physical role [*O'Neil and Hanks*, 1980; *Kerrich et al.*, 1984; *David et al.*, 1994; *Fournier*, 1996]. Specific hydrothermal water-rock interactions, including pressure solution, crack growth and healing, and clay formation through retrograde reactions, have been identified that could reduce or increase fault strength principally by modifying the cohesion strength and coefficient of friction of gouge [*Wintsch et al.*, 1995; *Moore et al.*, 1997]. Even though field investigations of exhumed faults have confirmed many of the above interactions [e.g. *Chester et al.*, 1992], the role of fluids in faulting processes is not universally accepted, because the results show large local variability and often are subject to multiple interpretations.

In order to improve our understanding of the physics and geochemistry of faulting at seismogenic depths, a group of scientists from the United States Geological Survey, National Laboratories and academia have proposed a scientific drilling program that includes a 10 km hole located in the San Andreas fault zone. As a preliminary part of this effort, we initiated a reconnaissance investigation of

the geochemistry of water and associated gases in thermal and mineralized springs and wells along the San Andreas fault system between the San Francisco Bay Area and San Bernardino, a region that includes the four selected candidate drilling segments (Figs. 1 and 2). The main objectives are to investigate: (1) the origin of water and solutes, and water-mineral interactions within and adjacent to the fault zone; (2) the origin and transport of volatile solutes (CO_2, CH_4, N_2, and noble gases) from mantle and deep crustal sources; and (3) the role of fluids on the dynamics of this fault. A literature search was conducted and a data base was assembled from several published reports [*Waring*, 1915, 1965; *Berkstresser*, 1968; *Barnes et al.*, 1973, 1975; *Majmundar*, 1984; *Thompson and White*, 1991; *Kharaka et al.*, 1988a; and others] and from unpublished files at the U.S. Geological Survey. Sampling of the important fluid sites was carried out because data on some of the critical chemical and isotope parameters, especially noble gases, were lacking from many of the sites previously sampled. The criteria for determining sampling targets were indications of deep fluid sources or extensive water-rock interactions, such as elevated discharge temperature or salinity, unusual solute chemistry, high gas discharges, and/or high flow rates.

In this report, we present the chemical and isotopic compositions of the water and gas samples obtained thus far, and we discuss the origin and geochemical evolution of fluids and solutes. Our results indicate that high fluxes of CO_2 of mantle and deep crustal origins may have important implications to the dynamics of the San Andreas fault system.

GEODYNAMIC MODELS OF THE SAN ANDREAS FAULT

The San Andreas fault system is a major transform plate boundary that has been the subject of intensive geological and geophysical studies beginning with the great San Francisco earthquake of 1906 [e.g., see review by *Wallace*, 1990]. Recent studies have lead to the development of sophisticated geodynamic models that describe the spatial and temporal behavior of this fault system [*Byerlee*, 1990, 1993; *Rice*, 1992; *Jones et al.*, 1994; *Jachens et al.*, 1995; *McLaughlin et al.*, 1996]. However, these models are incomplete primarily because of limited knowledge of fluid pressures, fluid-rock interactions and the role played by fluids in the dynamics of this fault, especially at seismogenic depths [*O'Neil and Hanks*, 1980; *Sibson*, 1981; 1992; *Kharaka et al.*, 1988a; *Sleep and Blanpied*, 1994; *Hickman et al.*, 1995; *Fournier*, 1996].

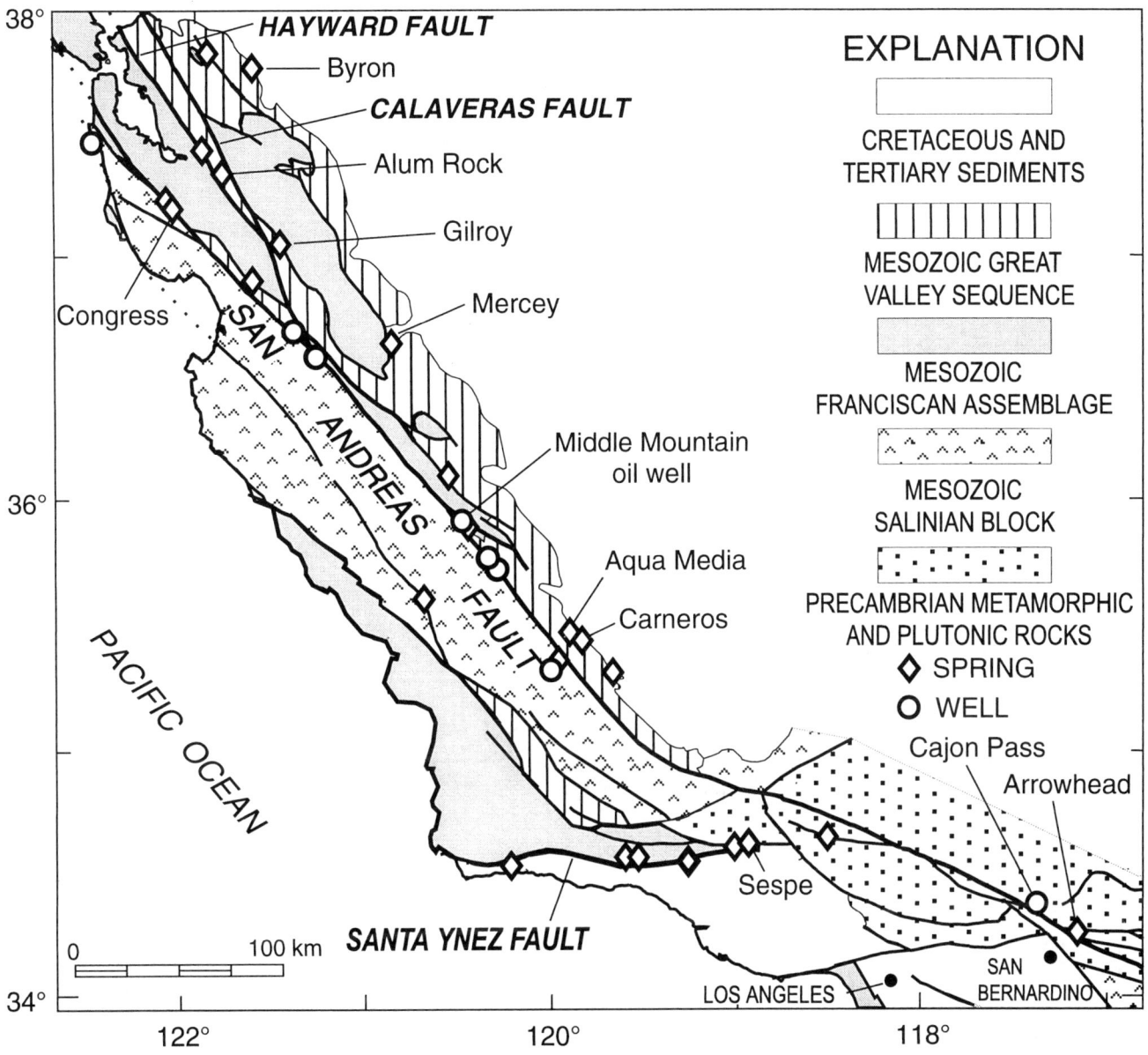

Figure 1. Simplified geologic map of the study area showing the locations for the sampled thermal and mineral springs and wells.

Most fault models accept the conclusion, based on field determinations of heat flow [*Brune et al.*, 1969; *Lachenbruch and Sass*; 1980; 1992; *Sass et al.*, 1997] and in situ stress orientations [*Zoback et al.*, 1987; *Mount and Suppe*, 1987; *Zoback and Healy*, 1992], that the San Andreas fault is anomalously weak, both in an absolute sense and relative to the adjoining crust [*Hickman*, 1991]. This fault weakness has been attributed to several factors, including the presence at depth of serpentines, clays or other authigenic minerals with low coefficient of friction [*Irwin and Barnes*, 1975; *Morrow et al.*, 1992; *Moore et al.*, 1997]. The most plausible explanation for the weakness, however, relates to the distribution of fluid pressures within the fault zone and in the surrounding crust [*Lachenbruch and Sass*, 1980; *Zoback et al.*, 1987; *Byerlee*, 1990].

Byerlee [1990; 1993] and *Rice* [1992] proposed models that require high fluid pressures in the fault zone, but normal hydrostatic pressures in the surrounding rocks, to satisfy both the heat flow and stress orientation constraints. The main differences in these two models relate to the origin and maintenance of the high pore pressures in the fault zone. In the continuous flow model of *Rice* [1992], the

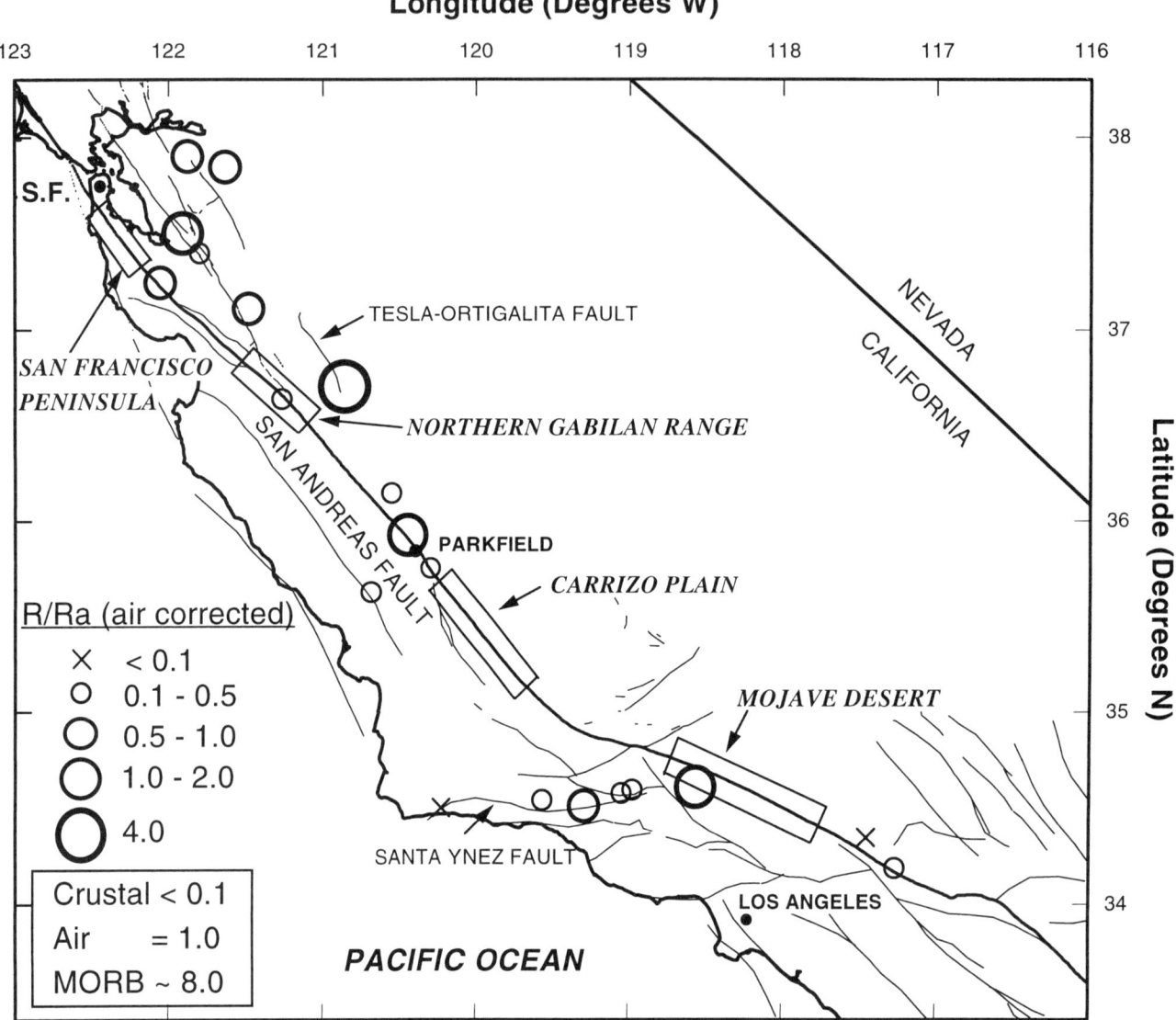

Figure 2. Distribution of ^{3}He/^{4}He values, normalized to air, measured in fluids from the study area, including the four candidate segments for siting the 10 km deep research well. Note the pervasive presence of a mantle component in the sampled gases.

high fluid pressures in the fault zone are maintained by flow of pressurized fluids originating in the root zone of the fault, including the mantle. *Byerlee* [1990; 1993] proposed an episodic model in which meteoric water from the adjacent country rock is trapped in the fault zone following an earthquake. The fluid pressure increases by compaction in a number of compartments sealed by mineral precipitation in the fault zone as the gouge is sheared. The pore pressure increase continues until a critical fluid pressure causes seal rupture, thereby triggering an earthquake and starting another earthquake cycle. Because mineral dissolution and precipitation are strongly influenced by grain

size (grains are broken during an earthquake), pore fluid pressures and mechanical stress, water-rock interactions must play an important role not only in the rheology of the fault zones, but also in the various aspects of the earthquake cycle [*O'Neil and Hanks*, 1980; *Sibson*, 1981; 1992; *Elias and Hajash*, 1992; *Kharaka et al.*, 1988a; *Sleep and Blanpied*, 1994; *Fournier*, 1996].

These fault models are difficult to verify because the required integrated geological, geophysical, and especially hydrologic data are not currently available [*Hickman et al.*, 1995]. However, there are several reports that discuss the distribution of fluid pore pressures [*Berry*, 1973; *Lico and*

Kharaka, 1985; *Johnson and McEvilly*, 1995] and the chemical composition and origin of waters [*Waring*, 1915, 1965; *Allen*, 1968; *White et al.*, 1973; *Barnes et al.*, 1975; *Kharaka et al.*, 1988a; *Donnelly-Nolan et al.*, 1993; *Davisson et al.*, 1994] in the Tertiary sedimentary section, in the Franciscan assemblage and the Great Valley sequence. The information in these reports, though regional in scope, has some general applicability to the San Andreas fault, especially in establishing the presence of high fluid pressures near this fault, in postulating mantle CO_2 fluxes into the fault zone [*Barnes et al.*, 1978; *Irwin and Barnes*, 1980] and in providing the chemical and isotopic signatures of waters that have reacted with different rock types. These data can also be used to speculate on the mechanochemical aspects of rock deformation, including pressure solution, crack growth and healing and compartment sealing [*Chester et al.*, 1992; *David et al.*, 1994; *Dewers and Hajash*, 1995].

The Cajon Pass research borehole, a 3.5 km well drilled about 3 km from the San Andreas fault, was the first attempt at an integrated investigation collecting relevant geological [*Silver*, 1988], geophysical [*Lachenbruch and Sass*, 1992; *Zoback and Healy*, 1992] and hydrogeochemical [*Evans et al.*, 1988a; *Kharaka et al.*, 1988a; *Torgerson and Clarke*, 1992] data necessary for the understanding of the geodynamic behavior of this fault. In 1994, four segments of the San Andreas fault (Mojave Desert, Carrizo Plain, northern Gabilan Range, and San Francisco Peninsula) were selected as candidate areas (Figure 2) for detailed study and site characterization for the new deep drilling program [*Hickman et al.*, 1995].

REGIONAL GEOLOGIC SETTING

The geology and tectonic evolution of California in general, and our study area in particular, are extremely varied and complex, but are now generally amenable to explanation by the conceptual framework of plate tectonics [*Atwater*, 1970; *Irwin*, 1990; *Jones et al.*, 1994; *Dickenson et al.*, 1996]. Tectonic interaction between continental American plate and oceanic Pacific and related plates has played a major role in the evolution of this region. This interaction has included plate convergence leading to the underthrusting of oceanic crust and the sweeping of island arcs and other crustal material into the zone of interaction, as well as lateral plate translation leading to fragmentation, slicing and movement of large crustal fragments [*Atwater*, 1970]. These processes have lead to the juxtaposition of "terranes" of variable lithology, metamorphic facies, origin and age [*Coney et al.*, 1980; *Irwin*, 1990; *Jones et al.*, 1994].

The San Andreas, which is a transform boundary fault between the North American and Pacific plates, has been the locus of tectonic activity in Californian for more than 30 Ma. The San Andreas fault system, as described by *Wallace* [1990], refers to the network of faults with predominantly right-lateral strike slip that collectively has accommodated most of the relative motion (300-350 km) between the North American and Pacific plates. The study area encompasses that section of the San Andreas fault system that extends from San Francisco southeastward through the central Coast Ranges and Transverse Ranges to Arrowhead Hot Springs east of San Bernardino. In this region, the main trace of the San Andreas fault extends for about 550 km, and combined with several major subsidiary faults and countless minor faults and splays defines a complex fault system 80 to 150 km wide.

A simplified regional geologic map of the study area shows the major faults of the San Andreas system and the distribution of principal tectonic blocks and lithologic units (Figure 1). Major subsidiary faults include the Calaveras and Hayward, that probably have accounted for a significant amount (up to 170 km) of the right-lateral offset between the plates [*McLaughlin et al.*, 1996]. The Tesla-Ortigalita fault, which is the likely flow path for the Mercey Hot Spring (Figure 1), bounds the east margin of the northern portion of the Diablo Range: It is an active fault system, dominantly strike-slip, although it is superimposed and often mingled with the Coast Range thrust [*LaForge and Lee*, 1982]. Farther south, the western Transverse Range is an east-west trending uplifted region which is tectonically dominated by left-lateral strike-slip along the Big Pine and Garlock faults, and by compression along the Santa Ynez fault which is considered to be a locally overturned, high angle reverse fault, and with an uplift estimated to be >3,000 m [*Dibblee*, 1966]. Numerous high flow thermal springs are present on Santa Ynez fault, but no major springs are located on Big Pine or Garlock faults.

The principal tectonic units in our study area are the Salinian block, the Franciscan assemblage and the Great Valley sequence with the Coast Range ophiolites at its base, and clastic Cenozoic sedimentary rocks of variable thickness overlie much of these basement units [*Norris and Webb*, 1990; *Irwin*, 1990]. A more detailed coverage of the lithotectonic units is presented below because, as will be demonstrated, these units control the chemical composition of fluids associated with the San Andreas fault system.

Salinian Block

The Salinian basement rocks occur between the San Andreas and Sur-Nacimiento fault zones throughout most of the Coast Ranges and consist of metamorphic rocks, and granitic plutons of Cretaceous age that formed in a magmatic arc system [*Norris and Webb*, 1990]. The metamorphic rocks commonly consist of gneiss, granofels, impure quartzite, and minor schist and marble. Plutonic rocks are mostly granite and tonalite, but range in composition to gabbro. Radiometric dates indicate that plutonic activity began 120-105 Ma, and the ages decrease southeastward. A thick sequence of Cenozoic sediments and volcanic rocks overlie much of the Salinian basement rocks, especially at Carrizo Plain, Salinas Valley and Half Moon Bay. Miocene rocks, consisting mainly of diatomites and belonging to the Monterey Formation comprise the bulk of these Cenozoic sediments.

Franciscan Assemblage

The Franciscan subduction complex forms the east wall of the San Andreas fault through the Coast Ranges, but is concealed in places by overlying sedimentary and volcanic rocks (Figure 1). The Franciscan assemblage is locally overlain by the Coast Range ophiolite and the Great Valley sequence, but is separated from them by the Coast Range thrust [*Bailey et al.*, 1970]. The contacts are well exposed in the Diablo antiform, which has four windows (Mt Diablo, Pacheco Pass, New Idria and Parkfield) of extensive Franciscan rocks. The Franciscan rocks are early Jurassic to Cretaceous in age and comprise a heterogeneous assemblage consisting of dismembered "eugeosynclinal" sequences of graywacke, shale, and lesser amounts of mafic volcanic rock, chert, and rare limestone. Fragments of serpentinite and tectonic pods of blueschist occur in melange zones that generally separate blocks of more coherent sequences. The origin of the Franciscan assemblage is problematic, paleomagnetic data suggest thousands of kilometers of northward travel [*Alvarez et al.*, 1980; *Gromme*, 1984].

Coast Range Ophiolite

The Coast Range ophiolite represents deformed and dismembered fragments of mafic to ultramafic oceanic crust and upper mantle on which much of the sedimentary rock of the Great Valley sequence was deposited [*Bailey et al.*, 1970]. The complete sequence, 3-5 km thick and present only in a few localities, consists of serpentinized harzburgite tectonite at the base, overlain by cumulate ultramafic and gabbroic rocks, passing upward into noncumulate gabbroic and related plutonic rocks, then into diabase dikes, and finally pillow lavas. Radiometric ages yield 170-150 Ma, indicating Middle to Late Jurassic [*Saleeby et al.*, 1984; *Mattinson and Hobson*, 1992]. Alternate origins of these rocks are the subject of excellent recent discussions [*Dickinson et al.*, 1996].

Great Valley Sequence

The Great Valley sequence comprises an enormous thickness (>15 km) of miogeosynclinal sedimentary rocks of late Jurassic to late Cretaceous age. The sequence consists of monotonous sections of interbedded marine mudstone, sandstone, and conglomerate that are markedly less deformed, more coherent and with greater lateral continuity than sedimentary sections of the Franciscan assemblage. The bulk of sediments, especially in the lower part of the sequence, are probably turbidites, deposited in submarine fans from rapid erosion of ancestral Klamath Mountains and Sierra Nevada. The Great Valley overlies the Coast Range Ophiolite, except where disrupted by faults. The sequence, on the east side of the Great Valley, is floored by mafic submarine arc rocks of the western Sierra Nevada [*Saleeby*, 1982; *Dickenson et al.*, 1996].

Cenozoic Volcanics

Tertiary volcanics occur in many localities in the Coast Ranges, notably the andesitic Quien Sabe volcanics of Miocene age in the center of the Diablo Range [*McLaughlin et al.*, 1996]. The age of these rocks and related volcanic activities generally decreases from south to north, being 25-22 Ma for Neenach (and equivalent, but displaced Pinnacle), 12-7 Ma for the Quien Sabe and 2.2-0.04 Ma for the Clear Lake volcanics [*McLaughlin et al.*, 1996, and references therein]. Models have been developed that relate this northward younging to the northward passage of the Mendocino triple junction from an initial contact location close to Los Angeles about 30 Ma to its present location at Point Arena. These models invoke crustal thinning and the introduction of magma into an asthenospheric slabless window beneath the North American plate in the wake of subducting lithosphere of the Pacific and related plates [*Fox et al.*, 1985; *Furlong et al.*, 1989; *McLaughlin et al.*, 1996].

FIELD AND LABORATORY METHODOLOGY

Water and gas samples were obtained from 41 selected thermal/mineral springs and relatively deep wells (Figure 1). A rough estimate of total site and individual spring water discharges was made, but a much more thorough and systematic effort is needed for this important parameter. About 20 samples were also obtained from local cold springs, shallow ground water wells and streams in the vicinity of some thermal waters, in order to determine the major chemical and stable isotope compositions of local meteoric and recharge waters.

Samples from springs were collected as close to the main orifice as possible and/or points with highest gas discharges. For diffuse springs, seeps and springs with water pools, temperature and conductivity surveys were carried out to determine optimal sampling points. Several discharge points and individual springs were sampled at many sites, but data for only one sample, with the least dilution by local meteoric water, are generally reported. Samples from wells were obtained at well heads after flowing the well for at least 30 minutes to allow for at least five bore volumes of flushing. Water from non-flowing wells was pumped from depths of up to 100 m with a Grundfos submersible pump; high pressure air was used in one well to pump water from a depth of ~200 m.

Water Samples

Methods used for the collection, preservation and field and laboratory analyses of water samples are generally those described in *Presser and Barnes* [1974] and *Lico et al.* [1982]. Field determinations were made for temperature, pH, specific conductance, alkalinity, H_2S and ammonia. Alkalinity was determined as total alkalinity by titration with 0.05 N sulfuric acid; H_2S was determined by the iodometric method; and ammonia was determined by specific ion electrode [*Lico et al.*, 1982].

Water samples were generally collected in a stainless steel pressure vessel and forced through 0.1 μm filter paper using argon gas. About 500 ml of the filtrate were used to prerinse all the sample containers. Samples for trace metal analysis were then collected in 250 ml polyethylene bottles, prewashed with a mixture of reagent grade nitric and sulfuric acids; water was acidified to pH ~1 with ultrex grade nitric acid (HNO_3). Samples for major cations and anions were filtered through 0.1μm or 0.45 μm and collected in 250 ml polyethylene bottles prewashed with 10% HNO_3 solutions for cations and distilled deionized water for anions; for cations, the sample was also acidified to pH ~1, but with ultrex grade HCl.

Samples for dissolved organic species were collected in pretreated brown glass bottles fitted with teflon inserts in the caps; mercuric chloride (40 mg/L Hg) was added as bactericide. The filtered samples were stored in a refrigerator until analysis. Samples for dissolved silica were filtered and then diluted 1:1 and 1:4 with distilled deionized water and stored in plastic bottles. Samples for water isotopes were not filtered and were collected in completely filled ~15 ml glass bottles with polyseal caps.

Samples for solute isotopes also require field treatment and preservation. Samples for stable and radiogenic C isotopes were prepared by the addition of ammoniacal $SrCl_2$ to ~15 ml or larger glass bottles filled with unfiltered water. Those for S isotopes were prepared by the addition of conc. $BaCl_2$ and a bactericide to filtered water in glass containers. Samples for Cl isotopes were filtered and stored in teflon container; those for Sr and B isotopes were treated like those for trace metals, but the pH was lowered to ~4 only.

The chemical compositions of water samples were determined in our USGS laboratory. Concentrations of Cl, SO_4, Br and other anions were obtained by ion chromatography as described in *Lico et al.* [1982]. Concentrations of other solutes were determined by inductively coupled plasma (ICP) spectrophotometry. Results generally have a precision of 2-5%.

Gas Samples

Methods for collection and analysis of gases are generally those described in *Evans et al.* [1981] and *Kennedy et al.* [1985]. Samples for both dissolved gas and a separate gas phase were obtained from sites with gas bubbles; otherwise only dissolved gas samples were collected. The exact sampling procedure was site specific, but the goal was to minimize air contamination and obtain gas data necessary to calculate the chemical and isotopic compositions and partial pressures of dissolved gases at subsurface conditions. In general, a 1.2 m stainless steel tube, ~1 cm diameter, was inserted into the discharge points and tygon tubing, connected to it, was used to obtain dissolved gas samples in a ~800 ml flow-through Pyrex cylinder, or a separate gas phase in a 40-250 ml preevacuated pyrex tube. Also to avoid air contamination, an inverted funnel was placed below water pools to capture gas bubbles for collection in preevacuated pyrex tubes and/or in a 1 cm diameter copper tube (~10 ml) crimp-sealed at both ends. For gas samples obtained from downhole samplers (e.g. Kuster sampler), the elaborate methodology and apparatus described in *Evans et al.* [1988b] was used.

Gas bulk compositions were obtained by gas chromatography, using Porapak Q and Linde Molecular Sieve 5Å as described in *Evans et al.* [1981]. Results generally have precision uncertainties of ±1%. Noble gases were analyzed using methods described in *Kennedy et al.* [1985] and *Higayon and Kennedy* [1992].

RESULTS AND DISCUSSION

Data from the literature and our rough estimates of water discharges indicate that high water flows are relatively few in the study area, and are especially rare on the trace of the San Andreas fault (Figs. 1 and 2). This lack of high discharge springs even in mountainous regions of significant relief is likely caused mainly by the low annual precipitation, which is generally <50 cm/yr [*Planert and Williams*, 1995]. Low percolation rates determined by regional geology, especially the presence of Salinian granitoids and other rocks with low porosity and permeability, also likely contribute to the rarity of high discharge springs.

In the case of the San Andreas fault, relatively high discharge is present only at Arrowhead Springs located at the southern end of the study area (Figure 1). The Arrowhead Hot Springs, located in and on the flanks of two canyons at the base of the San Bernardino Mountains (Figure 1), have the highest temperature (<91°C) and total water discharge (3-5 L/s, excluding thermal water from drilled wells). Arrowhead originally had about 30 individual springs in three main thermal areas, but the number is now reduced to about a dozen by development and drilling of thermal wells [*Waring*, 1915; *Burtell*, 1989]. Congress Springs, issuing CO_2-rich mineralized water from the Franciscan assemblage, has a moderate total discharge (~0.5 L/s), but the temperature is low (14-16°C), as is typical for soda springs.

Most of the high-discharge thermal springs are located on or near ancillary transform, normal and thrust faults on the margins of the Coast Ranges and the Transverse Ranges (Figure 1). The highest concentrations of high-discharge thermal springs in the study area are in the Transverse Ranges, on or near the E-W trending Big Pine, Santa Ynez and related faults. This region, with high precipitation and relatively low runoff and , thus high recharge [*Planert and Williams*, 1995], has more than a dozen thermal areas, with about half having total fluid discharges >5 L/s. The highest discharge and temperature are at Sespe Hot Springs, located ~25 km southwest of the San Andreas fault and composed of four groups of thermal springs with a high total discharge of ~50 L/s at temperatures ≤90°C.

There are three thermal and mineral areas (Gilroy, Alum Rock and Alameda) of moderate water discharges located on the Hayward and Calaveras faults (Figure 1). The highest temperature (42°C), but lowest discharge (~0.3 L/s) are obtained from Gilroy Hot Spring; the highest total discharge (total of 2-3 L/s), but lowest temperature (≤29°C) are obtained from 20-30 springs and seeps that comprise the mineralized area at Alum Rock (Figure 1).

The absence of major thermal discharges near most of the trace of the San Andreas fault in the study area likely reflects minor fluid circulation in the shallow crust. These results, although general, can be used as argument against the hypothesis that water circulation at shallow depths may be responsible for lack of a high heat flow anomaly over

the trace of this fault [*Lachenbruch and Sass*, 1980; 1992]. Our results instead support the conclusion of *Coyle and Zoback* [1988], whose data, yielding low permeability and essentially hydrostatic fluid pressures, indicated no significant fluid circulation at the Cajon Pass research well, and, thus also argue against the transport of significant amounts of frictional heat by water flow.

Isotopic Composition of Water

The δD and $\delta^{18}O$ values of water, though subject to several modifying processes, are the best parameters for indicating the origin of natural waters [for a review and references see *Kharaka and Thordsen*, 1992]. The isotopic values for the present study plot on a trend unrelated to the field ascribed to water of mantle or primary magmatic origin (δD = -40 to -80‰; $\delta^{18}O$ = 6 to 9‰), thus indicating no significant contribution from mixing with water from these sources (Figure 3). The majority of these values, especially of samples from granitoid rocks (e.g. Arrowhead thermal springs) and from all rocks in the Transverse Ranges, plot on or near the Global Meteoric Water Line, indicating that waters are predominantly of meteoric origin and have had a relatively low residence time in the granitoids or other local rocks; differences in these δD and $\delta^{18}O$ values reflect variable recharge elevations, distance from the ocean and longitude and latitude of the sites [*Kharaka et al.*, 1988a].

However, many samples, especially those obtained from the sedimentary rocks of the Great Valley sequence (e.g. from Parkfield area), show significant isotopic shifts from the Global Meteoric Water Line, that are similar to those obtained in meteoric or marine connate waters from petroleum fields in the Sacramento and San Joaquin Valleys [*Kharaka et al.*, 1973; *Kharaka and Carothers*, 1986; *Fisher and Boles*, 1990]. These isotopic shifts are also comparable to those measured in samples from thermal and mineral springs in the northern California Coast Ranges [*White et al.*, 1973; *Donnelly-Nolan et al.*, 1993; *Davisson et al.*, 1994]. Isotopic shifts in our samples do not correlate with the discharge temperatures, and generally do not correlate with calculated subsurface temperatures. A general increase in both the measured $\delta^{18}O$ values and the calculated $\delta^{18}O$ "excess" values (a measure of the horizontal distance to GMWL in Figure 3) with increased salinity is observed for samples with salinities higher than about 1,000 mg/L. Relatively small isotopic shifts (e.g. samples from Carrizo Plain, and part of the shift in the sample from Jack Ranch "Hwy 46" well, are likely due to evaporation effects (Figure 3). More significant shifts likely result mainly from isotopic exchange with minerals, especially

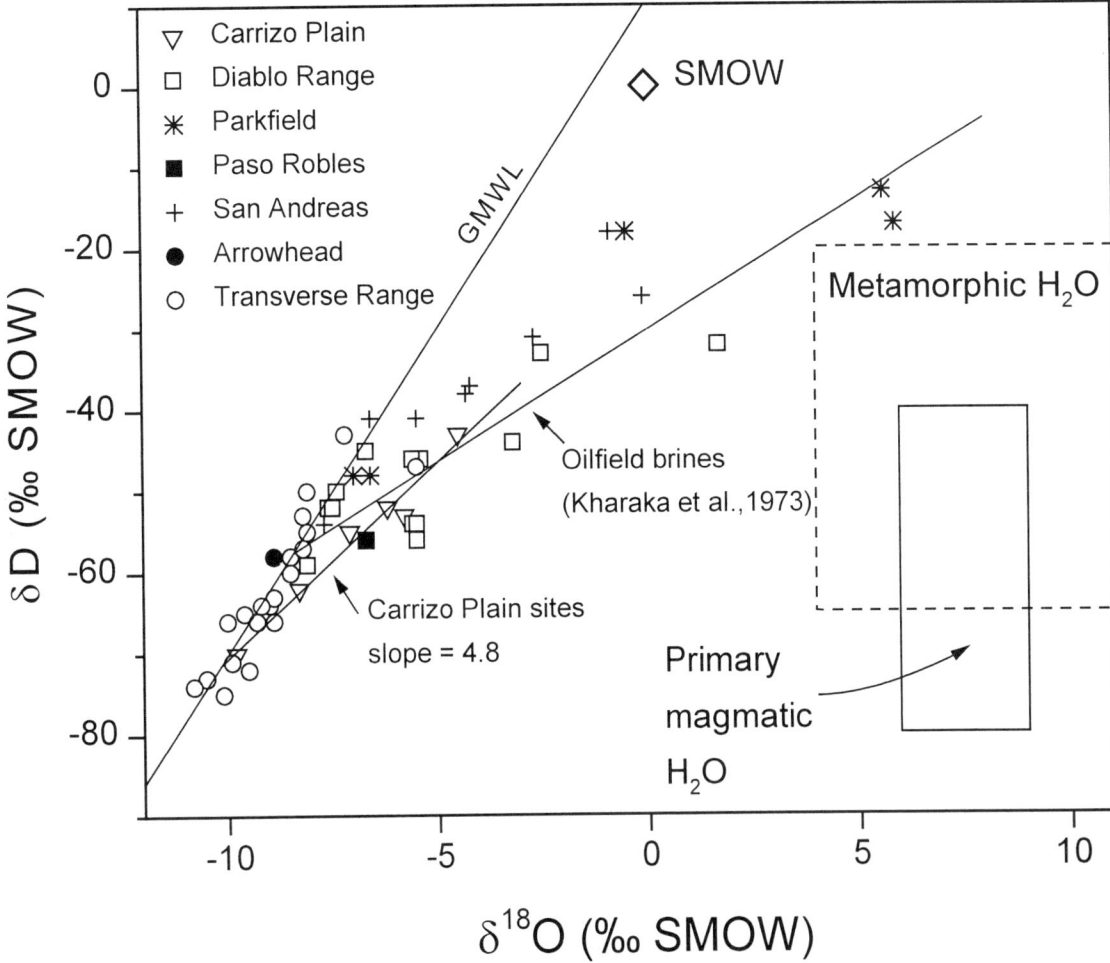

Figure 3. Isotopic composition of water from thermal and mineral springs and wells associated with the San Andreas fault system. The fields for metamorphic and primary magmatic waters are after *White et al.* [1973].

clays and carbonates and mixing of waters of different origin and evolutionary history [*Kharaka et al.*, 1973; *Fehn et al.*, 1992; *Kharaka and Thordsen*, 1992].

White [1957] and colleagues [*White et al.*, 1973] were the first to describe CO_2-rich waters from thermal and mineral springs from northern California Coast Ranges as "metamorphic water", attributing their isotopic and chemical compositions to dehydration and other metamorphic reactions involving rocks of the Franciscan assemblage. These waters show significant water isotopic shifts and are of the Na-HCO_3 type, with significant concentrations of B, NH_3 and H_2S. As pointed out by *Barnes et al.* [1975] and indicated from our data (Figure 3), many CO_2-rich waters from thermal and mineral springs from California Coast Ranges are clearly meteoric as their isotopes plot on the local meteoric water line and away from the field ascribed to

"metamorphic water". Recent detailed studies of CO_2-rich waters from the Clear Lake area of *White* [1957], have led to the conclusion that the waters are predominantly meteoric in origin and recharged < 80 k years ago; their isotopic shifts were attributed to exchange with rocks and multiple near-closed-system boiling in the Clear Lake magmatic system [*Fehn et al.*, 1992; *Donnelly-Nolan et al.*, 1993].

Tritium isotopes were also determined for most of the thermal and/or mineralized samples. Most samples have ^{3}H values (<0.1 tritium units) that indicate no modern meteoric water is present (Table 1). However, significant amounts of ^{3}H were found at Epsom Salt Spring (3.1 units), at Congress (2.0 units) and even at one of the high temperature (63°C) springs at Sespe (1.4 units). Mixing of variable amounts of modern meteoric water with much older ascending water at shallow depths is expected in

Table 1. Chemical and Isotopic Compositions of Water from Selected Springs and Wells

Sample No. Site	94SAF-01 Arrowhead Hot Sp	94SAF-13 Carneros Sp	94SAF-14 Aqua Media Sp	94SAF-17 Middle Mountain oil well	94SAF-29 Congress Sp	94SAF-38 Gilroy Hot Sp	94SAF-31 Mercey Hot Sp	94SAF-52 Sespe Hot Sp	95SAF-10 Byron Hot Sp
T(°C)	87	22	25	17	14	42	45	90	34
pH	8.55	7.44	7.30	6.51	6.39	6.78	8.86	8.26	6.80
Na	203	72.5	441	8900	1490	276	828	333	4870
K	6.5	2.4	2.9	97	81	5.2	5.6	14	62
Mg	0.1	19	123	93	237	117	0.4	0.2	163
Ca	11	80	246	125	210	11	41	22	549
Sr	0.39	1.00	3.06	9.40	2.91	0.27	0.41	0.80	20.4
Li	0.17	0.04	0.09	4.1	1.5	0.16	0.07	0.75	0.15
Cl	48	21	195	12700	1304	120	1300	285	8570
SO_4	316	207	1470	1.1	4.7	2.1	5.1	294	0.65
HCO_3	78	204	224	2400	3730	1220	60	69	504
Br	0.1	0.1	0.6	41	3.5	0.3	4.0	0.8	33
F	6.3	0.2	1.2	<0.2	2	0.1	<0.3	13.1	0.3
NH_4^+	<0.1	<0.1	<0.1	168	11	0.4	4.3	0.8	39
SiO_2	74	27	14	40	84	105	73	94	26
B	2	0	5	265	90	15	14	13	29
TDS	756	642	2750	26100	7720	1950	2400	1210	15000
δD	-58	-62	-52	-13	-37	-45	-54	-75	-33
$\delta^{18}O$	-8.9	-8.3	-6.2	5.6	-4.2	-6.7	-5.6	-10.1	-2.5
^3H	0	1.3	2.0	0	2.0	0	0	0.12	0.43
$\delta^{13}C$	-19.9	-13.9	-12.2	23.3	-7.2	-10.8	-10.8	-16.4	0.42

Chemical concentrations reported in mg/L; isotopic compositions in permil; ^3H in tritium units (TU).

these highly fractured and altered rocks, but quantification of this mixing process was not attempted, because it requires multiple sampling at different seasons [*King et al.*, 1994].

Chemical Composition of Water

The chemical composition of water (Table 1) is highly variable, with salinity generally being that of fresh water (<1,000 mg/L, total dissolved solids) or brackish water (salinity 1,000-10,000 mg/L, TDS). Seven samples have salinities higher than that for brackish water; the highest values are obtained for three samples from the sedimentary section at Parkfield (salinity 26,000-47,000 mg/L). The major cations are Na, Ca and Mg; the major anions are Cl, HCO_3 and SO_4. We made numerous plots depicting various combinations of cations and anions. Our preference is for pattern diagrams depicting the equivalent proportions of the major cations and anions, with data on salinity and pH (Figs. 4-7).

The chemical composition of water is determined primarily by the origin of water and the minerals with which it reacts in its flow path; thus local geology, especially the enclosing rock types, plays the dominant role. In general, water samples from shallow sedimentary rocks and granitoids are meteoric in origin and have low salinity (<1,000 mg/L) and a mixed major cation and anion composition (Figure 4). Waters originating in deeper sections of the sedimentary Great Valley sequence have the highest salinity (up to ~50,000 mg/L) and are dominantly of the Na-Cl type (Figure 5); samples from the Franciscan assemblage have a low to moderate salinity (500-8,000 mg/L), are Na-HCO_3 type, but contain significant concentrations of Mg, B, NH_3 and dissolved CO_2 (Figure 6). Finally, samples originating from deeper parts of the Salinian block granitoids have relatively low salinity (<3,000 mg/L) and high pH, and are of the Na-SO_4 type (Figure 7).

The evolution of meteoric or marine connate waters to Na-Cl (Figure 5) and Na-Ca-Cl type waters in sedimentary rocks, especially those present in petroleum basins, is reasonably well known, from extensive study over the last 50 years [*De Sitter*, 1947; *Fisher and Boles*, 1990; *Kharaka and Thordsen*, 1992; *Hanor*, 1994; *Davisson et al.*, 1994; *Land*, 1995]. In short, Na and Cl are ultimately derived

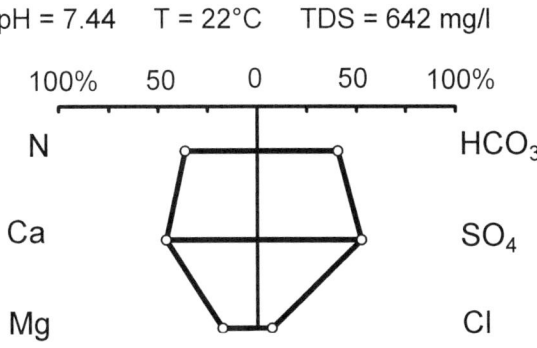

94SAF-13 Carneros Sp

pH = 7.44   T = 22°C   TDS = 642 mg/l

Figure 4. Pattern diagram for a low salinity and mixed cation and anion composition water that is typical of samples obtained from springs with shallow circulation depths.

from a marine source; Ca values, which commonly increase with increasing temperatures in the basin, are generally obtained from albitization of plagioclase feldspar and/or dolomitization of calcite. Increased concentrations of Ca (up to 50,000 mg/L in the central Mississippi Salt Dome basin), and to a lesser extent, of Sr and Ba, with increased temperatures ultimately lead to the precipitation of sulfate and carbonate minerals, resulting in the loss of most of SO_4 and HCO_3 from oil field brines [e.g., *Kharaka and Thordsen*, 1992].

The chemical composition and evolution of Na-HCO_3 type waters (Figure 6) having relatively high concentrations of Mg, B, NH_3 and dissolved CO_2 and H_2S obtained from thermal and soda springs located in California Coast Ranges, especially those issuing from Franciscan rocks, have also received a great deal of discussion [*White*, 1957; *Barnes*, 1970; *Barnes et al.*, 1973; *White et al.*, 1973; *Irwin and Barnes*, 1975; 1980; *Donnelly-Nolan et al.*, 1993; *Davisson et al.*, 1994]. *White* [1957] was the first to describe these fluids as "metamorphic water", attributing their chemical and isotopic compositions to dehydration and other metamorphic reactions involving rocks of the Franciscan assemblage. Water-rock interactions in the Franciscan assemblage, regionally metamorphosed to laumontite, pumpellite, greenschist and lower grade facies, are responsible for the relatively high Na (from albite), Mg (from serpentinite) and B (from a magmatic source or clay minerals) [e.g., *Barnes et al.*, 1973; *Donnelly-Nolan et al.*, 1993]. As we point out in the discussion of gases below, our CO_2-rich water samples from Franciscan rocks have $\delta^{13}C$ values of -6 to -11‰ , which combined with the isotopic composition of associated noble gases, indicate a

mixture of mantle and organic sources for HCO_3 and dissolved CO_2. The isotopic data preclude metamorphism of marine limestone, which is a rare constituent of the Franciscan assemblage, as an important source of HCO_3 and dissolved CO_2. Organic matter is probably also the main source of the relatively high concentrations of NH_3, Br, I and dissolved H_2S in these waters [*Fehn et al.*, 1992; *Donnelly-Nolan et al.*, 1993].

The Na-SO_4 type waters of relatively low salinity (<3,000 mg/L) and high pH (Figure 7) obtained from Salinian block and other granitoids are discussed in more detail below, because they are less well studied than the other water types. The chemical composition of water in our samples is similar to that of water obtained from thermal springs and intermediate temperature (50-150°C) wells in granitic rocks from Western United States and other areas worldwide [*Feth et al.*, 1964; *McCulloh et al.*, 1981; *Nordstrom and Olsson*, 1987]. However, these waters are totally different from water from shield areas in North America and the USSR where salinities may exceed 300,000 mg/L, and the water is of the Ca-Na-Cl type [*Frape et al.*, 1984]. Water encountered at shallower depths in granitic rocks generally has low pH and salinity (<300 mg/L), but has relatively high SiO_2 and is of a mixed cation-HCO_3 type. The mixed cations and silica are formed by the incongruent dissolution of aluminosilicates, and bicarbonate is mainly of soil-organic origin.

The detailed chemical composition of water, gases and minerals obtained from Arrowhead Hot Springs and the nearby Cajon Pass research well [*Kharaka et al.*, 1988a] provide a good data set to investigate the path of water

94SAF-17 Middle Mountain oil well

pH = 6.51   T = 17°C   TDS = 26,100 mg/l

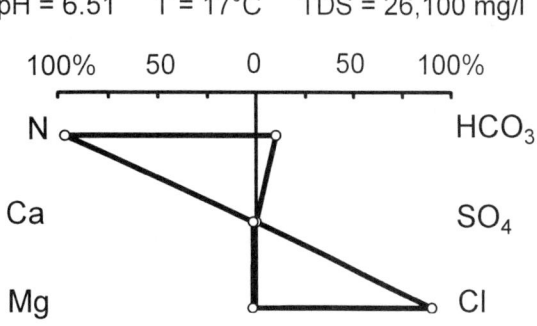

Figure 5. Pattern diagram for a Na-Cl type water that is generally obtained from sedimentary sections in the San Andreas fault system.

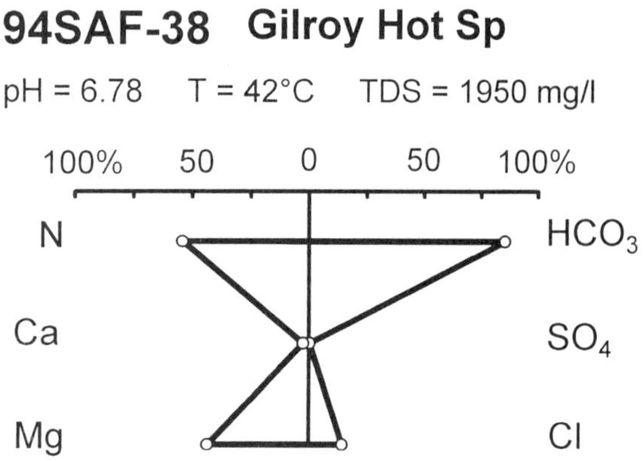

94SAF-38 Gilroy Hot Sp

pH = 6.78   T = 42°C   TDS = 1950 mg/l

Figure 6. Pattern diagram for a CO_2-rich water generally obtained from the Franciscan assemblage sections in the San Andreas fault system. Note the relatively high concentrations of Na, Mg and HCO3.

evolution from the mixed cation-HCO_3 type, generally obtained from shallow granitic rocks or sandstones, to Na-SO_4 type water encountered at the well and Arrowhead. We used the geochemical code SOLMINEQ.88 [*Kharaka et al.*, 1988b] to compute the saturation states of those minerals that most likely control the chemical composition of these waters. Computations show that the shallow groundwaters in the area and formation water in the well are highly undersaturated with respect to feldspar (albite-anorthite), K-feldspar, and other major aluminosilicates comprising the granitic rock; it is generally supersaturated with respect to kaolinite and Mg-chlorite. These results indicate congruent and incongruent dissolution of feldspar and other aluminosilicates leading to high pH values and increased salinity. The reaction with oligoclase, for example, may lead to precipitation of kaolinite as in reaction:

$$5Ca_{0.2}Na_{0.8}Al_{1.2}Si_{2.8}O_8 + 6H^+ + 19H_2O \rightarrow$$
$$Ca^{++} + 4Na^+ + 3Al_2Si_2O_5(OH)_4 + 8H_4SiO_4^0.$$

The precipitated kaolinite probably reacts with dissolved Mg (from dissolution of mafic minerals) to form Mg-chlorite, as in the reaction:

$$Al_2Si_2O_5(OH)_4 + 5Mg^{++} + 5H_2O + H_4SiO_4^0 \rightarrow$$
$$Mg_5Al_2Si_3O_{10}(OH)_8 + 10H^+.$$

The formation water in the well and at Arrowhead is generally saturated with calcite, laumontite, heulandite and fluorite, indicating that Ca leached from silicates is precipitated in the fractures. Calcite, laumontite, stilbite (equivalent to heulandite) and chlorite are the dominant

authigenic phases reported in fractures at this well [*Silver et al.*, 1988]. Removal of Mg and Ca thus leaves Na as the dominant cation in formation water. Precipitation of calcite lowers the concentration of bicarbonate in these waters.

The relatively high concentrations of SO_4 in these waters likely result mainly from oxidation of pyrite and other reduced sulfur minerals present in granitic rocks [*Kharaka et al.*, 1988a; *Anderson et al.*, 1988]. At shallow depth and wherever dissolved oxygen is present the reaction, mediated by bacteria, is:

$$2FeS_2 + 7O_2 + 2H_2O \rightarrow 2Fe^{++} + 4SO_4^{--} + 4H^+.$$

Other oxidizing agents may become more important at depth. The reaction with NO_3, for example, may cause formation of NH_3 as in:

$$4FeS_2 + 7NO_3^- + 11H_2O + 6H^+ \rightarrow 4Fe^{++} + 8SO_4^{--} + 7NH_4^+.$$

The concentration of Fe in these waters is generally below the detection limit of ≈ 0.05 mg/L. A portion of the dissolved Fe produced from the above and other reactions probably enters chlorite, but a large portion is precipitated as Fe-oxyhydroxides because of the high pH of water. Some Cl is added to formation water from fluid inclusions [*Nordstrom and Olsson*, 1987], but the amount at these temperatures is small and the water is mainly of a Na-SO_4 type.

Chemical Geothermometry

The chemical composition of spring water in general and the concentrations of dissolved SiO_2 and cations in

94SAF-01 Arrowhead Hot Sp

pH = 8.55   T = 87°C   TDS = 756 mg/l

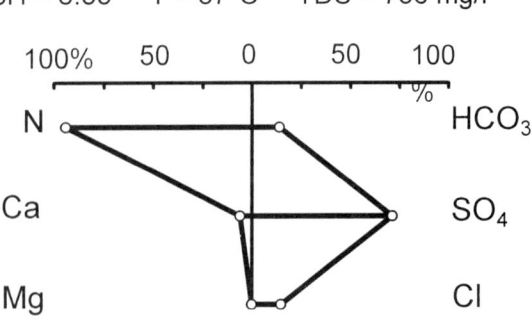

Figure 7. Pattern diagram for a Na-SO_4 type water of low salinity and high pH that is generally obtained from thermal springs issuing from Salinian block granitoids.

particular, have been used in chemical geothermometry, pioneered by Fournier and colleagues [e.g., *Fournier et al.*, 1974], to estimate the equilibrium subsurface temperature and maximum water circulation depths. We used the equations reported in *Kharaka and Mariner* [1989] and coded in a modified version of SOLMINEQ [*Kharaka et al.*, 1988b], to compute 12 different temperature values for each water sample. The assumptions and limitations of chemical geothermometry, applied to waters from thermal and mineral springs, are discussed by *Fournier et al.* [1974] and *Kharaka and Mariner* [1989]. The most reliable estimates with this approach are obtained for thermal springs with large water discharges, subsurface temperatures >70°C, and when concordant (within ±10°C) values are obtained from the quartz, Mg-Li, Mg-corrected Na-K-Ca and Na-Li geothermometers. For this reconnaissance study, we made no corrections for any chemical composition changes resulting from water-rock interactions in the ascending flow path, which could be important in the results from soda springs or thermal springs of low discharge. We also assumed no mixing with water from shallower zones, but such mixing usually leads to lower computed temperature values [*Kharaka and Mariner*, 1989].

Results show (Table 2) that about half of our samples give subsurface temperatures >80°C, with a maximum computed value of 150°C for the main source of the Sespe Hot Springs. The values selected are for individual discharges or wells. Where a group of springs is present in a thermal area, subsurface temperature was computed for each sample collected, and the highest temperature is reported (Table 2). Water circulation depths (Table 2) were calculated assuming a regional temperature gradient of 25°C/km and a mean annual surface temperature of 18°C [Colin Williams, 1997, personal communication]. Results give a maximum circulation depth of 5.3 km for the Sespe Hot Springs. A maximum circulation depth >6 km could be calculated for Sespe assuming a lower local temperature gradient of 20-24°C/km, based on values for several nearby petroleum wells [*McCulloh et al.*, 1981; *De Rito et al.*, 1989]. Results give shallow to moderate reservoir depths of up to 6 km, indicating that meteoric water, recharged at high elevations in the San Andreas system, percolates to depths of 3-6 km. Where this heated water encounters a permeable flow path (e.g., a fault zone), it changes direction and flows relatively rapidly to emerge as a thermal or mineral spring. A comparable depth of <5 km is indicated for a few of our samples, as well as those reported by *Davisson et al.* [1994], that originate from high pressure pore water squeezed tectonically from shales and siltstones of the Great Valley sequence.

Table 2. Calculated Subsurface Temperatures and Circulation Depths From Chemical Geothermometry[a]

Sample No.	Site	Selected T, (°C)	Circulation Depth[b], km
94SAF-01	Arrowhead Hot Sp	120	4.1
94SAF-02	Warm Sp	100	3.3
94SAF-06	Aqua Caliente Hot Sp	110	3.7
94SAF-07	Little Caliente Hot Sp	100	3.3
94SAF-08	Gaviota Hot Sp	90	2.9
94SAF-09	Paso Robles Mineral Hot Sp	110	3.7
94SAF-10	Coalinga Mineral Sp	100	3.3
94SAF-17	Middle Mountain oil well	100	3.3
94SAF-18	Middle Mountain tar seep	120	4.1
94SAF-27	Epsom Salt Sp, Mt Diablo	130	4.5
94SAF-29	Congress Sp	130	4.5
94SAF-31	Mercey Hot Sp	100	3.3
94SAF-33	Stone Canyon Well	80	2.5
94SAF-37	UN Pescadero Cr. Sp #2	130	4.5
94SAF-38	Gilroy Hot Sp	110	3.7
94SAF-52	Sespe Hot Sp	150	5.3
95SAF-104	Matilija Hot Sp	90	2.9

[a]Table includes only samples with subsurface temperature ≥ 80°C.
[b]Geothermal gradient is 25°C/km; mean annual air temperature is 18°C.

Chemical and Isotopic Composition of Gases

The gas compositions (Table 3) are also controlled mainly by geology and rock types, with samples from the Salinian block granitoids being composed mainly of N_2 (principally, but not exclusively of atmospheric origin) and minor CH_4 and CO_2. Samples from the Great Valley sequence are dominated by CH_4 of thermogenic origin ($\delta^{13}C$ = -36 to -50‰). Gas samples from Franciscan rocks are CO_2-rich but also have relatively high concentrations of CH_4 and N_2. The $\delta^{13}C$ values of bicarbonate and CO_2 range from -20 to +23‰, spanning the main known sources of carbon, including organic, marine carbonate and mantle, and known processes, including mixing and isotopic exchange with carbonate minerals [*Carothers and Kharaka*, 1980]. Samples obtained from CO_2-rich springs in Franciscan rocks have $\delta^{13}C$ range of -6 to -11‰ (Figure 8), which is closer to that expected from a mantle source [$\delta^{13}C$ = -4 to -8‰, *Marty and Jambon*, 1987; *Trull et al.*, 1993].

The fluid samples are enriched in total He relative to its concentration in air saturated water; $^4He/^{36}Ar$ enrichment factors, relative to 10°C air-saturated water, range from ~1 to 4,300. The He isotopic compositions (Figure 2) have Ra values from 0.06 to 4.0 (Ra is the $^3He/^4He$ normalized to ratio in air), indicating a range from crustal values (<0.1 Ra) derived from α-decay of U and Th in the rocks, to high enrichment in ^3He of mantle (~8 Ra in MORB) origin

Table 3. Chemical and Isotopic Compositions of Gas Discharges from Selected Springs

Site	Arrowhead	Alum Rock	Congress	Mercey	Gilroy	Sespe	Byron
Sample No.	94SAF-01	94SAF-22	94SAF-29	94SAF-31	94SAF-38	94SAF-52	95SAF-10
He	0.185	0.0001	<0.0002	0.0071	0.0023	0.0344	0.0015
H_2	<0.001	0.0006	0.0004	0.0106	0.0021	0.0002	0.0181
Ar	1.66	0.0163	0.0207	0.219	0.246	1.37	0.0239
O_2	0.670	<0.0005	0.0161	0.221	0.0538	0.0640	0.0361
N_2	96.2	3.69	0.738	15.8	28.6	92.5	4.03
CH_4	0.423	12.2	0.0064	84.6	31.3	4.20	92.9
CO_2	0.325	83.4	98.3	0.0203	40.3	1.55	4.48
C_2H_6	0.0018	0.0229	<0.0002	0.0468	0.0383	0.0344	0.0481
H_2S	<0.0005	0.0975	<0.0005	<0.0005	<0.0005	0.0150	<0.0005
Total	99.5	99.4	99.1	100.9	100.5	99.8	101.5
$\delta^{13}C$ (CO_2)	---	-9.96	-11.0	---	-14.3	-18.6	-3.81
$\delta^{13}C$ (CH_4)	---	-36.7	---	-40.8	-37.4	-18.2	-54.9
$\delta^{15}N$ (N_2)	---	-0.71	---	0.23	0.05	---	---
R/Ra	0.41	0.34	0.85	4.1	0.99	0.42	0.53

Chemical compositions reported in mole percent, free-water basis; isotopic compositions reported in permil; ---, no data.

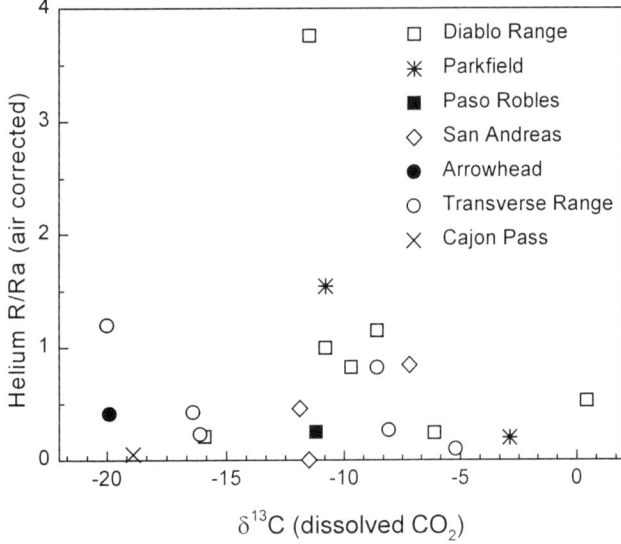

Figure 8. Isotopic composition of total dissolved inorganic C and He gas in fluids from the San Andreas fault system. Note that mantle gases have $\delta^{13}C$ values of -4 to -8‰ and an $^3He/^4He$ value of 8 Ra.

[*Mamyrin and Tolstikhin*, 1984]. In *Kennedy et al.* [1997], we discussed the weak correlation between high $^3He/^4He$ ratios and the proximity to the trace of the San Andreas fault; we also explored, but concluded against the possibility, that the high ratios are derived from Tertiary volcanics, either directly or via a staining process. The highest $^3He/^4He$ ratio (~4 Ra), for example, is obtained in fluids from Mercey Hot Springs located on the Tesla-Ortigalita

fault (Figs. 1 and 2). Our model calculations [*Kennedy et al.*, 1997] indicate that this high ratio could not be derived from the 7.4-11.6 Ma [*McLaughlin et al.*, 1996] Quien Sabe volcanics. This conclusion is supported by the chemical data for water from this site that show a Great Valley chemical signature, similar to Middle Mountain oil well (Figure 4), but with TDS = 2,400 mg/L; the chemical data show no indication of reaction with the nearby andesitic volcanics.

Active magmatic systems generally yield Ra values ≥6 [*Kennedy et al.*, 1985; *Sorey et al.*, 1993], and these or very young volcanic rocks are clearly responsible for the high $^3He/^4He$ ratios measured in gases obtained from sites in California, including ~8 Ra from the Gulf of California (active system) and Clear Lake area (0.01-2.1 Ma); values of up to 2.8 Ra were measured in gases from petroleum fields in the Sacramento Valley, with higher values obtained from gas fields close to the 1-2 Ma Sutter Butte volcanics [*Poreda et al.*, 1986; *Jenden et al.*, 1988; *Poreda and Craig*, 1989]. Our main conclusion, however, is that the high $^3He/^4He$ ratios indicate a pervasive and currently active mantle degassing throughout the San Andreas fault system that does not correlate with rock provenance, with water types, or with the age or distance of nearby volcanic rocks.

Fluid-San Andreas Fault Interactions

Our model estimations, discussed in *Kennedy et al.* [1997], give the flux of ^3He in the Phillips-Varian well at Parkfield to be ~2x10^{-15} mol/cm^2-year. Extrapolating this rate to the entire San Andreas fault system (~1500 km length by 200 km width), as described by *Wallace* [1990], yields a modest ~6 mol/year. More important is the fact that mantle ^{3}He is coupled to other more abundant volatiles, especially CO_2 and possibly water. Mid-oceanic ridge basalts originating in the upper mantle, for instance, have a relatively constant $CO_2/^3$He ratio of ~10^{10}, which is independent of vesicularity and total gas abundance [*Marty and Jambon*, 1987; *Trull et al.*, 1993; *Kennedy et al.*, 1997]. Using this ratio and the ^{3}He flux at Parkfield, yields a very high mantle CO_2 flux of ~10^{11} mol/year for the entire San Andreas system. The total CO_2 flux, of mantle and other sources, is expected to be even higher as calculations show that the measured total C (H_2CO_3, HCO_3^-, CO_3^{--}, and other C species) in our samples is higher, at times by >1000, than that supportable by the measured mantle ^{3}He (see discussion below). Because the density, specific heat and viscosity of CO_2 are comparable to those of water at seismogenic depths [e.g., *Duan et al.*, 1996], CO_2 flux into the fault zone will generate high pore pressures, thus lowering the effective stress and weakening the fault as originally suggested by *Irwin and Barnes* [1980], as calculated by *Bredehoeft and Ingebritsen* [1990] and as postulated in the *Rice* [1992] model. *Bredehoeft and Ingebritsen* [1990] modified a computer package, developed for simulating geothermal systems [*Bodvarsson*, 1982], to show that global CO_2 fluxes, focussed along seismically active areas, could generate high pressures in the crust in time scales of 10^4-10^6 years.

Generation of high fluid pressures by CO_2 flux into the San Andreas fault was modeled in this study with the same simulation package and parameters used by *Bredehoeft and Ingebritsen* [1990], but with temperature, pressure and CO_2 flux data obtained from our study. Results indicate that for reasonable geologic and hydrologic parameters operating at a 10 km depth, fluid pressures could increase from hydrostatic (~74 MPa) to lithostatic (~250 MPa) values in a short period of only ~200 years using the highest total CO_2 flux of mantle and other sources (Figure 9). A total CO_2 flux, that is lower by a factor of 10 from the maximum, yields a period of ~2000 years. Because the stress relieved by an earthquake is probably of the order of ~10 MPa [*Lachenbruch and Sass*, 1992], the fluid pressure at the start of an earthquake cycle could be close to lithostatic and time periods required to increase that to a fluid pressure that would cause fault rupture and trigger an earthquake is probably of the order of 1/10 of the values shown in Figure 9. These preliminary model results indicate that deep CO_2 flux alone could account for the generation of high fluid pressures that may be responsible for triggering earthquakes in parts of the San Andreas fault system. The model times required

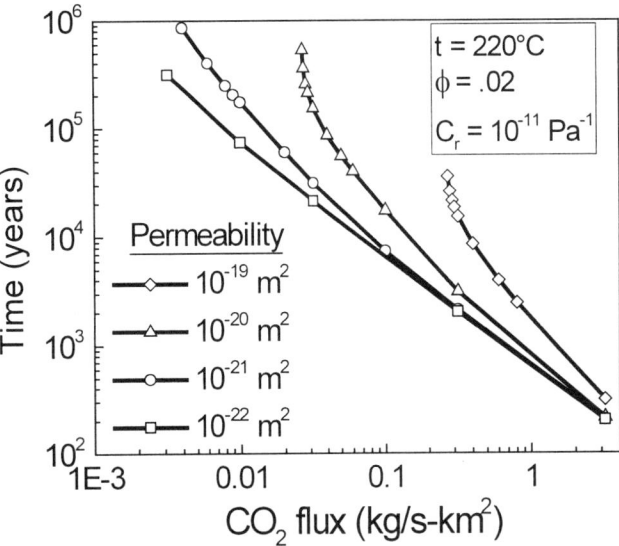

Figure 9. Computed times required to generate lithostatic fluid pressures from hydrostatic values by CO_2 flux at a 10 km depth. Values used for rock porosity (ϕ) and compressibility (C_r) are indicated.

to increase fluid pressures to lithostatic values are comparable to those of earthquake recurrence.

The high fluxes of total CO_2 used in our modeling, also add considerable amounts of advective heat and significantly increase the subsurface temperatures close to the fault. This result is contrary to field evidence that shows no increase of heat flow over the trace of the San Andreas fault [*Brune et al.*, 1969; *Lachenbruch and Sass*, 1980; *Sass et al.*, 1997]. The simulation package is being modified to refine calculations of the advective heat transport by CO_2 flux, without significantly changing the periods calculated to generate high fluid pressures indicated on Figure 9. Model modifications being tested involve: (1) heat loss by CO_2 expansion due to expected pressure drops following an earthquake; and (2) focussing the CO_2 flux into a narrower width of the fault zone than the 1 km width used in our simulations and those by *Bredehoeft and Ingebritsen* [1990].

Addition of CO_2 may also play an important role in water-rock interactions that impact the mechanical behavior of the San Andreas fault. Thermochemical data indicate that CO_2, present mainly as H_2CO_3 at high temperatures, becomes much more reactive (as HCO_3^- and H^+) with enclosing rocks at temperatures ≤300°C [e.g., *Bischoff and Rosenbauer*, 1996]. Addition of CO_2 to subsurface waters generally results in a lowering of pH values, in redissolution of carbonate minerals and in an increase of dissolution reaction rates. In addition, low pH solutions in the presence of dissolved Mg could result in the formation of chlorite or biotite, and thus cause "fault softening." Increased fluid pressure and "fault softening" reactions could lead to earthquake triggering. On the other hand, loss of CO_2 from water following a fluid pressure drop induced by an earthquake, would result in higher pH and in precipitation of calcite and other minerals, and would shift water-rock interactions to "fault hardening" reactions [*Wintsch et al.*, 1995]. Mineral precipitation would lead to fault crack healing, to sealing of fluid flow paths and to the start of a new earthquake cycle.

The total fluid-rock interactions in the seismogenic zone are probably more complex than those described above. Calculations, already mentioned, show that the measured total C in our samples is higher, at times by >1000, than that supportable by the measured mantle ^3He. These high $C/^3He$ could be due to selective He loss from some fluids prior to sampling, and/or to the addition of CO_2 from non-mantle sources. The source of the additional C is probably mainly from decarbonation of organic matter as indicated by the low $\delta^{13}C$ (up to -20‰) values, but some may be derived from the rare marine carbonate units ($\delta^{13}C$ ~0‰). Unlike ^3He, however, ^{13}C isotopes do not provide an unambiguous tracer for the source of dissolved C because of mixing of C from different sources and isotopic exchange between dissolved C and carbonate minerals [*Carothers and Kharaka*, 1980; *Trull et al.*, 1993].

The role of any water associated with mantle ^3He is also problematic. Our simulations assumed the fluid to be pure CO_2, which has properties (density, viscosity, specific heat) that are comparable to those of water at the conditions of simulations [e.g., *Duan et al.*, 1996]. The presence of water would be equally as effective as that of pure CO_2, but we assume an absence of a water phase at depths >10 km based on the measured δD and $\delta^{18}O$ values (Figure 2), indicating that water sampled is meteoric or connate in origin with no significant mantle component. Also, meteoric water circulation, as indicated by chemical geothermometry, ranges to a depth of only 6 km. Meteoric or connate water may penetrate deeper in this fault system, but probably not deeper than ~15 km, the "rheological transition zone", that separates brittle upper crust from ductile lower crust [*Nesbitt and Muehlenbachs*, 1989; *Thompson and Connolly*, 1992; *Lachenbruch and Sass*, 1992]. Finally, thermochemical data indicate that any water present with mantle He and CO_2, likely would rapidly become incorporated in clays, serpentine and other hydrous minerals, rendering lower crust in non subducting and non magmatic continental regions 'dry' [*Yardley*, 1985; *Frost and Bucher*, 1994]. These results indicate that meteoric water would be dominant only at shallower depths, and, likely would also play an important role in the earthquake cycle as postulated by the episodic model of *Byerlee* [1993].

CONCLUSIONS

In order to investigate the geochemistry and the role of fluids in the dynamics of the San Andreas fault system, we have carried out detailed chemical and isotopic analysis of water and associated gases from (41) thermal and saline springs and a few wells located along or near this fault system between San Francisco and Los Angeles. The important results are listed below.

(1) The δD and $\delta^{18}O$ values establish that waters are predominantly of meteoric origin, with no significant mantle or metamorphic components.

(2) The chemical compositions of water and gases are controlled mainly by the enclosing rock types.

(3) Application of chemical geothermometry gives reservoir temperatures of 80-150°C, indicating shallow to moderate circulation depths of up to 6 km for the water.

(4) In contrast to the above results, the compositions and isotope abundances of noble gases, especially the $^3He/^4He$ and $\delta^{13}C$ values of HCO_3, indicate a significant (up to 50%) mantle component for volatiles.

(5) Numerical simulations indicate that the obtained high CO_2 fluxes, of mantle and deep crustal origins, are sufficient to generate lithostatic fluid pressures in time scales comparable to those of earthquake cycles. These CO_2 fluxes into and out of the fault zone could also play an important role in mineral dissolution and precipitation reactions in the compartmentalized fluid pressure regimes and, thus, effect the dynamics of this fault.

The presence of a mantle component in the sampled fluids is clearly shown by the $^3He/^4He$ results, but the physical-chemical processes that focus and transport mantle fluids in this system are largely speculative. Several assumptions, all reasonable in our opinion, are made in extrapolating the CO_2 fluxes from spring measurements to areal coverages required for the numerical simulations. These results, thus, carry large uncertainties, and a detailed areal study of CO_2 fluxes, including determinations of gas transport in the unsaturated/soil zone, is planned to provide more reliable values for fluid fluxes. A deep drilling program, also planned for this fault, would allow for direct observation of fluids and rocks within and adjacent to the fault at seismogenic depths. Deep drilling and measurements of areal CO_2 fluxes would thus greatly improve our understanding of the role of fluids in the dynamics of this important plate boundary fault.

Acknowledgements. We thank our colleagues James Bischoff and Robert Mariner for a thorough review of an earlier version of this manuscript. We are grateful also to Jim Boles, Jeff Hanor and Peter Mozley for helpful comments and suggested modifications of the copy submitted for this Special Paper.

REFERENCES

Allen, C. R., The tectonic environments of seismically active and inactive areas along the San Andreas fault system, in *Proceedings of Conference on Geologic Problems of San Andreas Fault System*, edited by W.R. Dickenson and A. Grantz, pp. 70-82, Stanford University Publications in the Geological Sciences, vol. 11, Stanford, Calif., 1968.

Alvarez, W., D. V. Kent, I. Premoli-Silva, R. A. Schweikert, and R. A. Larson, Franciscan Complex limestone deposited at 17° south paleolatitude, *Geol. Soc. Am. Bull., 91,* 476-484, 1980.

Anderson R. N., R. E. Dove, C. Broglia, and others, Elemental and mineralogical analyses using geochemical logs from the Cajon Pass Scientific Drill hole, California, and their preliminary comparison with core analyses, *Geophys. Res. Lett., 15,* 969-972, 1988.

Awater, T., Implications of plate tectonics for the Cenozoic tectonic evolution of western North America, *Geol. Soc. Am. Bull., 81,* 3513-3536, 1970.

Bailey, E. H., M. C. Blake, and D. L. Jones, On-land Mesozoic oceanic crust in California Coast Ranges, *U.S. Geol. Surv. Prof. Pap. 700-C,* C70-C81, 1970.

Barnes, I., Metamorphic waters from the Pacific Tectonic belt of the west coast of the United States: *Science, 168,* 973-975, 1970.

Barnes, I, W. P. Irwin, and H. A. Gibson, Geologic map showing springs rich in carbon dioxide or chloride in California, *U.S. Geol. Surv. Water Resour. Invest. Open-File Map 75-34,* 1975.

Barnes, I., W. P. Irwin, and D. E. White, Global distribution of carbon dioxide discharges, and major zones of seismicity, *U.S. Geol. Surv. Water Resour. Invest. Open-File Rep. 78-39,* 12pp., 1978.

Barnes, I., M. E. Hinkle, J. B. Rapp, C. Heropoulous, and W. M. Vaughn, Chemical composition of naturally occuring fluids in relation to mercury deposits in part of north-central California, *U.S. Geol. Surv. Bull. 1382-A,* A1-A19, 1973.

Berkstesser, C. F., Jr., Data for springs in the southern Coast, Transverse, and Peninsular Ranges of California: *U.S. Geol. Surv. Open-File Rep.,* 1968.

Berry, F. A. F., High fluid potentials in California Coast Ranges and their tectonic significance, *AAPG Bull., 57,* 1219-1249, 1973.

Bischoff, J. L., and R. J. Rosenbauer, The alteration of rhyolite in CO_2 charged water at 200 and 350^oC: The unreactivity of CO_2 at higher temperature, *Geochim. Cosmochim. Acta, 60,* 3859-3867, 1996.

Bodvarsson, G. S., Mathematical modeling of geothermal systems, Ph.D. dissertation, Berkeley, University of California, 353pp., 1982.

Bredehoeft, J. D., and S. E. Ingebritsen, Degassing of carbon dioxide as a possible source of high pore pressure in the crust, in *The Role of Fluids in Crustal Processes*, National Academy Press, Washington D.C., pp. 158-164, 1990.

Brune, J. N., T. L. Henyey, and R. F. Roy, Heat flow, stress, and the rate of slip along the San Andreas fault, California, *J. Geophys. Res., 74,* 3821-3827, 1969.

Byerlee, J. D., Friction, overpressure and fault normal compression, *Geophys. Res. Lett., 17,* 2109-2112, 1990.

Byerlee, J. D., A model for episodic flow of high pressure water in fault zones before earthquakes, *Geology, 21,* 303-306, 1993.

Burtell, S. G., Geochemical Investigations at Arrowhead Springs, San Bernardino and Along the San Andreas Fault in Southern California, M. Sc. Thesis, Pittsburgh, University of Pittsburgh, 169pp., 1989.

Carothers, W. W., and Y. K. Kharaka, Stable carbon isotopes of HCO_3^- in oil-field waters – Implications for the origin of CO_2, *Geochimica et Cosmochimica Acta,* 44, 323-332, 1980.

Chester, F. M., J. P. Evans, and R. L. Beigel, Internal structure and weakening mechanisms of the San Andreas Fault, *J. Geophys. Res.,* 98, 771-786, 1992.

Coney, P. J., D. L. Jones, and J. W. H. Monger, Cordilleran suspect terranes, *Nature,* 288, 329-333, 1980.

Coyle, B. J., and M. D. Zoback, In situ permeability and fluid pressure measurements at 2 km depth at the Cajon Pass research well, *Geophys. Res. Lett.,* 15, 1027-1032, 1988.

David, C., T. F. Wong, W. Zhu, and J. Zhang, Laboratory measurements of compaction-induced permeability change in porous rocks: Implications for the generation and maintenance of pore pressure excess in the crust, *Pure Appl. Geophys., 143,* 425-456, 1994.

Davisson, M. L., T. S. Presser, and R. E. Criss, Geochemistry of tectonically expelled fluids from the northern Coast ranges, Rumsey Hills, California, USA, *Geochim. Cosmochim. Acta,* 58, 1687-1699, 1994.

De Rito, R. F., A. H. Lachenbruch, T. H. Moses Jr., and R. J. Munroe, Heat flow and thermotectonic problems of the central Ventura basin, southern California, *J. Geophys. Res.,* 94, 681-699, 1989.

De Sitter, L. U., Diagenesis of oil-field brines, *AAPG Bull., 31,* 2030-2040, 1947.

Dewers, T., and A. Hajash, Rate laws for water assisted compaction and stress-induced water-rock interaction in sandstones: *J. Geophys. Res.,* 100, 13093-13112, 1995.

Dibblee, T. W., Jr., Geology of the Central Santa Ynez Mountains, Santa Barbara County, California, *Calif. Div. Mines Geol. Bull., 186,* 1966.

Dickinson, W. R., C. A. Hopson, and J. B. Saleeby, Alternate origins of the Coast Range Ophiolite (California): Introduction and Implications, *GSA Today,* 6, 1-10, 1996.

Donnelly-Nolan, J. M., M. G. Burns, F. E. Goff, E. K. Peters, and J. M. Thompson, The Geysers-Clear Lake area, California: Thermal waters, mineralization, volcanism, and geothermal potential: *Econ. Geol.,* 88, 301-316, 1993.

Duan, Z., N. Møller, and J. H. Weare, A general equation of state for supercritical fluid mixtures and molecular dynamics simulation of mixture PVTX properties, *Geochim. Cosmochim. Acta*, 60, 1209-1216, 1996.

Elias, B., and A. Hajash, Changes in quartz solubility due to effective stress: An experimental investigation of pressure solution, *Geology, 20*, 451-454, 1992.

Evans, W. C., N. G. Banks, and L. D. White, Analyses of gas samples from the summit crater, *U.S. Geol. Surv. Prof. Pap. 1250*, 227-231, 1981.

Evans, W. C., L. D. White, and Y. K. Kharaka, Dissolved gases in the Dosecc Cajon Pass Well: First Year Results, *Geophys. Res. Lett., 15-9*, 1041-1044, 1988a.

Evans, W. C., L. D. White, and J. B. Rapp, Geochemistry of Some Gases in Hydrothermal Fluids from the Southern Juan de Fuca Ridge, *J. Geol. Res., 93*, 15,305-15,313, 1988b.

Fehn, U., E. K. Peters, S. T. Fitzpatrick, P. W. Kubik, P. Sharma, R. T. D. Teng, H. E. Gove, and D. Elmore, ^{129}I and ^{36}Cl concentrations in waters of the eastern Clear Lake area, California: Residence times and source ages of hydrothermal fluids, *Geochim. Cosmochim. Acta, 56*, 2069-2079, 1992.

Feth, J. H., C. E. Roberson, and W. L. Polzer, Sources of mineral constituents in water from granitic rocks, Sierra Nevada, California and Nevada, *U.S. Geol. Surv. Water Supply Pap. 1535-I*, 70pp., 1964.

Fisher, J. B., and J. R. Boles, Water-rock interaction in Tertiary sandstones, San Joaquin basin, California, U.S.A.: Diagenetic controls on water composition, *Chemical Geology, 82*, 83-101, 1990.

Fournier, R. O., Compressive and tensile failure at high fluid pressure where preexisting fractures have cohesive strength, with application to the San Andreas fault, *J. Geophys. Res., 101*, 25499-25509, 1996.

Fournier, R. O., D. E. White, and A. H. Truesdell, Geochemical indicators of subsurface temperatures: Part 1. Basic assumptions, *J. Res. U.S. Geol. Surv., 2*, 259-262, 1974.

Fox, K. F. Jr., R. J. Fleck, G. H. Curtis, and C. E. Meyer, Implications of the northwestwardly younger age of the volcanics of west central California, *Geol. Soc. Am. Bull., 96*, 647-654, 1985.

Frape, S. K., P. Fritz, and R. H. McNutt, The role of water-rock interaction in the chemical evolution of groundwaters from the Canadian Shield, *Geochim. Cosmochim. Acta, 48*, 1617-1627, 1984.

Frost, B. R., and K. Bucher, Is water responsible for geophysical anomalies in the deep continental crust? A petrological perspective, *Tectonophysics, 21*, 293-309, 1994.

Fyfe, W. S., N. J. Price, and A. B. Thompson, *Fluids in the Earth Crust*, 383pp., Elsevier, New York, 1978.

Furlong, K. P., W. D. Hugo, and G. Zandt, Geometry and evolution of the San Andreas fault zone in northern California: *J. Geophys. Res., 94*, 3100-3110, 1989.

Gromme, C. S., Paleomagnetism of Franciscan basalt, Marin County, California, in *Franciscan Geology of Northern California*, edited by M. C. Blake Jr., pp. 113-119, Society of Economic Paleontologists and Mineralogists, Pacific Section, vol. 43, Los Angeles, 1984.

Hacker, B. R., Diagenesis and fault valve seismicity of crustal faults: *J. Geol. Res., 102*, 24459-24467, 1997.

Hanor, J. S., Physical and chemical controls on the composition of waters in sedimentary basins, *Marine and Petroleum Geology, 11*, 31-45, 1994.

Hickman, S., Stress in the lithosphere and the strength of active faults: U.S. National Report International Union Geodetic Geophysics, 1987-1990, *Rev. Geophys., 29*, 759-775, 1991.

Hickman, S. H., R. Sibson, and R. Bruhn, Introduction to special section: Mechanical involvement of fluids in faulting, *J. Geophys. Res., 100*, 12,831-12,840, 1995.

Hiyagon, H., and B. M. Kennedy, Noble gases in CH_4-rich gas fields, Alberta, Canada, *Geochim. Cosmochim. Acta, 56*, 1569-1589, 1992.

Hubbert, M. K., and W. W. Rubey, Role of fluid pressure in mechanics of overthrust faulting, *Geol. Soc. Am. Bull., 70*, 115-205, 1959.

Irwin, W. P., Geology and plate-tectonic development, in "The San Andreas Fault System, California", edited by R. E. Wallace, *U.S. Geol. Surv. Prof. Pap. 1515*, pp. 61-80, 1990.

Irwin, W. P., and I. Barnes, Effect of geologic structure and metamorphic fluids on seismic behavior of the San Andreas fault system in central and northern California, *Geology, 3*, 714-716, 1975.

Irwin, W. P., and I. Barnes, Tectonic relations of carbon dioxide discharges and earthquakes, *J. Geophys. Res., 85*, 3115-3121, 1980.

Jachens, R. C., A. Griscom, and C. W. Roberts, Regional extent of Great Valley basement west of the Great Valley, California: Implications for extensive tectonic wedging in the California Coast Ranges, *J. Geophys. Res., 100*, 12,769-12,790, 1995.

Jenden, P. D., I. R. Kaplan, R. J. Poreda, and H. Craig, Origin of nitrogen-rich natural gases in the California Great Valley: evidence from helium, carbon, and nitrogen isotope ratios, *Geochim. Cosmochim. Acta, 52*, 851-861, 1988.

Johnson, P. A., and T. V. McEvilly, Parkfield seismicity: Fluid-driven?, *J. Geophys. Res., 100*, 12,937-12,950, 1995.

Jones, D. L., R. Graymer, C. Wang, T. V. McEvilly, and A. Lomax, Neogene transpressive evolution of the California Coast Ranges, *Tectonics, 13*, 561-574, 1994.

Kennedy, B. M., M. A. Lynch, S. P. Smith, and J. H. Reynolds, Intensive sampling of noble gases in fluids at Yellowstone: I. Early overview of the data: regional patterns, *Geochim. Cosmochim. Acta, 49*, 1251-1261, 1985.

Kennedy, B. M., Y. K. Kharaka, W. C. Evans, A. Ellwood, D. J. DePaolo, J. Thordsen, G. Ambats, and R. H. Mariner, Mantle fluids in the San Andreas fault system, California, *Science, 278*, 1278-1281, 1997.

Kerrich, R., T. E. La Tour, and L. Willmore, Fluid participation in deep fault zones: Evidence from geological, geochemical, and ^{18}O/^{16}O relations, *J. Geophys. Res., 89*, 4331-4343, 1984.

Kharaka, Y. K., and W. W. Carothers, Oxygen and hydrogen stable isotope geochemistry of deep basin brines, in *Handbook of Environmental Isotope Geochemistry: Vol. 2*, edited by P. Fritz, and J. Ch. Fontes, pp. 305-360, 1986.

Kharaka, Y. K., and R. H. Mariner, Chemical geothermometers and their application to formation waters from sedimentary basins, in *Thermal History of Sedimentary Basins: Methods and Case Histories*, edited by N. D. Naeser, and T. H. McCulloh, pp. 99-117, Springer-Verlag, Berlin, 1989.

Kharaka, Y. K., and J. J. Thordsen, Stable isotope geochemistry and origin of water in sedimentary basins, in *Isotope Signatures and Sedimentary Records*, edited by N. Clauer, and S. Chaudhuri, pp. 411-466, Springer-Verlag., Berlin, 1992.

Kharaka, Y. K., F. A. F. Berry, and I. Friedman, Isotopic composition of oil field brines from Kettleman North Dome oil field, *Geochim. Cosmochim. Acta, 37*, 1899-1908, 1973.

Kharaka, Y. K., G. Ambats, W. C. Evans, and A. F. White, Geochemistry of water at Cajon Pass, California: preliminary results, *Geophys. Res. Lett., 15*, 1037-1040, 1988a.

Kharaka Y. K., W. D. Gunter, P. K. Aggarwal, E. H. Perkins, and J. D. DeBraal, SOLMINEQ 88: A computer program for geochemical modeling of water-rock interactions, *U.S. Geological Survey, Water Resources Investigation Report 88-4227*, 420pp., 1988b.

King, C. Y., T. S. Presser, W. C. Evans, and L. D. White, In search of earthquake-related hydrologic and chemical changes along Hayward fault, *Appl. Geochemistry, 9*, 83-91, 1994.

Lachenbruch, A. H., and J. H. Sass, Heat flow and energetics of the San Andreas fault zone, *J. Geophys. Res., 85*, 6185-6223, 1980.

Lachenbruch, A. H., and J. H. Sass, Heat flow form Cajon Pass, fault strength and tectonic implications, *J. Geophys. Res., 97*, 4995-5015, 1992.

LaForge, R., and W. H. K. Lee, Seismicity and tectonics of the Ortigalita fault and southeast Diablo Range, California, in "Conference on Earthquake Hazards in the Eastern San Francisco Bay Area, Hayward Calif., 1982, Proceedings", edited by E. W. Hart, S. E. Hirschfeld, and S. S. Schulz, *Calif. Div. Mines Geol. Spec. Pub. 62*, 93-101, 1982.

Land, L. S., Na-Ca-Cl saline formation waters, Frio Formation (Oligocene), south Texas, USA: Product of diagenesis, *Geochim. Cosmochim. Acta, 59*, 2163-2174, 1995.

Lico, M. S., Y. K. Kharaka, W. W. Carothers, and V. A. Wright, Methods for collection and analysis of geopressured geothermal and oil-field waters, *U.S. Geol. Surv. Water Supply Pap. 2194*, 21pp., 1982.

Lico, M. S., and Y. K. Kharaka, Subsurface pressure and temperature distributions in Sacramento basin, California, *Selected Papers of Pacific Section of the American Association of Petroleum Geologists 1983 Meeting, vol. 1,* 57-77, 1985.

Majmundar, H. H., Technical map of the geothermal resources of California, *Calif. Div. Mines Geol., Geologic Data Map No. 5*, 45pp., 1984.

Mamyrin, B. A., and I. N. Tolstikhin, *Helium Isotopes in Nature*, Elsevier, 360pp., 1984.

Marty, B., and A. Jambon, $C/^3He$ in volatile fluxes from the solid Earth: implications for carbon geodynamics, *Earth Planet. Science Lett.*, 83, 6-26, 1987.

Mattinson, J. M., and C. A. Hopson, U/Pb ages of the Coast Range Ophiolite: A critical reevaluation based on new high-precision Pb/Pb ages, *AAPG Bull. 76*, 425, 1992.

McCulloh, T. H., V. A. Frizzell Jr., R. J. Stewart, and I. Barnes, Precipitation of laumontite with quartz thenardite, and gypsum at Sespe Hot Springs, Western Transverse Ranges, California, *Clays and Clay Minerals, 29*, 353-364, 1981.

McLaughlin, R. J., W. V. Sliter, D. H. Sorg, P. C. Russell, and A. M. Sarna-Wojcicki, Large-scale right-slip displacement on the East San Francisco Bay Region fault system, California: implications for location of late Miocene to Pliocene Pacific plate boundary, *Tectonics, 15*, 1-18, 1996.

Moore, D. E., D. A. Lockner, M. Shengli, R. Summers, and J. D. Byerlee, Strengths of serpentinite gouges at elevated temperatures, *J. Geophys. Res., 102*, 14,787-14,801, 1997.

Morrow, C. A., B. Radney, and J. Byerlee, Frictional strength and the effective pressure law of montmorillonite and illite clays, in *Earthquake Mechanics and the Transport Properties of Rocks*, edited by B. Evans, and T. F. Wong, Academic Press, London, 69-88, 1992.

Mount, V. S., and J. Suppe, State of stress near the San Andreas fault: Implications for wrench tectonics, *Geology, 15*, 1143-1146, 1987.

Nordstrom, D. K., and T. Olsson, Fluid inclusions as a source of dissolved salts in deep granitic groundwaters, in "Saline Water and Gases in Crystalline Rocks", edited by P. Fritz, and S. K. Frape, *Geol. Assoc. Canada Spec. Pap. 33*, 111-119, 1987.

Norris, R. M., and R. W. Webb, *Geology of California*, John Wiley and Sons, Inc., New York, 541pp., 1990.

O'Neil, J. R., and T. C. Hanks, Geochemical evidence for water-rock interaction along the San Andreas and Garlock faults of California, *J. Geophys. Res., 85*, 6286-6292, 1980.

Planert, M., and J. S. Williams, Groundwater Atlas of the United States, Segment 1, California, Nevada, *U.S. Geol. Surv. Hydrologic Investigations Atlas 730-B*, B1-B28, 1995.

Poreda, R., and H. Craig, Helium isotope ratios in circum-Pacific volcanic arcs, *Nature, 338*, 473-478, 1989.

Poreda, R. J., P. D. Jenden, I. R. Kaplan, and H. Craig, Mantle helium in Sacramento basin natural gas wells, *Geochim. Cosmochim. Acta*, 50, 2847-2853 1986.

Presser, T. S., and I. Barnes, Special techniques for determining the chemical properties of geothermal waters, *U.S. Geol. Surv. Water Res. Invest. Rep. 22-74*, 11pp., 1974.

Rice, J. R., Fault stress states, pore pressure distributions, and the weakness of the San Andreas fault, in *Fault Mechanics and Transport Properties of Rocks*, edited by B. Evans, and T. –F. Wong T.-F., Academic Press, San Diego, pp. 475-503, 1992.

Robert, F., A. Boullier, and K. Firdaous, Gold-quartz veins in metamorphic terranes and their bearing on the role of fluids on faulting, *J. Geophys. Res., 100*, 12,861-12,880, 1995.

Saleeby, J. B., Polygenetic ophiolite belt of the California Sierra Nevada: Geochronological and tectonostratigraphic development, *J. Geophys. Res. 87*, 1803-1824, 1982.

Saleeby, J. B., M. C. Blake, and R. G. Coleman, Pb/U zircon ages on thrust plates of west-central Klamath Mountains and Coast Ranges, northern California and southern Oregon, *Am. Geophys. Union Trans., 65*, 1147, 1984.

Sass, J. H., C. F. Williams, A. H. Lachenbruch, S. P. Galanis, Jr., and F. V. Grubb, Thermal regime of the San Andreas fault near Parkfield, California, *J. Geophys. Res., 102*, 27,575-27,585, 1997.

Sibson, R. H., Fluid flow accompanying faulting: Field evidence and models, in Earthquake Prediction: An International Review, edited by D. W. Simpson and R. G. Richards, *Am. Geophys. Union, Maurice Ewing Series, 4*, 593-603, 1981.

Sibson, R. H., Earthquake faulting, induced fluid flow, and fault-hosted gold-quartz mineralization, *Proceedings of the International Conference on Basement Tectonics, 8*, 603-641, 1988.

Sibson, R. H., Implications of fault-valve behavior for rupture nucleation and recurrence, *Tectonophysics, 211*, 283-293, 1992.

Silver, L. T., E. W. James, and B. W. Chappell, Petrological and geochemical investigation at Cajon Pass deep drillhole, *Geophys. Res. Lett., 15*, 961-964, 1988.

Sleep, N. H., and M. L. Blanpied, Ductile creep and compaction: A mechanism for transiently increasing fluid pressure in mostly sealed fault zones, *Pure Appl. Geophys., 143*, 9-40, 1994.

Sorey, M. L., B. M. Kennedy, W. C. Evans, C. D. Farrar, and G. A. Sumnicht, Helium isotope and gas discharge variations associated with crustal unrest in Long Valley caldera, California, 1989-1992, *J. Geophys. Res., 98*, 15,871-15,889, 1993.

Thompson, J. M., and L. D. White, Chemical Analyses of Water from the GTA-1 Well, Parkfield, California, and Other Nearby Spring and Well Waters, *U.S. Geol. Surv. Open-File Rep. 91-0003*, 27pp., 1991.

Torgerson, T., and W. B. Clarke, Geochemical constraints on formation fluid gases, hydrothermal heat flux, and crustal mass transport mechanisms at Cajon Pass, *J. Geophys. Res., 97*, 5031-5038, 1992.

Trull, T., S. Nadeau, F. Pineau, M. Polve, and M. Javoy, C-He systematics in hotspot xenoliths: implications for mantle carbon contents and C recycling, *Earth Planet. Science Lett., 118*, 43-64, 1993.

Wallace, R. E., General features, in The San Andreas Fault System, edited by R. E. Wallace, *U.S. Geol. Surv. Prof. Pap. 1515*, 3-12, 1990.

Waring, G. A., Springs of California, *U.S. Geol. Surv. Water Supply Pap. 338*, 410pp., 1915.

Waring, G. A., Thermal Springs of the United States and Other Countries of the World – A Summary, *U.S. Geol. Surv. Prof. Pap. 492*, 383pp., 1965.

White, D. E., Magmatic, connate, and metamorphic waters, *Geol. Soc. Am. Bull., 68*, 1659-1682, 1957.

White, D. E., I. Barnes, and J. R. O'Neil, Thermal and mineral waters of nonmeteoric origin, California Coast Ranges, *Geol. Soc. Am. Bull., 84*, 547-560, 1973.

Wintsch, R. P., R. Christofferson, and A. K. Kronenberg, Fluid-rock reaction weakening of fault zones, *J. Geophys. Res., 100*, 13,021-13,032., 1995.

Yardley, B. W. D., Is there water in the deep continental crust? *Nature, 323*, 111, 1986.

Zoback, M. D., M. L. Zoback, V. S. Mount, and others, New evidence on the state of stress of the San Andreas fault system, *Science, 238*, 1105-1111, 1987.

Zoback, M. D., and J. H. Healy, In situ stress measurements to 3.5 km depth in the Cajon Pass scientific research borehole: implications for the mechanics of crustal faulting, *J. Geophys. Res., 97*, 5039-5057, 1992.

Y. Kharaka, J. Thordsen, and W. Evans, U.S. Geological Survey, Water Resources Division, 345 Middlefield Road, MS 427, Menlo Park, CA 94025. ykharaka@usgs.gov jthordsn@usgs.gov; wcevans@usgs.gov.

M. Kennedy, Lawrence Berkeley National Laboratory, Center for Isotope Geochemistry, MS 70A-3363, Berkeley, CA 94720. bmkennedy@lbl.gov

Solute-Sieving-Induced Calcite Precipitation on Pulverized Quartz Sand: Experimental Results and Implications for the Membrane Behavior of Fault Gouge

T.M. Whitworth

New Mexico Bureau of Mines and Mineral Resources, Socorro, New Mexico

W.C. Haneberg[1]

New Mexico Bureau of Mines and Mineral Resources, Albuquerque, New Mexico

P.S. Mozley and L.B. Goodwin

Department of Earth and Environmental Science, New Mexico Tech, Socorro, New Mexico

We report on a preliminary experiment with an analog for fault gouge composed of clay-sized quartz particles. An undersaturated calcium carbonate solution was forced through a layer of clay-sized quartz particles. Calcite crystals, identified by secondary electron imaging and energy dispersive X-ray analysis, formed on the gouge. Calcite precipitation was not due to changes in temperature or pH related to ion exchange. A rise in Ca and bicarbonate concentrations on the high pressure side of the membrane is the most likely cause of calcite precipitation. Solute-sieving by clay-rich sediments has been previously described. Our results suggest that the solute-sieving properties of fault gouge should be considered as a possible mechanism for the selective cementation of some faults even when the gouge has a low clay mineral content. The formation of calcite cements in one fault zone in central New Mexico may be explained by solute-sieving by fault gouge. This zone is strongly preferentially cemented by calcite relative to the surrounding materials, yet oriented concretions within the cemented zone suggest that the groundwater flow was nearly perpendicular to the fault at the time of cementation.

---

[1]Now at Haneberg Geoscience, Port Orchard, Washington

INTRODUCTION

Faults can act as barriers, conduits, or complex barrier-conduit systems with respect to fluid flow [*Caine et al.*, 1995]. Cementation can greatly reduce fault-zone permeability [*Knipe*, 1993], which can in turn reduce the ability of a fault zone to act as a conduit for flow. In extreme cases, cemented faults may act as capillary or absolute barriers to flow, resulting in compartmentalization of groundwater basins and hydrocarbon reservoirs [*Maclay and Small*, 1983; *Harding and Tuminas*, 1989; and *Edwards, et al.*, 1993]. Faults in unconsolidated sediments may act as barriers or partial barriers to groundwater flow for several reasons [*Davis and DeWeist*, 1966]: grain-size reduction of wall rock material; realignment of elongated clasts to reduce fault-normal permeability; juxtaposition of saturated strata of differing permeability; and cementation of the fault zone. *Knipe* [1993] recast these mechanisms in terms of permeability reduction modes along faults: 1) porosity collapse seals, which include permeability reduction due to both grain crushing and realignment; 2) pore-filling cement seals; and 3) juxtaposition seals.

Some fault zones are preferentially cemented with respect to the surrounding rocks [e. g. *Flournoy and Ferrell*, 1980; *Burley et al.*, 1989; *Knipe*, 1993; *Mozley and Goodwin*, 1995]. This cementation is generally thought to result from preferential flow of the cementing fluids within the fault zone [*Mozley and Goodwin*, 1995]. In this paper we suggest another possible mechanism for the preferential cementation of some fault zones—solute-sieving in which the fault gouge acts as a semipermeable membrane, even in the absence of clay minerals.

GEOLOGIC MEMBRANES

A semi-permeable membrane is defined as a material that will permit the passage of some molecules but not others [*Noggle*, 1984]. A perfect semi-permeable membrane would completely stop the passage of certain molecular species. Such perfect membranes probably do not exist in nature [*Fritz*, 1986]. Therefore, a better working definition for a semi-permeable geological membrane is any lithology which retards one solution component more effectively than another.

Studies have experimentally confirmed the ability of clays (composed of clay minerals) to act as semi-permeable membranes [*Whitworth and Fritz*, 1994]. Faults in sand and sandstone commonly contain cataclastically deformed quartz and feldspar grains [*Aydin*, 1978; *Antonellini and Aydin*, 1994, 1995; *Goodwin and Haneberg*, 1996], and clay-sized particles, although not necessarily clay minerals, are common in fault gouge. Our experiment tests the ability of clay-sized particles of quartz, which lack the significant electrical double-layer properties common to many clay minerals, to act as a membrane. If clay-sized quartz particles are capable of acting as a membrane, then many fault gouges, which commonly contain both clay-sized particles and clay minerals, should be capable of functioning as membranes.

The flow of solution and solute through membranes is described by:

$$J_v = L_p(\Delta P - \sigma \Delta \pi) \tag{1}$$

and

$$J_s = \bar{c}_s(1 - \sigma)J_v + \omega \Delta \pi \tag{2}$$

Where J_v = solution flux (m/s) through the membrane, L_p = water permeation coefficient (m^3/Pa·s), ΔP = pressure difference across the membrane (Pa), σ = reflection coefficient (dimensionless), $\Delta \pi$ = theoretical osmotic pressure difference across the membrane (Pa), J_s = solute flux (mole/m^2·s) through the membrane, ω = solute permeation coefficient (mole/N·s), and $\bar{c}_s$ = average solute concentration in mole/cm^3 across the membrane where $\bar{c}_s = (c_o + c_e)/2$, c_o = concentration at the high-pressure membrane face (M) and c_e = effluent concentration [M; *Kedem and Katchalsky*, 1962].

The equation for $\Delta \pi$ for a dilute solute [*Fritz*, 1986] is

$$\Delta \pi = \nu RT(c_o - c_e) \tag{3}$$

Where ν is a factor that corrects for the number of particles due to ion formation. For example, since NaCl forms two ions in solution, Na$^+$ and Cl$^-$, then for NaCl, $\nu = 2$. However, for CaCl$_2$, which forms one Ca^{++} ion and two Cl$^-$ ions for each molecule of CaCl$_2$, $\nu = 3$. In Equation 3, R is the gas constant (0.082 bar/mole·K) and T is the temperature in K.

Fritz [1986] suggested that three of the phenomenological coefficients of *Kedem and Katchalsky* [1962]—σ, ω, and L_p—are useful in describing the behavior of non-ideal, geologic membrane systems. *Fritz and Marine* [1983] state that σ is important because it is a measure of the osmotic efficiency of a membrane. Thus, a membrane with a $\sigma = 0.90$ would exhibit 90 percent of the theoretically predicted osmotic pressure. Values of σ range from zero to one. If $\sigma = 0$, there is no membrane effect. In this case equation 1 reduces to $J_v = L_p \Delta P$, which is equivalent to a special case of Darcy's Law for zero elevation head gradient. If $\sigma = 1$, the membrane is ideal and no solute can pass through the membrane. The value of σ for non-ideal geological membranes must be greater than zero, but less than one. *Fritz and Marine* [1983] calculated values of σ for a series of six experiments using different NaCl solutions and bentonite clay membranes compacted to different porosities. The determined values of σ ranged from 0.04 to 0.89.

The solute permeation coefficient ω describes the diffusion of solute through the membrane. For ideal membranes, ω = 0 because no solute can pass through the membrane. For non-ideal geologic membranes ω should be greater than zero. *Elrick et al.* [1976] measured ω for a Na-bentonite slurry with 90 percent porosity and obtained a value of 3 x 10^{-10} mole/N·s. *Fritz and Marine* [1983] suggested that for more compacted bentonites, the value of ω should be considerably lower than 3 x 10^{-10} mole/N·s.

The water permeation coefficient (L_p) is a measure of the membrane's permeability to water and is thus a form of permeability measurement.

METHODS

To test the efficacy of clay-sized quartz particles as a semi-permeable membrane, a calcite-free, quartz sand (Ottawa Sand) was ground in a percussion mill. X-ray analysis shows this sand to contain no clay minerals and to be almost 100% quartz. To assure that the sand contained no calcite, it was soaked in dilute hydrochloric acid, and then rinsed with deionized water several times. The 0.4 μm and smaller particles were separated and then sedimented in an acrylic experimental cell in deionized water to form an artificial, 0.2-cm thick, fault gouge. The permeability of this ground quartz was determined at 0.25 millidarcies by measuring the volume of deionized water which passed through it at a known pressure in a given time. Published fault gouge permeability values typically range from a high of approximately 10 millidarcies to a low of approximately 1 x 10^{-5} millidarcies or less [*Morrow et al.*, 1981; *Morrow et al.*, 1984; *Antonellini and Aydin*, 1994]. Thus, the synthetic material used in this experiment had a high-range permeability for fault gouge.

The experimental hyperfiltration cell is the same design used by *Fritz and Whitworth* [1994] and is capable of operating at a maximum working pressure of 6.89 x 10^6 Pa (1000 psi). An ISCO Model 500D syringe pump was used to force the solutions through the clay-sized quartz particles at a constant solution flux (Fig. 1). Once the synthetic fault gouge was sedimented in place and its permeability to deionized water measured, the syringe pump was flushed with calcium carbonate stock solution at 22% saturation (pH = 7.82) four times, and the deionized water was decanted from the cell and replaced with the stock solution. The stock solution was then forced through the two-mm thick

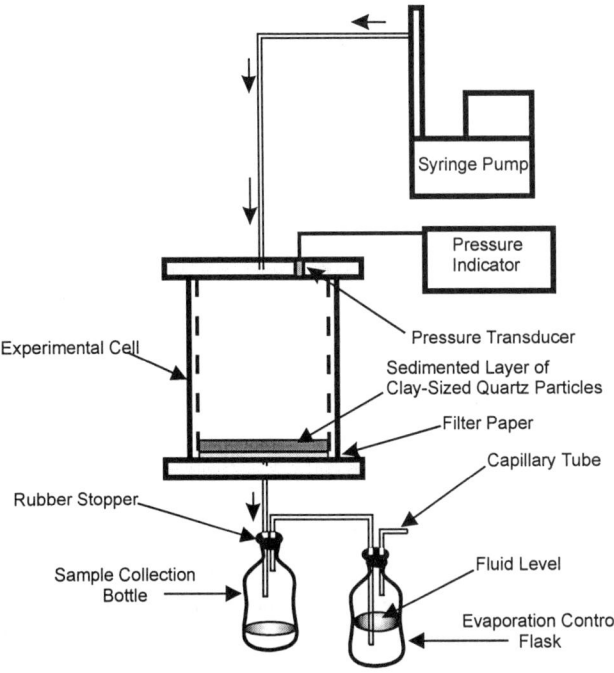

Figure 1. Experimental setup used in this study. This is the same configuration used by Fritz and Whitworth [1994].

layer of crushed quartz at a flow rate of 500 ml/hour. During the experiment, effluent samples were periodically collected and analyzed. The pH was measured for each effluent aliquot. Alkalinities were determined by titration, and calcium concentrations were determined by ion chromatography. At the end of the experiment, the cell solution was decanted and analyzed and the quartz filter cake was repeatedly washed in ethanol to displace water, dried, and then examined by secondary electron imaging and energy dispersive X-ray analysis.

RESULTS AND DISCUSSION

In the initial effluent samples, calcium concentrations were reduced by as much as 44% compared to that of the input solution (Fig. 2a). When solute-sieving occurs, a portion of the solute is retarded by the membrane, with the result that the solute concentration on the high-pressure side of the membrane increases and forms a concentration polarization layer (CPL) (Fig. 3). Typically, as time passes, assuming that no precipitation or other chemical reactions occur in the CPL, the concentration of the effluent increases until a steady state equilibrium is reached in which the effluent concentration is equal to that of the input concentration c_i

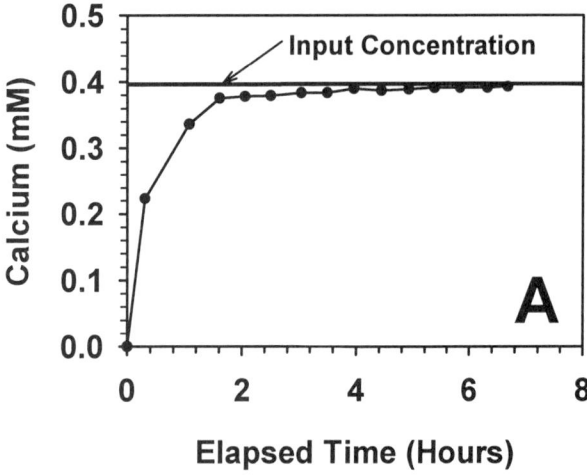 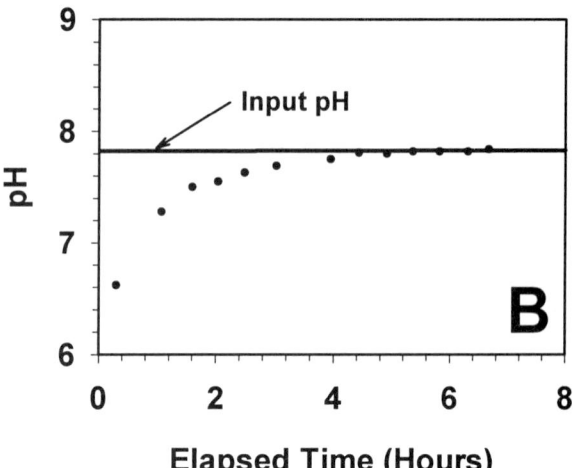

Figure 2. Plot of calcium concentration in the effluent versus time of collection (a). The black horizontal line represents the input concentration. The two-standard-deviation precision of these analyses is 0.4 percent. Note that the initial effluent concentrations are significantly less than the input concentration. This is attributed to solute-sieving by the synthetic fault gouge. The graph on the right (b) is a plot of effluent pH versus collection time. The horizontal black line represents the pH of the input solution. Note that the initial effluent samples are more acidic than the input solution. This is attributed to exchange of calcium for hydronium ions on the exchange sites of the crushed sand. The hydronium ions were adsorbed onto the particles when first exposed to deionized water and then desorbed when exposed to calcium ions. Mass balance calculations indicate that only a small fraction of one percent of the calcium ions were involved. Therefore, ion exchange effects are insignificant with respect to calcium concentrations.

[*Fritz*, 1986; Fig. 3]. Our experimental data (Fig. 2a) suggest this relationship, confirming the possibility that solute-sieving was occurring.

Alternatively, the observed reduction in effluent concentrations might have been due to exchange of calcium for hydronium ions adsorbed to the newly-formed, negatively-charged exchange sites on the crushed quartz particles. A simple beaker experiment with crushed Ottawa sand demonstrated that when exposed to deionized water (as our fault gouge proxy was before it was exposed to the undersaturated calcium carbonate solution) the pH of the stirred deionized water/crushed sand mixture (8.68) is more alkaline than the original deionized water (6.71). This suggests that hydronium ions are adsorbed by the newly formed exchange sites. The initial effluent samples had more acidic pH values than either the stock solution (represented by the horizontal black line in Fig. 2b) or later effluent samples. This suggests that some calcium ions were replacing hydronium ions on exchange sites in the synthetic gouge. Mass balance calculations demonstrated that the calcium for hydronium ion exchange during the experiment was only a small fraction of one percent and thus negligible.

Temperature or pH of a solution can influence calcite precipitation. Temperature was held constant during our experiment and did not influence precipitation. In addition,

desorption of hydronium ions occupying the exchange sites on the quartz grains could only lower the pH of solution. Therefore, ion exchange at the higher-pressure surface of the clay-sized quartz layer would induce calcite dissolution rather than precipitation. Therefore, the small (10-100 μm) calcite crystals observed on the surface of the gouge (Fig. 4) are thought to have precipitated as a result of the solute-sieving properties of the synthetic fault gouge.

We now attempt to determine values of σ, ω, and L_p for our experiment, so that our results can be compared to other published geologic membrane experiments [*Fritz and Marine*, 1983; *Fritz and Whitworth*, 1994; *Whitworth and Fritz*, 1994]. L_p is calculated from the experimental results using

$$L_P = J_v / L_p \qquad (4)$$

where the measured J_v (for deionized water) was 6.85×10^{-1} m/s, ΔP was 5.654×10^5 Pa, and L_p is 1.21×10^{-6} m³/Pa·s.

Calculating experimental values of σ and ω is usually done by either the steady state or transient solutions presented by *Fritz and Whitworth* [1994]; however, their methods require accurate knowledge of the concentration within the cell at the end of the experiment. The width of the CPL is approximately equivalent to $10D/J_v$, where D is

$$\omega = D / RT\Delta x\zeta \qquad (7)$$

$$c_e = c_i \qquad (8)$$

where Δx is membrane thickness in m and ζ is tortuosity, the ratio of the actual path length through the membrane to the membrane thickness (Note: the definition of tortuosity used here is slightly different than the commonly used definition where tortuosity equals the square of the ratio of the straight line distance to the actual path length as defined by *Bear* [1972]), into Equations 1 and 2 we obtain

$$c_o = \frac{-c_i(J_v\Delta x\zeta(1+\sigma)+2Dv)}{J_v\Delta x\zeta(\sigma-1)-2Dv} \qquad (9)$$

And by substituting Equations 3 and 8 into Equation 1 we obtain

$$J_v = L_p(\Delta P - \sigma vRT(c_o - c_i)) \qquad (10)$$

Equations 9 and 10 form a pair of simultaneous equations with two unknowns (σ and c_o). When solved for c_o, the solution is a polynomial with two roots—one always negative and one always positive.

$$
\begin{aligned}
c_o = &\frac{1}{2L_pvRT(J_v\Delta x - 2vD)} \cdot \\
&(-2L_pvRTJ_vc_i\Delta x\zeta - 4L_pv^2RTDc_i \\
&+ \Delta x\zeta J_v L_p\Delta P - \Delta x^{1/2}\zeta^{1/2} \\
&(-8\zeta\Delta xc_iJ_v^3TRvL_p + 8\zeta\Delta xc_iJ_v^2TRvL_p^2 \\
&+ \zeta\Delta xJ_v^4 - 2\zeta\Delta x\Delta PL_pJ_v^3 \\
&+ \zeta\Delta x\Delta P^2L_p^2J_v^2 - 16J_v^2c_iDTRv^2L_p^2)^{1/2})
\end{aligned} \qquad (11)
$$

Because negative values of c_o have no physical meaning, the positive root (Eqn. 11) is chosen. Equation 11 reproduces within 2.4 percent the values of c_o calculated by *Fritz and Whitworth* [1994] for their LiCl experiment by their transient solution; a difference within experimental error.

This system of equations is not sensitive to values of ζ, which can vary as much as one order of magnitude and change the calculated value of c_o by less than 0.02 percent. Therefore, we can use a representative value for ζ determined from previous studies. Values of measured ζ for the Bison mudstone and for the Queenston shale are from $5.0 \le \zeta \le 6.7$ and from $5.0 \le \zeta \le 10$, respectively [*Barone et al.*, 1992; *Barone et al.*, 1990] for an average value of ζ of about 7.0. We will assume a value of $\zeta = 7.0$ for our experimental membrane.

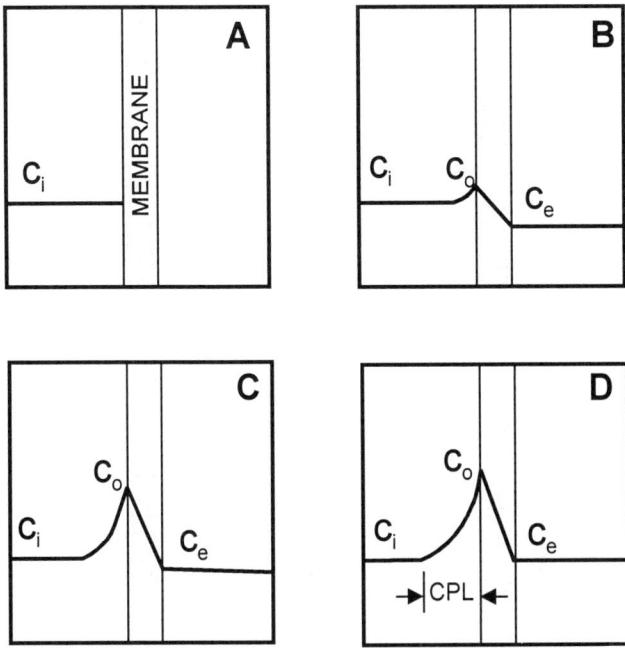

Figure 3. Conceptual development of the concentration polarization layer. The initial conditions (a) are such that the solute is all on the high pressure side of the membrane and the pore fluids within the membrane contain no solute. Some time after solute flux through the membrane begins (b), the concentration at the high pressure membrane face c_o increases because some of the solute is rejected by the membrane. Some solute also begins to pass through the membrane so that the effluent now contains some solute as well. Even later (c) c_o has increased further as has the effluent concentration c_e. If no precipitation or other chemical reactions are occurring, at steady-state (d) the input concentration c_i is now equal to c_e and the value of c_o is constant (Redrawn from Fritz and Marine, 1983).

the diffusion coefficient in m^2/s [*Fritz and Whitworth*, 1994]. For our experiment, the width of the CPL was approximately 0.01 cm. Consequently, the overall concentration increase in the 12.7 cm long cell was less than the two standard deviation precision of the analytical measurements ($\pm$ 0.4% for Ca^{++} analyses, and $\pm$ 2.5% for alkalinity). Therefore, we will develop a steady state solution which does not require *a priori* knowledge of the average concentration in the experimental cell and thus is more applicable to modeling membrane effects in real aquifers.

By substituting Equation 3 and the following steady state relationships

$$\overline{c}_s = (c_o + c_e)/2 \qquad (5)$$

$$J_s = J_vc_e \qquad (6)$$

Figure 4a. Secondary electron image of calcite crystals on synthetic fault gouge membrane.

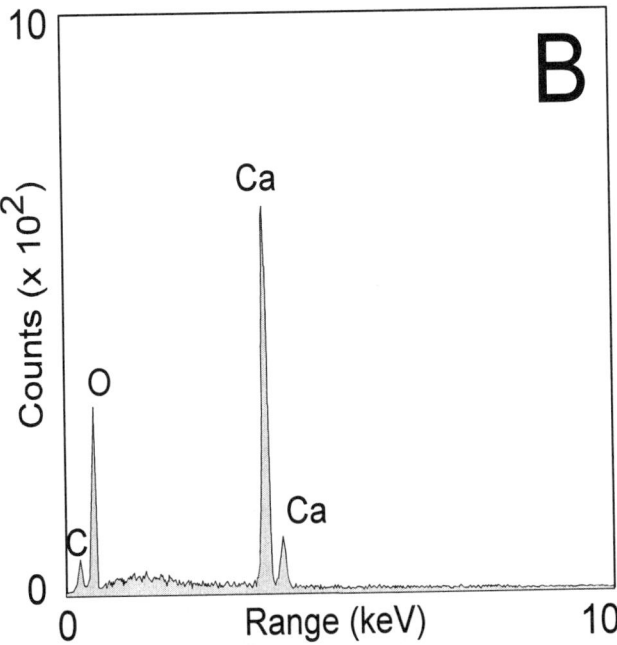

Figure 4b. Energy dispersive scan of crystal shown in figure 4a that shows the composition to be calcite.

Using Equation 11 with $\zeta = 7.0$, $D = 8.5 \times 10^{-10}$ m²/s as calculated from the Nernst relation [*Lide*, 1990], $v = 2$, $J_v = 6.85 \times 10^{-1}$ m/s, $\Delta P = 8.825 \times 10^5$ Pa, $\Delta x = 0.02$ m, and $c_i = 3.85 \times 10^{-4}$ M, we calculate that the steady state value of c_o for the synthetic fault gouge experiment was 6.51×10^{-2} M. Thus, theoretical calculations suggest that calcium carbonate was extremely oversaturated at the membrane face (log IAP/KT = 3.00), or would have been if no precipitation occurred. All saturation calculations were done with the geochemical reaction model PHREEQE [*Parkhurst et al.*, 1993].

We then substitute Equation 3 into Equation 1, substitute c_i for c_e and solve for σ

$$\sigma = L_p \Delta P - J_v / L_p vRT(c_o - c_i) \qquad (12)$$

The steady state value of the reflection coefficient for our experiment is then calculated from Equation 12 as 0.99, demonstrating that our crushed quartz layer is an efficient membrane. The steady state value of ω is then calculated by Equation 7 to be 2.5×10^{-11} mole/N·s. These values are calculated based on the assumption that there is no precipitation, so they are not exact. However, although calcite precipitation rates depend on the state of saturation, calcite precipitation kinetics are typically quite slow [*Appelo and Postma*, 1993] with respect to the length of our experiment,

so we suspect error due to calcite precipitation may not be large. The results of these calculations suggest the material used in our experiment, if similar to fault gouge, may cause calcite cementation on the high-pressure side of fault gouge.

An important question is what head differences are required across the fault gouge to cause solute-sieving? If high head differences are required, yet are uncommon across faults, then solute-sieving effects along faults may be rare. But, if the required head differences are low, then solute-sieving may be a common cementation mechanism along faults in the saturated zone, especially in systems where faults offset the water table to generate head differences.

To examine this question, we need to derive two equations and make a few reasonable assumptions. To begin we substitute Equations 3, 4, 5, 6, 7, and 8 into Equation 2 to derive the following steady state equation for J_v

$$J_v = \frac{2c_o Dv - 2c_i Dv}{-c_o \Delta x \zeta + c_o \sigma \Delta x \zeta + c_i \Delta x \zeta + c_i \sigma \Delta x \zeta} \qquad (13)$$

Now by solving Equation 12 for ΔP

$$\Delta P = \frac{J_v + L_p \sigma vRT(c_o - c_i)}{L_p} \qquad (14)$$

we will be able to calculate the head difference across the membrane required to achieve some arbitrary, steady state value of c_o. Consider the example where $\Delta x = 1$ cm, $\sigma = 0.25$ (representing a relatively inefficient membrane), the water is 78 percent saturated with calcium carbonate ($c_i = 3.0 \times 10^{-3}$ M) at a pH of 7.00, and $L_p = 1.2 \times 10^{-10}$ m³/N·s, a mid-range value for fault gouge. We want c_o to equal 1.2 c_i, or 3.6×10^{-3} M, which yields a 107 percent calcite saturation at the membrane face. The calculated ΔP necessary to produce our desired c_o is 776 Pa or an equivalent head across the membrane of only 0.064 m. In semi-arid areas, where shallow groundwater is often close to calcite saturation [*Mozley et al.*, 1995], even a small fault displacement might generate sufficient head difference across the fault to begin the solute-sieving process. Under conditions of lesser calcite saturation, greater head differences would obviously be required. If we consider another example using the same parameters as above, except that the value of c_i is 1.6×10^{-3} M (13 percent saturated), and we want our c_o to be 3.0 times c_i, we find that the required head difference across the membrane is 0.322 m, which produces a calcite saturation of 176 percent at the membrane face. Our preliminary conclusion is that high head differences across faults are not required to cause solute-sieving by fault gouge and resultant

Figure 5. Elongate concretions in the Santa Anna fault zone in central New Mexico indicate that groundwater flow was subhorizontal and normal to the fault at the time of calcite precipitation. The orientation of concretions in the fault zone is subparallel to that of elongate concretions immediately outside the fault zone in undeformed lower Santa Fe Group sands. The arrow indicates the inferred direction of groundwater flow.

cementation due to calcite precipitation. Of course, a natural head difference is maintained across many faults because they act as barriers to flow.

In order for CPL development to occur as described in Figure 3, there must be flow across the fault through membrane-functioning gouge. At least one fault zone in central New Mexico provides field evidence that cross-fault flows existed at the time of calcite precipitation. This zone is strongly preferentially cemented by calcite, yet oriented concretions within the zone indicate that the flow was nearly perpendicular to the fault at the time of cementation (Fig. 5; See *Mozley and Goodwin* [1995] for a discussion of the significance of such concretions). In addition, another study [*Edwards et al.,* 1993] found that some faults have asymmetric cementation. Asymmetric cementation is compatible with cementation due to solute-sieving during cross-fault flow because, under these conditions, only one side of

the gouge membrane will exhibit supersaturated calcite concentrations.

However, flow is primarily parallel to faults in many instances [e.g. *Knipe*, 1993]. If the gouge is capable of acting as a membrane a CPL will form, although it will be diminished to some extent by the fault-parallel component of flow. Commercial reverse osmosis desalination units are designed so that the membrane is swept by high velocity, turbulent flow. In these units, CPL development is minimized but not completely eliminated [*Mariñas and Selleck,* 1992]. Consequently, it is unlikely that fault-parallel groundwater flow could completely eliminate the CPL and it might be possible for some calcite to precipitate within the CPL adjacent to the gouge, even during fault parallel flow. As long as there is a pressure differential across the fault, there will be a component of cross-fault flow, even if much of the flow is parallel to the fault. Even a minute amount of membrane-induced calcite precipitation could result in strong preferential cementation of a fault zone by providing nucleation sites for later cementation (i.e., calcite precipitation typically will occur preferentially on a preexisting carbonate substrate). This is particularly important in siliclastic sediments that do not contain significant carbonate detritus.

CONCLUSIONS

This experimental study suggests that fault gouges containing little or no clay minerals may have membrane properties, and that some calcite cementation associated with faults may be due to the solute-sieving properties of the fault gouge. Example calculations suggest that head differences across the fault gouge of a few centimeters to a few tens of centimeters may be sufficient to drive this process.

Acknowledgments.   We are indebted to Dennis Romero and Gina DeRosa for their assistance in the lab and to Mike Spilde for operating the microprobe and obtaining the SEM images. The manuscript was considerably improved by the comments and suggestions of Joris Gieskes and an anonymous reviewer. We are grateful to John Hawley and David Love for showing us numerous cemented faults in outcrop and for many informative discussions on this topic.

REFERENCES

Allan, U. S. Model for hydrocarbon migration and entrapment within faulted structures, *Am. Assoc. Petrol. Geol. Bull., 73*, 803-811, 1989.

Antonellini, M. and Aydin, A. Effect of faulting on fluid flow in porous sandstones: petrophysical properties, *Am. Assoc. Petrol. Geol. Bull., 78*, 355-377, 1994

Antonellini, M. and A. Aydin. Effect of faulting on fluid flow in porous sandstones: geometry and spatial distribution, *Am. Assoc. Petrol. Geol. Bull., 79*, 642-671, 1995.

Appelo, C. A. J., and Postma, D. *Geochemistry, groundwater, and pollution*, A. A. Balkema, Rotterdam, 536 p., 1993.

Aydin, A. Small faults formed as deformation bands in sandstone, *Pure Appl. Geoph., 116*, 913-930, 1978.

Barone, F. S., Rowe, R. K., and Quigley, R. M. Estimation of chloride diffusion coefficient and tortuosity factor for mudstone, *J. Geotech. Engrg., 118*, 1031-1047, 1992.

Barone, F. S., Rowe, R. K., and Quigley, R. M. Laboratory determination of chloride diffusion coefficient in intact shale, *Canadian Geotech. J., 27.*, 177-184, 1990.

Bear, J. *Dynamics of Fluids in Porous Media*, Dover Publications, New York, 1972.

Burley, S.D., Mullis, J., and Matter, A. Timing of diagenesis in the Tartan Reservoir (UK North Sea): Constraints from combined cathodoluminescence microscopy and fluid inclusion studies, *Mar. and Petrol. Geol., 6*, pp. 98-120, 1989.

Caine, J.S., J.P Evans, and C.B. Forster. Fault zone architecture and permeability structure, *Geology, 24*, 1025-1028, 1996.

Dahlberg, E. C. *Applied hydrodynamics in petroleum exploration*, Springer Verlag, New York, 1994.

Davis, S. N. and R.J.M. DeWiest. *Hydrogeology*, John Wiley & Sons, Inc., New York, 1966.

Edwards, H.E., A.D. Becker, and J A Howell. Compartmentalization of an aeolian sandstone by structural heterogeneities: Permo-Triassic Hopeman Sandstone, Moray Firth, Scotland, in *Characterization of Fluvial and Aeolian Reservoirs*, edited by C.P. North and D.J. Prosser, pp. 339-365, Geological Society of America Special Publication No 73, 1993.

Elrick, D. E., D.E. Smiles, N. Baumgartner, and P.H. Groenvelt. Coupling phenomena in saturated homo-ionic montmorillionite: I. Experimental, *J. Soil Sci. Soc. Am., 40*, 490-491, 1976.

Flournoy, L.A., and R.E. Ferrell. Geopressure and diagenetic modifications of porosity in the Lirette field area, Terrebonne Parish, Louisiana, *Gulf Coast Geol. Assoc. Trans., 30*, 341-345, 1980.

Fritz, S. J. Ideality of clay membranes in osmotic processes: a review, *Clays and Clay Minerals, 34*, 214-223, 1986.

Fritz, S. J. and C.D. Eady. Hyperfiltration-induced precipitation of calcite, *Geochim. Cosmochim. Acta, 49*, 761-768, 1985.

Fritz, S. J. and I.W. Marine. Experimental support for a predictive osmotic model of clay membranes, *Geochim. Cosmochim. Acta, 47*, 1515-1522, 1983.

Fritz, S. J. and T.M. Whitworth. Hyperfiltration-induced fractionation of lithium isotopes: ramifications relating to representativeness of aquifer sampling, *Water Resour. Res., 30*, 225-235, 1994.

Goodwin, L.B. and W.C. Haneberg. Deformational fabrics and inferred permeability of faulted sands from the Rio Grande rift, New Mexico, *Geol. Soc. Am. Abstr. Prog., 28*, A-255, 1996.

Harding, T. P. and A. C. Tuminas, Structural interpretation of hydrocarbon traps sealed by basement normal blocks and at stable flank of foredeep basins and at rift basins, *Am. Assoc. Petrol. Geol. Bull., 73*, 812-840, 1989.

Katchalsky, A. and P.F. Curran. *Biophysics*, Harvard University Press, Cambridge, 1965.

Kedem, O. and A. Katchalsky. A physical interpretation of the phenomenological coefficients of membrane permeability, *J. Gen. Physiol., 45.*, 143-179, 1962.

Knipe, R.J., The influence of fault zone processes and diagenesis on fluid flow, in *Diagenesis and Basin Development*, edited by A.D. Horbury and A.G. Robinson, pp. 135-151, Am. Assoc. Petrol. Geol. Studies in Geology No. 36, 1993.

Lide, D. R. editor., *CRC Handbook of Chemistry and Physics*, CRC Press, Inc., 1990.

Maclay, R. W. and T. A. Small. Hydrostratigraphic subdivisions and fault barriers of the Edwards aquifer, south-central Texas, U.S.A., *J. Hydrol., 61*, 127-146, 1983.

Marinas, B. L. and R. E. Selleck. Reverse osmosis treatment of multicomponent electrolyte solutions, *J. Membrane Sci., 72*, 211-229, 1992.

Morrow, C. A., L.Q. Shi, and J. Byerlee. Permeability of fault gouge under confining pressure and shear stress, *J. Geophys. Res., 89*, 3193-3200, 1984.

Morrow, C., L.Q. Shi, and J. Byerlee. Permeability and strength of San Andreas fault gouge under high pressure, *Geoph. Res. Lett., 8*, pp. 325-328, 1981.

Mozley, P.S., and L.B. Goodwin. Patterns of cementation along a Cenozoic normal fault: A record of paleoflow orientations, *Geology*, 23, 539-542, 1995.

Mozley, P. S., J. Beckner, and T.M. Whitworth. Spatial distribution of calcite cement in the Santa Fe Group, Albuquerque Basin, NM: Implications for groundwater resources, *New Mexico Geol.*, 17, 88-93, 1995.

Noggle, J. H. *Physical Chemistry*, Little, Brown & Company Limited, 1984.

Parkhurst, D. L., D.C. Thorstenson, and L.N. Plummer, *PHREEQE, a geochemical reaction model based on an ion pairing aqueous model*, Institute for Ground-Water Research and Education, Colorado School of Mines, Golden CO., 1993.

Whitworth, T. M. and S.J. Fritz. Electrolyte-induced solute permeability effects in compacted smectite membranes, *Appl. Geochem.*, 9, 533-546, 1994.

T.M. Whitworth, New Mexico Bureau of Mines and Mineral Resources, New Mexico Tech, 801 Leroy Place, Socorro NM 87801-4796 (e-mail: mikew@nmt.edu)

W.C. Haneberg, Haneberg Geoscience, 10411 SE Olympiad Drive, Port Orchard WA 98366, (e-mail: bill@haneberg.com)

P.S. Mozley, Department of Earth and Environmental Science, New Mexico Tech, 801 Leroy Place, Socorro NM 87801 (e-mail: mozley@nmt.edu)

Laurel B. Goodwin, Department of Earth and Environmental Science, New Mexico Tech, 801 Leroy Place, Socorro NM 87801 (e-mail: lgoodwin@nmt.edu)

Flow-Path Textures and Mineralogy in Tuffs of the Unsaturated Zone

Schön Levy, Steve Chipera, and Giday WoldeGabriel

EES-1, Los Alamos National Laboratory, Los Alamos, New Mexico

June Fabryka-Martin and Jeffrey Roach

CST-7, Los Alamos National Laboratory, Los Alamos, New Mexico

Donald Sweetkind

U. S. Geological Survey, Denver, Colorado

The high concentration of chlorine-36 (^{36}Cl) produced by above-ground nuclear tests (bomb-pulse) provides a fortuitous tracer for infiltration during the last 50 years, and is used to detect fast flow in the unsaturated zone at Yucca Mountain, Nevada, a thick deposit of welded and nonwelded tuffs. Evidence of fast flow as much as 300 m into the mountain has been found in several zones in a 7.7-km tunnel. Many zones are associated with faults that provide continuous fracture flow paths from the surface. In the Sundance fault zone, water with the bomb-pulse signature has moved into subsidiary fractures and breccia zones. We found no highly distinctive mineralogic associations of fault and fracture samples containing bomb-pulse ^{36}Cl. Bomb-pulse sites are slightly more likely to have calcite deposits than are non-bomb-pulse sites. Most other mineralogic and textural associations of fast-flow paths reflect the structural processes leading to locally enhanced permeability rather than the effects of ground-water percolation. Water movement through the rock was investigated by isotopic analysis of paired samples representing breccia zones and fractured wall rock bounding the breccia zones. Where bomb-pulse ^{36}Cl is present, the waters in bounding fractures and intergranular pores of the fast pathways are not in equilibrium with respect to the isotopic signal. In structural domains that have experienced extensional deformation, fluid flow within a breccia is equivalent to matrix flow in a particulate rock, whereas true fracture flow occurs along the boundaries of a breccia zone. Where shearing predominated over extension, the boundary

Faults and Subsurface Fluid Flow in the Shallow Crust
Geophysical Monograph 113
This paper is not subject to U.S. copyright
Published in 1999 by the American Geophysical Union

between wall rock and breccia is rough and irregular with a tight wall-rock/breccia contact. The absence of a gap between the breccia and the wall rock helps maintain fluid flow within the breccia instead of along the wall-rock/breccia boundary, leading to higher ^{36}Cl/Cl values in the breccia than in the wall rock.

INTRODUCTION

The faulted and fractured tuffs of Yucca Mountain, Nevada, have been the subject of intensive study because the mountain is a potential site for a high-level nuclear waste repository. In such a repository, situated in the unsaturated zone about 300 m below the ground surface, the nuclear waste packages would be subject to interaction with percolating ground water. Bounding values must be set on the percolation rates, and fast flow paths, preferential flow paths, and diversionary structures must be adequately represented in the conceptual model of groundwater flow. Surface studies and surface-based drill-hole studies have been supplemented by integrated hydrologic, structural, and mineralogic/textural research conducted in an underground complex, the Exploratory Studies Facility (ESF, Figure 1). This 7.7-km tunnel as much as 300 m beneath the surface of Yucca Mountain provides opportunities to study the role of faults and fractures in unsaturated-zone fluid flow, as well as the mineralogic associations of these transmissive features. In particular, we would like to investigate whether fast-flow paths have distinctive mineralogic or textural attributes. By documenting links between the development of fault and fracture flow paths, the secondary minerals deposited along those paths, and the hydrologic behavior of the paths, we should improve our ability to interpret fracture mineralogy as an indicator of fluid-flow regimes.

Chlorine-36 investigations of infiltration are part of the hydrologic characterization program at Yucca Mountain. The introduction of bomb-pulse ^{36}Cl into infiltrating water provides a fortuitous tracer whose ^{36}Cl/Cl isotopic value is significantly higher than the pre-1950's values of natural infiltration and percolating ground water. Detection of bomb-pulse ^{36}Cl in the subsurface is the primary documentation for the existence of a fast fluid pathway, defined for this study as a transmissive feature that has received at least a small component of water that traveled from the surface in the last ~50 years or less. Most such pathways in the ESF are fractures, shears, breccia zones, or faults with direct or indirect connections to major throughgoing faults. Because active fast pathways have received percolation from the surface within the past ~50 years, and because the isotopic composition of the percolation they receive has changed significantly over this short period due to input of bomb-pulse ^{36}Cl, isotopic disequilibrium within the fluid pathway is readily de-tectable. In areas of the ESF where ground-water travel times may be of the order of 10^3 to 10^4 years, the changes of isotopic composition in the infiltrating water moving through these rocks probably occurred so gradually that isotopic disequilibrium between waters in breccia and in wall rock, for example, would not be detectable by a comparison of their ^{36}Cl/Cl values.

GEOLOGIC SETTING OF YUCCA MOUNTAIN

Yucca Mountain, located within the Southwest Nevada Volcanic Field, is a 1200-m-thick accumulation of silicic tuffs ranging in age from 14 to 11 Ma [*Sawyer et al.*, 1994]. The mountain is made up of interlayered deposits of densely welded and nonwelded tuffs, with highly fractured welded tuffs exposed at the surface. Located within the extensional Basin and Range Province, the study site is cut by block-bounding normal faults with tens to hundreds of meters of vertical displacement and by intrablock faults with as much as tens of meters of displacement.

The lithostratigraphy of Yucca Mountain, based on surface mapping and drill-hole data [*Buesch et al.*, 1996], is shown in Table 1. The silicic tuffs exposed in the ESF and comprising the overlying rock column include the mostly welded Tiva Canyon Tuff, a sequence of five mostly non-welded formal and informal units, and the mostly welded Topopah Spring Tuff, in order of increasing age and depth. The PTn hydrogeologic unit, used in conceptual and numerical models of site hydrology, consists of the mostly nonwelded units plus immediately underlying and overlying nonwelded rocks of the lower Tiva Canyon and upper Topopah Spring Tuffs, respectively.

History of Secondary-Mineral Deposition

Deposition of secondary minerals in and adjacent to faults, fractures and other void spaces began very soon after the ash flows were deposited and has continued into the Quaternary Period. The general chronology of secondary-mineral deposition has been established by many studies and continues to be refined [*e.g., Levy and O'Neil*, 1989; *Cowan et al.*, 1993; *Carlos et al.*, 1995; *Levy et al.*, 1996; *Whelan et al.*, 1998; *Paces et al.*, 1998; *Neymark et al.*, 1998]. This brief summary emphasizes aspects of mineral deposition that are helpful for documenting the development of transmissive features.

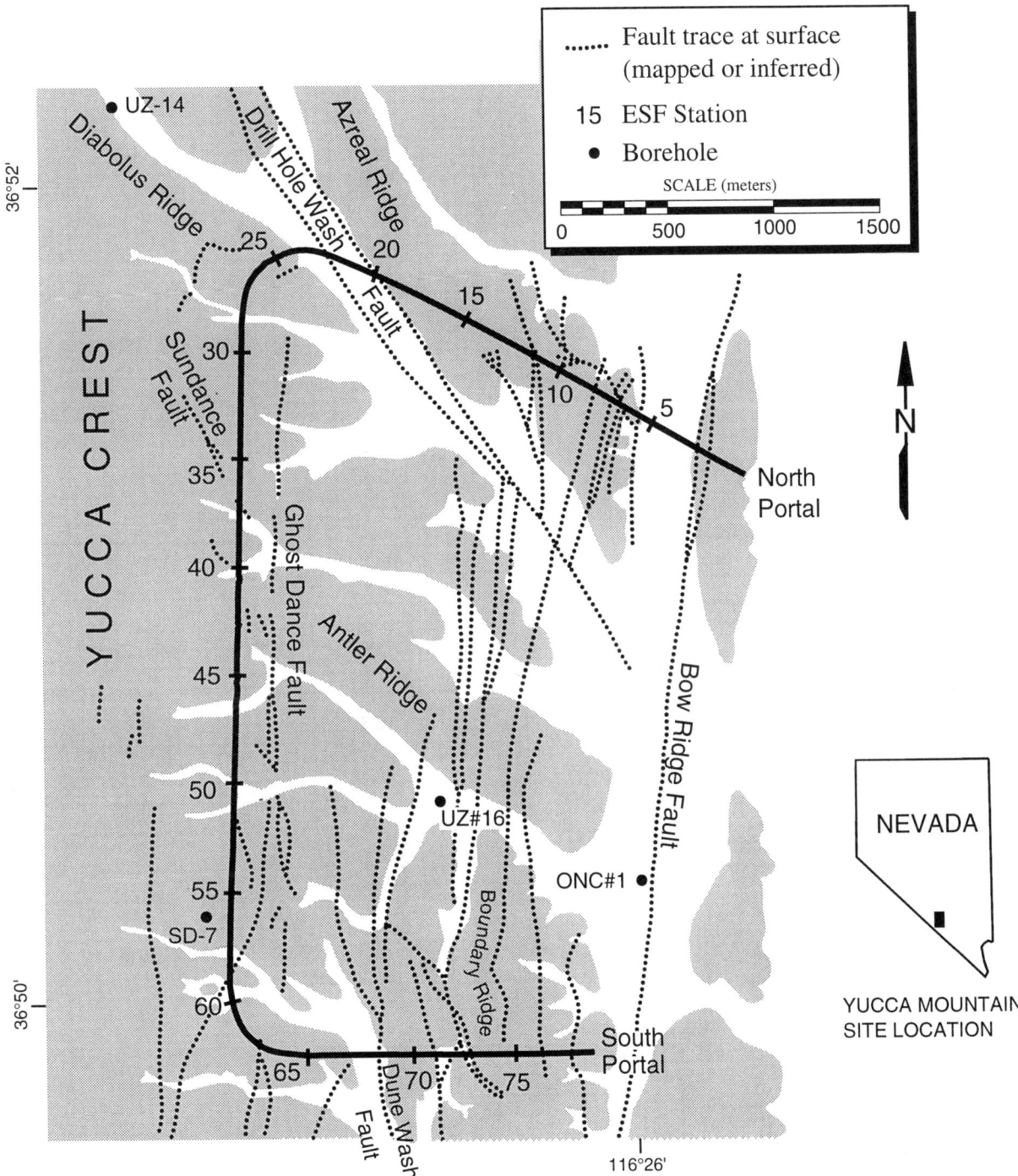

Figure 1. Plan view of ESF tunnel, showing its relationship to selected boreholes and faults mapped at the surface. Fault locations are based on *Day et al.* [1998]; not all faults on that map are shown here. Gray-stipple pattern represents pyroclastic bedrock; unshaded areas are Quaternary alluvial cover. ESF stationing is in hundreds of meters from the North Portal.

Table 1. Lithostratigraphy of the upper part of the Yucca Mountain unsaturated zone.

Lithostratigraphy[a]	Age (Ma)[b]	Hydrogeologic Unit[c]
TIMBER MOUNTAIN GROUP (Tm)		
Rainier Mesa Tuff (Tmr)	11.6	
PAINTBRUSH GROUP		
Tiva Canyon Tuff (Tpc)	12.7	UO
crystal-rich member (Tpcr)		
vitric zone (Tpcrv)		
nonwelded (Tpcrv3)		
moderately welded (Tpcrv2)		
densely welded (Tpcrv1)		
nonlithophysal zone (Tpcrn)		
lithophysal zone (Tpcrl)		
crystal-poor member (Tpcp)		TCw
upper lithophysal zone (Tpcpul)		
middle nonlithophysal zone (Tpcpmn)		
lower lithophysal zone (Tpcpll)		
lower nonlithophysal zone (Tpcpln)		
hackly (Tpcplnh)		
columnar (Tpcplnc)		
vitric zone (Tpcpv)		
densely welded (Tpcpv3)		
moderately welded (Tpcpv2)		
nonwelded (Tpcpv1)		
Pre-Tiva Canyon Tuff bedded tuff (Tpbt4)		
Yucca Mountain Tuff (Tpy)		
Pre-Yucca Mountain Tuff bedded tuff (Tpbt3)		PTn
Pah Canyon Tuff (Tpp)		
Pre-Pah Canyon Tuff bedded tuff (Tpbt2)		
Topopah Spring Tuff (Tpt)	12.8	
crystal-rich member (Tptr)		
vitric zone (Tptrv)		
nonwelded (Tptrv3)		
moderately welded (Tptrv2)		
densely welded (Tptrv1)		
nonlithophysal zone (Tptrn)		
lithophysal zone (Tptrl)		
crystal-poor member (Tptp)		
upper lithophysal zone (Tptpul)		
middle nonlithophysal zone (Tptpmn)		TSw
lower lithophysal zone (Tptpll)		
hackly fractured (Tptpllh)		
lower nonlithophysal zone (Tptpln)		
hackly fractured (Tptplnh)		
columnar (Tptplnc)		
vitric zone (Tptpv)		
densely welded (Tptpv3)		
moderately welded (Tptpv2)		
nonwelded (Tptpv1)		CHn

[a]Group and formation names, shown in bold type, are formal stratigraphic nomenclature after *Sawyer et al.* [1994]. Member, zone, and subzone designations and unit abbreviations follow informal usage of *Buesch et al.* [1996].

[b]Ages are from *Sawyer et al.* [1994].

[c]Hydrogeologic units based on *Montazer and Wilson* [1984] and *Ortiz et al.* [1985]: UO = unconsolidated overburden; TCw = Tiva Canyon welded; PTn = Paintbrush Tuff nonwelded; TSw = Topopah Spring welded; CHn = Calico Hills nonwelded.

Vapor-phase crystallization in fractures and lithophysal cavities (gas pockets) is a high-temperature process that occurs early in the cooling history of an ash flow deposit. The vapor-phase mineral products tridymite, cristobalite, and alkali feldspar are most prominent in the upper portions of the Tiva Canyon and Topopah Spring Tuffs. Late-stage syngenetic minerals that formed at ambient or near-ambient temperatures include quartz, chalcedony, cristobalite, and opal-CT. Some calcite and opal-A also formed under these conditions [*Levy*, 1993; *Whelan et al.*, 1998]. Many other minerals likely crystallized during the late stages of cooling; among these are smectite, mordenite, heulandite-clinoptilolite, potassium feldspar, apatite, and zircon [*Levy et al.*, 1996].

For the past 11 million years since the decline of major volcanic activity near Yucca Mountain, the rocks at and above the ESF level have remained in the unsaturated zone [*Levy*, 1991; *Paces et al.*, 1998]. These rocks did not experience the pervasive diagenetic or hydrothermal alteration that affected rocks at greater depth. During at least the last eight to ten million years, secondary-mineral deposition of calcite and opal was the main mineralogic modification in the shallow unsaturated zone [*Vaniman and Chipera*, 1995; *Whelan et al.*, 1998; *Paces et al.*, 1998; *Neymark et al.*, 1998]. Many geochemical and isotopic attributes of the calcites reflect soil-zone processes and interactions with infiltrating water that subsequently deposited the minerals [*Whelan and Stuckless*, 1992].

PRINCIPLES OF CHLORINE-36 HYDROLOGIC STUDIES

The interpretation of chlorine isotopic data is simple in principle and complex in application. The use of ^{36}Cl as a tracer and as a means of dating ground water is based on the production of this cosmogenic isotope in the atmosphere and its incorporation into infiltrating water. Once isolated from the atmosphere, the ^{36}Cl content of the ground water gradually decreases as a result of radioactive decay. Chloride extracted from the accessible pore spaces of subsurface samples is representative of the water that moved through the rock. Water ages, representing estimates of ground-water travel time from the surface to the underground sample sites, may be calculated from the decay constant for ^{36}Cl, the sample $^{36}Cl/Cl$, the initial atmospheric $^{36}Cl/Cl$, and the secular equilibrium $^{36}Cl/Cl$ resulting from *in situ* production by the subsurface neutron flux (from the decay of uranium and thorium isotopes) according to the standard equation of *Bentley et al.* [1986]. The most important uncertainties affecting interpretation of the isotopic data are temporal variations in the production and deposition rates of cosmogenic ^{36}Cl relative to the deposition of stable chloride on the ground surface.

The production rate of cosmogenic ^{36}Cl is controlled by variations in the earth's magnetic field strength, and the deposition of the cosmogenic isotope is influenced by climatic factors. The deposition of stable chloride on the ground surface at a given location, which affects the $^{36}Cl/Cl$ value independent of variations in cosmogenic ^{36}Cl production, is largely a function of climate and distance from the sea. The compound effect of climatic change includes varying eolian deposition of salt derived from salt flats, dry lake beds, and salt lakes and changing patterns and frequency of storms from the Pacific, carrying salt of marine origin. A theoretical reconstruction of ^{36}Cl production rates for the last million years and a ~40-ky record of $^{36}Cl/Cl$ variations from regional packrat middens, dated by the ^{14}C method, both suggest that $^{36}Cl/Cl$ ratios were higher throughout most of the Pleistocene than during the last few thousand years [*Plummer et al.*, 1997].

Detection of Fast Flow Paths

Waters entering the subsurface during the last ~50 years contain high concentrations of ^{36}Cl relative to natural background values. The elevated values are traceable primarily to global fallout from more than 70 above-ground nuclear tests conducted between 1952 and 1958 [*Glasstone*, 1962]. This input provides a fortuitous tracer to identify infiltrating water of very recent origin. ESF samples with $^{36}Cl/Cl$ values indicating a component of bomb-pulse ^{36}Cl are the basis for identifying fast pathways in the subsurface. The selection of a lower bounding value to identify samples containing bomb-pulse ^{36}Cl is based on in-progress statistical analysis of the sample population.

Conceptual Model of Fast-Flow Paths

The presence of bomb-pulse ^{36}Cl at certain locations in the ESF records travel of waters through the entire PTn hydrogeologic unit above those locations in 50 years or less, indicating that some flow followed structural pathways and largely bypassed the nonwelded rock matrix of the PTn-equivalent lithologic units. Our conceptual model of fast flow paths in the unsaturated zone at Yucca Mountain requires three conditions for bomb-pulse ^{36}Cl to reach the sampled depths within 50 years:

1) *A continuous fracture path must extend from the surface to the sampled depth.* This condition is necessary because transport simulations allowing only matrix flow through the PTn hydrogeologic unit produce fluid travel times that are too long and therefore inconsistent with the bomb-pulse $^{36}Cl/Cl$ values from fault zones. Fracture paths, predominantly cooling joints, exist in most of the welded portions of the Tiva Canyon and Topopah Spring

Tuffs [*Anna and Wallman*, 1997], but these units are separated from each other by the much less fractured PTn hydrogeologic unit. Therefore, the presence of faults that cut the PTn hydrogeologic unit and increase its fracture conductivity is required to satisfy the condition of a continuous fracture pathway.

2)   *The magnitude of surface infiltration must be sufficiently high to initiate and sustain at least a small component of fracture flow along the connected-fracture path.* Based on estimates of net infiltration derived from neutron logging, soil mapping, hydrologic-properties measurements, and watershed modeling, the fracture component of flux is estimated to range from 1 to 15 mm/yr [*Henning and Rickertsen, 1998*].

3)   *The residence time of water in the surface alluvium must be less than 50 years.* Travel times through alluvial cover must be sufficiently rapid for infiltrating waters to reach the soil/bedrock interface without being lost through evapotranspiration. At Yucca Mountain, this generally requires alluvial thicknesses less than about 3 m [*Flint and Flint, 1995*].

COLLECTION AND ANALYSIS OF SAMPLES

Geologic samples were selected to provide a systematic representation of the bedrock and to include transmissive features such as faults, fractures, and breccia zones. Bedrock samples were collected at 100-m or 200-m intervals along the tunnel. A large variety of features was sampled to test for the existence of fast fluid pathways. ESF sample locations are measured in meters inward from the north portal (Figure 1). For example, a location 115 m inward is designated as Station 1+15. Depths from the ground surface to the ESF range from about 40 m at Station 2+00 to about 300 m at Station 34+00.

Basic Sample Preparation and Isotopic Analysis

Chloride was extracted from one- to five-kg rock samples by leaching in an equal mass of deionized water for 48 hours. The Cl of interest is on the outer surfaces of particles or fractures. Poorly cohesive material was leached without further comminution, but other samples were crushed to 1- to 2-cm size fragments prior to leaching. An aliquot of the leachate was analyzed for Cl and Br by ion chromatography to estimate the contribution of Cl from construction water traced with LiBr. The remaining leachate was decanted, acidified to promote settling of particulates, and filtered. Known quantities of isotopically pure ^{35}Cl were added to samples with low Cl concentrations. Silver nitrate was added to the leachates to precipitate silver chloride, AgCl. The AgCl was purified of S by multiple cycles of dissolution in ammonium hydroxide, addition of barium nitrate to precipitate barium sulfate, followed by reprecipitation of the AgCl with nitric acid. The purified AgCl precipitates were sent to the Purdue Rare Isotope Measurement (PRIME) Laboratory for Cl isotopic analysis by accelerator mass spectrometry.

Isotopic compositions of samples were corrected for contamination by construction water, which has been isotopically characterized and labeled with lithium bromide to achieve a known Br/Cl ratio. Water ages, representing estimates of maximum ground-water travel time from the surface to the underground sample sites, were calculated from the corrected data assuming a maximum initial ^{36}Cl/Cl value of 1250×10^{-15} according to the standard equation of *Bentley et al.* [1986]. Table 2 lists the corrected ^{36}Cl/Cl values for all mineralogically characterized samples. The complete data set of isotopically analyzed samples is graphically depicted in *Fabryka-Martin et al.* [1998].

Special Preparation of Paired Breccia/Wall-Rock Samples

In addition to detecting variations in the ^{36}Cl/Cl ratio associated with the processes affecting the isotopic ratios of infiltrating water or with large structures that promote fast flow, the isotopic data potentially could be used to investigate fluid flow on a smaller scale. The kilogram-size samples required for isotopic analysis preclude the use of microsampling to investigate the fine-scale spatial distribution of bomb-pulse ^{36}Cl. Small breccia zones, up to a few tens of centimeters wide, may offer an opportunity to detect isotopic differences between the breccia and adjacent wall rock as possible indicators of local factors that affect water flow. These structural features, commonly bounded by tectonically modified cooling-joint surfaces, are abundant and widespread in the middle nonlithophysal zone of the Topopah Spring Tuff and in other units, as well.

We have made preliminary attempts to detect isotopic differences between breccias and adjacent wall rock by collecting and analyzing pairs of texturally distinct materials from breccia sites, including faults. In many locations, the distinction between breccia and wall rock is ambiguous. The existence of variably gradational boundaries between breccia and broken wall rock may require slightly different sample-selection criteria and processing for each site. The basic collection protocol was to collect separately bagged sample pairs at each site to represent texturally distinct materials, such as breccia and the adjacent rock or cooling-joint surface or multiple generations of fault fillings. In order to quantify and standardize the distinction between breccia and wall rock, some of the

Table 2. Lithostratigraphic and structural settings, ^{36}Cl/Cl values, and secondary mineralogy of ESF sample sites.

Sample[a]	Station	Lithostratigraphic unit[b]	Sampled feature	Corrected[c] ^{36}Cl/Cl $\times10^{-15}$	Calcite	Opal[d]	Clay/Mord. (note e)	Feldspar±cr.silica ±Fe-Ti oxides[f]	Transported Particulates[g]	Mn minerals	Other mineral(s) (note h)
E001	1+98	Tpcpll?	fault breccia	518	-	-	-	-	-	-	-
E008	1+99.8	Tpcp	fault breccia	**2138**	●	-	●	●	●	●	-
E009	1+99.8	Tpcp	fault breccia	**2444**	-	-	●	●	●	●	-
E010	1+99.8	Tpcp	fault breccia	720	-	-	●	-	●	●	-
E011	1+99.8	Tpcp	fault breccia	**2378**	●	●	?	-	●	●	●
E012	1+99.8	Tpcp	fault breccia	**2398**	-	-	●	●	●	●	-
E007	2+03	Tmbt1	bedrock	519	-	-	-	-	-	-	-
E163	4+94	Tpcpul	syst./bedrock	485	-	-	●	-	-	●	-
E073	5+04	Tpcpul	fracture	468	-	-	-	-	-	●	-
E164	7+00	Tpcpul	syst./bedrock	571	-	-	-	-	-	-	-
E165	7+70	Tpcpln/Tpcpv	subunit contact	496	-	●	●	?	-	●	?
E166	7+70	Tpcpln/Tpcpv	subunit contact	484	-	-	●	?	-	●	-
E167	7+70	Tpcpv2	subunit contact	427	-	-	●	-	-	●	-
E188	8+26.5	Tpcpv1	bedrock	766	-	-	●	-	-	-	-
E189	8+26.5	Tpcpv1/Tpbt4	contact	625	-	-	●	-	●	-	●
E190	8+26.5	Tpbt4	bedrock	647	-	-	-	-	-	?	●
E244	8+38	Tpbt3	fault	488	-	-	-	-	-	●	-
E168	8+59	Tpcpv1	above contact	1096	●	-	●	-	-	●	-
E169	8+59	Tpcpv1/Tpbt4	contact/fault	1096	-	-	●	-	-	-	-
E170	8+59	Tpbt4	bedrock/fault	635	-	-	-	-	-	-	-
E191	8+75	Tpbt3	subunit	904	-	●	-	-	-	-	●
E192	8+75	Tpbt3	subunit contact	698	-	●	●	-	-	-	●
E171	8+90	Tpp	below contact	**1335**	-	-	●	-	-	●	●
E172	8+90	Tpbt3/Tpp	contact	637	-	-	●	-	-	-	●
E173	8+90	Tpbt3	above contact	806	-	-	?	-	-	●	●
E174	9+00	Tpp	syst./bedrock	660	-	-	-	-	-	-	-
E126	10+34	Tpbt2	fault	633	●	●	●	-	-	●	●
E130	10+41	Tpbt2	fault	773	-	-	●	-	-	-	●
E197	10+62.5	Tptrv1	subunit contact	**1452**	-	-	●	●	-	-	-
E027	11+00	Tptrv1	syst./bedrock	1076	-	-	-	●	-	-	-
E213	12+36.5	Tptrm	broken rock	719	-	-	-	●	-	●	-
E028	12+44	Tptrm	cooling jts.	**2637**	-	-	-	●	●	●	●
E214	12+44	Tptrm	cooling jts.	750	-	-	-	●	-	●	-
E215	12+49	Tptrm	cooling jts.	668	-	-	-	●	-	-	-
E029	13+00	Tptrm	syst./bedrock	640	-	-	●	-	-	-	-
E030	13+67	Tptrm	cooling jts.	**1621**	-	-	-	●	-	-	-
E031	14+00	Tptrm	shear zone	**2398**	-	-	-	●	-	-	-

Table 2. Continued. Lithostratigraphic and Structural Settings, [36]Cl/Cl Values, and Secondary Mineralogy of ESF Sample Sites

Sample[a]	Station	Lithostrati-graphic unit	Sampled feature[b]	Corrected[c] [36]Cl/Cl ×10^{-15}	Calcite	Opal[d]	Clay/Mord. (note e)	Feldspar±cr.silica ±Fe-Ti oxides[f]	Transported Particulates[g]	Mn minerals	Other mineral(s) (note h)
E033	14+41	Tptrn	fault breccia	876	•	•		•			
E035	15+05	Tptrn	fracture	628	•	•		•		•	
E036	16+12	Tptrn	cooling jt.	382	•	•	•	•		•	•
E037	16+19	Tptrn	fracture	982							
E038	17+00	Tptrn	syst./bedrock	714			•	•	•	•	
E040	18+96	Tptpul	broken rock	**1642**	(•)[j]			•			
E041	19+00	Tptpul	syst./bedrock	746	•			•		•	
E042	19+31	Tptpul	breccia	**3019**	•		•	•		•	
E044	19+42	Tptpul	breccia	**2290**	•			•			•
E216	20+71	Tptpul	fractures	842				•		•	
E045	21+00	Tptpul	syst./bedrock	799				•			
E046	22+71	Tptpul	fractures	862				•			
E047	23+00	Tptpul	syst./bedrock	663				•			
E050	24+40	Tptpul	fault breccia	**2579**	•			•		•	
E020	24+68	Tptpul	fracture	814	•	•		•		•	
E217	26+19	Tptpul	cooling joints	522			•	•		•	
E218	26+36	Tptpul	fractured rock	603			•	•		•	
E219	26+46	Tptpul	fractured rock	578			•	•		•	
E052	26+79	Tptpul/tpmm	shear zone	**2036**	•			•			
E220	26+79	Tptpul/tpmm	fract. bedrock	565				•		•	
E058	27+66	Tptpmm	fault breccia	458				•	•	•	
E141	29+00	Tptpmm	syst./bedrock	922	•			•			
E142	29+21	Tptpmm	fracture	583				•			
E143	29+65	Tptpmm	fault breccia	1077	•		•	•		•	
E144	29+73	Tptpmm	cooling jt.	815	•						
E149	31+64	Tptpmm	cooling jt.	631	•	•	•	•		•	
E150	33+00	Tptpmm	syst./fr. bedrock	**1341**						•	
E151	33+16	Tptpmm	lith. cavity	529	•	•		•		•	
E152	34+28	Tptpmm	fract. bedrock	**4105**	•		•	•		•	
E153	34+32	Tptpmm	cooling jts.	**3291**	•		•	•	•	•	
E154	34+71	Tptpmm	cooling jts.	**3767**	(•)[j]				•	•	
E154	34+71	Tptpmm	breccia	803	(•)[j]				•	•	
E155	35+00	Tptpmm	syst./bedrock	1013			•	•		•	
E156	35+00	Tptpmm	broken rock	626				•		•	
E157	35+03	Tptpmn	cooling jts.	**(1339)**	•					•	

Table 2. Continued. Lithostratigraphic and Structural Settings, $^{36}Cl/Cl$ Values, and Secondary Mineralogy of ESF Sample Sites

Sample[a]	Station	Lithostratigraphic unit[b]	Sampled feature[b]	Corrected[c] $^{36}Cl/Cl$ ×10^{-15}	Calcite	Opal[d]	Clay/Mord. (note e)	Feldspar±cr.silica ±Fe-Ti oxides[f]	Transported Particulates[g]	Mn minerals	Other mineral(s) (note h)
E158	35+08	Tptpmn	cooling jts.	(2605)	●	-	●	●	-	●	-
E160	35+45	Tptpmn	cooling jts.	3329	(●)j	-	●	●	●	●	-
E161	35+58	Tptpmn	cooling jt.	(2141)	●	-	●	●	●	●	-
E175	35+93	Tptpmn	fault breccia	(2840)	(●)j	-	●	●	-	●	-
E177	37+00	Tptpmn	syst./bedrock	484	-	-	-	-	-	●	-
E178	37+60	Tptpmn	cool. jt./breccia	471	●	-	●	●	●	●	-
E179	37+68	Tptpmn	cool. jt./breccia	363	-	-	-	●	-	●	-
E179	37+68	Tptpmn	broken wall rock	397	-	-	●	●	-	●	-
E182	38+79	Tptpmn	cool. jt./breccia	379	-	-	-	●	-	●	-
E183	38+95	Tptpmn	cool. jt./breccia	745	-	-	●	●	-	●	-
E184	39+00	Tptpmn	syst./fractures	536	-	-	-	●	-	●	-
E185	39+39	Tptpmn	frct./lith. cavity	897	(●)j	-	●	●	-	●	-
E186	39+47	Tptpmn	cool. jt./breccia	561	●	-	-	●	-	●	-
E187	39+61	Tptpmn	cool. jt./breccia	540	-	-	●	●	-	●	-
E221	41+00	Tptpmn	syst./cool. jts.	773	-	-	-	●	-	●	-
E198	41+65	Tptpmn	cool. jts./br. rock	291	(●)j	-	●	●	-	●	-
E222	42+55	Tptpmn	cool. jts./shear	531	-	-	-	●	-	●	-
E199	43+00	Tptpmn	syst./bedrock	1042	-	-	-	●	-	●	-
E201	43+63	Tptpmn	cooling jts.	1974	-	-	●	●	-	●	-
E202	44+20	Tptpmn	cooling jts.	3463	-	-	●	●	●	●	-
E203	44+21	Tptpmn	cooling joints	849	-	-	-	●	-	●	-
E204	44+22	Tptpmn	cooling joints	772	-	-	●	●	-	●	-
E205	45+00	Tptpmn	syst./cool. jts.	1514	(●)j	-	-	●	-	●	-
E207	45+79	Tptpmn	fractured rock	593	-	-	●	●	-	●	-
E223	47+00	Tptpmn	syst./ frct. rock	734	-	-	-	●	-	●	-
E224	49+00	Tptpmn	syst./cool. jts.	499	-	-	●	●	-	●	-
E226	49+56	Tptpmn	cool. jts./breccia	456	●	-	●	●	-	●	-
E227	49+89	Tptpmn	cool. jts./breccia	497	●	-	-	●	-	●	-
E230	51+00	Tptpmn	syst./fract. rock	555	-	-	-	●	-	●	-
E231	51+07	Tptpmn	cooling joints	709	(●)j	-	●	●	-	●	-
E233	51+73	Tptpmn	fractures	647	(●)j	-	●	●	-	●	-
E234	52+43	Tptpmn	cooling joints	367	(●)j	-	●	●	-	●	●
E236	53+00	Tptpmn	syst./fract. rock	417	-	-	-	●	-	●	-
E237	53+61	Tptpmn	broken rock	539	-	-	-	●	-	●	-
E238	54+20	Tptpmn	cool. jts./breccia	727	-	-	●	●	-	●	-

Table 2. Continued. Lithostratigraphic and Structural Settings, ^{36}Cl/Cl Values, and Secondary Mineralogy of ESF Sample Sites

Sample[a]	Station	Lithostratigraphic unit	Sampled feature[b]	Corrected[c] ^{36}Cl/Cl ×10^{-15}	Calcite	Opal[d]	Clay/Mord. (note e)	Feldspar±cr.silica ±Fe-Ti oxides[f]	Transported Particulates[g]	Mn minerals	Other mineral(s) (note h)
E239	55+00	Tptpmn	syst./fract. rock	464				●		●	
E241	56+85	Tptpmn	cool. jts./breccia	778	(●)[j]			●		●	
E242	56+93	Tptpmn	cool. jt./breccia	1117	(●)[j]			●		●	
E252	57+00	Tptpmn	syst./fractures	388	●			●		●	
E255	58+77	Tptpmn	below frcts.	140				●		●	
E256	59+00	Tptpll	syst./fract. rock	361			●	●		●	
E290	59+98	Tptpmn	syst./fract. rock	205		?	●	●		●	
E258	61+92	Tptpmn	fractures	276			●	●		●	
E260	62+05	Tptpmn	fault	261		?		●		●	
E266	63+26	Tptpmn	fractures	486		?		●		●	
E267	63+30	Tptpmn	fault	427	(●)[j]			●		●	
E270	63+81	Tptpul	fractured rock	439			●	●		●	
E271	64+00	Tptpul	fractured rock	467	●		●	●		●	
E272	64+34	Tptpul	broken rock	467				●		●	
E274	64+93	Tptpul	fracture/breccia	491	●	●		●		●	
E275	65+00	Tptrl	syst./fractures	443	●	●				●	
E268	65+20	Tptrl	fractures/breccia	468	(●)[j]			●		●	
E277	65+80	Tptrm	fracture zone	424	(●)[j]	(●)[j]		●		●	
E279	66+15	Tptrm	fault breccia	402		●		●	●	●	●
E280	66+40	Tptrm	fault breccia	238		●		●		●	
E281	67+00	Tpbt3	systematic	453				●			
E283	67+27	Tpcpv2/1	thin bed	470			●	●		●	
E284	67+35	Tpcplnc/pv2	subunit contact	502			●	●		●	●
E284	67+35	Tpcpv2	subunit contact	509			●	●		●	●
E289	67+61	Tpcpv1	fault margin	589			●	●		●	●
E285	67+73	Tpcpv2/1	thin bed	468			●	●		●	
E286	67+87	Tpcpv	fault/graben	475	●		●	●	●	●	
E287	67+87	Tptr	fault/graben	517			●	●		●	
E288	67+90	Tptpul	broken rock/fault	557				●		●	
E298	68+00	Tptpul	syst./bedrock	606				●		●	
E292	69+00	Tptrm	syst./cool. jt.	414				●		●	
E293	69+14.5	Tptrm	cool. jt./breccia	454				●		●	
E294	69+32.5	Tptrm	cool. jt./breccia	473				●		●	
E295	69+41.7	Tptrm	cool. jt./breccia	476	●	●	●	●		●	
E299	69+47	Tptrm	syst./shear zone	441				●		●	
E300	69+68	Tptrm	fracts./fault(?)	354	(●)[j]			●		●	

Table 2. Continued. Lithostratigraphic and Structural Settings, ^{36}Cl/Cl Values, and Secondary Mineralogy of ESF Sample Sites

Sample[a]	Station	Lithostratigraphic unit	Sampled feature[b]	Corrected[c] ^{36}Cl/Cl ×10^{-15}	Calcite	Opal[d]	Clay/Mord. (note e)	Feldspar±cr.silica ±Fe-Ti oxides[f]	Transported Particulates[g]	Mn minerals	Other mineral(s) (note h)
E301	69+95.8	Tptrn	fault breccia	224	•	•	•	•	-	•	-
E302	70+19	Tpbt2	fault breccia	327	-	•	•	•	-	-	•
E303	70+36	Tpbt2	fault	439	-	-	-	-	-	-	•
E304	70+50	Tpbt2	syst./bedrock	491	-	-	-	-	-	•	-
E306	70+66	Tptpmn	fault breccia	499	-	-	•	•	-	•	-
E309	71+41	Tptpmn	fault	445	-	-	•	•	-	•	-
E310	71+50	Tptpmn	syst./frct. rock	441	-	-	•	•	-	•	-
E312	72+69	Tptpul	fault breccia	463	-	-	-	•	-	•	-
E313	73+48	Tptrl	syst./shear	367	-	-	-	-	-	•	-
E314	74+43	Tptrn	fault	*341*	(•)[j]	(•)[j]	(•)[j]	-	-	•	-
E315	74+49	Tptrv3	syst./bedrock	435	-	-	•	•	-	•	-
E317	75+09	Tpcpv1	subunit contact	402	-	-	-	-	-	-	-
E320	75+20	Tpcpv2	fract./breccia	457	(•)[j]	•	•	•	-	•	•
E322	75+47.5	Tpcplnc	syst./frct. rock	*318*	-	-	•	•	•	•	-
E325	76+30	Tpcpln	fault wall rock	380	•	•	-	-	-	•	-
E327	76+50	Tpcpln	syst./frct. rock	281	•	-	•	-	-	•	-
E329	77+10	Tpcpln	cool. jts./fault	394	-	-	-	•	-	•	-
E335	77+19	Tpcpmn	fract. rock	*186*	-	-	•	•	-	•	-

[a]Samples were divided into separate splits for isotopic and mineralogic analysis. Mineralogic data were also recorded for the sample sites. In cases where more than one split of a sample was measured for chlorine isotopic ratios, there are separate entries for each analyzed split or the value reported in this table is the highest value obtained.

[b]Abbreviations: br. rock = broken rock, cooling jt., cool. jt. = cooling joint; fr. (or fract.) bedrock = fractured bedrock; frct(s). = fracture(s) or fractured; syst. = systematic sample collected at fixed intervals.

[c]Isotopic ratios are corrected for the presence of construction water. Values in boldface (^{36}Cl/Cl values > 1250 × 10^{-15}) denote samples containing an unambiguous component of bomb-pulse chlorine inferred to be less than ~50 years old. Italicized values below 350 may denote samples with longer water residence times than elsewhere.

[d]As used here, opal is transparent, colorless to light-colored, and fluoresces yellow-green in short-wave ultraviolet light. X-ray diffraction analysis of selected samples indicates opal-A. The silica in sample E126 has no discernible morphology. The opal in sample E192 (and some of the opal in E191 and E295) has a morphology suggestive of opal-CT. A translucent character is suggestive of opal-CT (E274, E295, E327). Queried entries "?" (E260, E290) refer to glassy, silica-rich coatings usually associated with microbreccia.

[e]This category includes clay and/or mordenite. The clay is predominantly smectite, but may include palygorskite.

[f]This category includes minerals inferred to be of early to late syngenetic origin. Reported occurrences in this category are limited to minerals in growth position on the rock surfaces. Cr. silica = crystalline silica, including quartz, chalcedony, cristobalite, tridymite, opal-CT.

[g]This category includes physically transported particulates, mostly silt- and sand-size material. Deposits of clay-size material are not included here.

[h]This category includes fluorite (E012, E036, E126?, E234), zeolite (E130, E165?, E189?, E191, E192, E216, E284, E297), hematite (other than specular hematite of vapor-phase origin; E126, E189, E190, E304), chabazite, kenyaite, and moganite (E320), and unidentified minerals.

[i]"-" indicates a mineral or textural component was not observed; "•" indicates a mineral or component was observed; "?" indicates uncertain identification.

[j]Parentheses indicate that calcite (or opal or clay) was not present in the aliquot for mineralogic characterization but was observed at the collection site in fractures or voids (E040, E154, E160, E186, E207, E242, E277) or in fractures adjacent to the fault (E175, E267) or in the same fault (E314).

paired samples were individually sieved to remove material larger or smaller than a certain size. The finer fractions were processed for analysis without further size reduction. For the coarser fractions, the larger rock fragments were individually crushed with a steel plate and hammer to a maximum size of ~2 cm, then processed for analysis. The texturally distinct materials of each pair were processed and analyzed separately. Table 3 summarizes the textural/structural setting, processing, and isotopic compositions of each sample pair.

Mineralogic and Textural Analysis

Subsamples designated for mineralogic and petrologic study were examined by binocular microscope at up to 500× magnification. Contamination of the rock materials by human Cl was avoided by the use of new examination gloves for each sample. Mineral identifications are based on color, morphology, hardness, comparison to known materials, limited X-ray diffraction (XRD) analysis, and evidence of fluorescence under ultraviolet light. Some ambiguities of mineral identification are unresolved; for example, clay and mordenite (a zeolite) in minute deposits cannot be distinguished without electron microscopy or other techniques (e.g., XRD). The mineralogy of representative samples was verified by XRD analysis. XRD data for selected samples were obtained on an automated Siemens D-500 diffractometer using Cu-Kα radiation, incident- and diffracted-beam Soller slits, and a Kevex (SiLi) solid-state detector. Data were collected from 2 to 50° 2θ and count times ≥ 2.0 s per step. Quantitative analyses employed the internal standard method of *Chung* [1974a,b] using 1.0-μm corundum as the internal standard.

Detailed textural descriptions of two breccias that represent contrasting structural domains, included here, are integral to the investigation of paired breccia and wall-rock samples. Textural analysis documents the syngenetic attributes of the bedrock, including degree of welding, presence of devitrification, and development of lithophysal porosity. The geometric and mineralogic characteristics of fracture walls, such as planarity, smoothness, and presence of bleached margins typical of cooling joints, are also recorded. Textures of tectonic modification include polishing and pulverization of fracture asperities. Breccias are characterized by sizes and distributions of fragments, the existence of fragment size segregation or zonation within a breccia deposit, and lithologic variation among fragments.

The mineralogy of breccia cements helps reconstruct the relative timing of deformation and permeability modification. A well characterized sequence of syngenetic alteration and mineral deposition is used to reconstruct the early chronology of fracturing and brecciation. Post-syngenetic deposition of calcite and opal-A by percolating water locally preserved the fabrics and modified the permeabilities of breccia deposits.

ISOTOPIC RESULTS

The $^{36}Cl/Cl$ results for mineralogically characterized samples are listed in Table 2. Bomb-pulse ^{36}Cl is present in breccia samples from the block-bounding Bow Ridge fault (E008, E011). Additional bomb-pulse samples in the interval from Station 2+00 to Station 18+00 (E171, E028, E030, and E031, plus three additional sample sites not characterized for mineralogy) have a general spatial association with numerous faults in the hanging (western) wall of the Bow Ridge fault. Three bomb-pulse samples (E040, E042, and E044) coincide with the northeastern splay of the intrablock Drill Hole Wash fault. The locations of two bomb-pulse samples (E050, E052) below Diabolus Ridge (Figure 1) may be related to a shallowly dipping thrust fault that cuts the PTn hydrogeologic unit above the ESF.

Breccia from a trace of the Sundance fault, an intrablock fault, contains bomb-pulse ^{36}Cl (E175), and the numerous bomb-pulse samples from within 300 m north of the fault intersection in the ESF may reflect diversion of fast flow from the fault zone into subsidiary local transmissive features such as cooling joints and breccia zones. Three samples from between Stations 43+00 and 45+00, near the boundary of a zone of abundant cooling joints, also have elevated ^{36}Cl values. There is no identified fault in the ESF at this location, nor do the sample localities correspond to any fault mapped at the surface. As yet, the presence of bomb-pulse ^{36}Cl at these localities remains unexplained.

DISTRIBUTIONS OF SECONDARY MINERALS

Calcite

Examination of the subset of analyzed samples for which mineralogic data are available shows that calcite is present at slightly more than half of the sample sites that have received infiltration during the last ~50 years (Table 2), with 15 of 26 bomb-pulse values obtained for samples associated with calcite. In comparison, calcite is present in 44 of the 135 samples in which no unambiguous bomb-pulse signal is present. Between Stations 17+00 and 36+00, calcite is present at 20 of the 28 feature-based sample sites, 14 of which are also bomb-pulse sites. This interval includes the Drill Hole Wash and Sundance fault zones and may be influenced hydrologically by a thrust

Table 3. ^{36}CL/CL ratios of paired breccia/wall rock samples.

Sample Pair	Station	Lithostratigraphic Unit	Structural Feature	Sample Constituents	Corrected ^{36}Cl/Cl $\times 10^{-15}$ $\pm 1\sigma$
E030-1 E030-2	13+67	Tptrn	Broken rock and fine-grained breccia at intersection of two high-angle cooling joints	Breccia < 0.5 cm Broken bedrock/breccia > 0.5 cm	697 ± 35 1643 ± 85
E154-1 E154-3	34+71	Tptpmn	1- to 4-cm-wide breccia zone between high-angle cooling-joint faces in fractured wall rock	Breccia Fractured wall rock	803 ± 42 3794 ± 120
E158-1 E158-3	35+08	Tptpmn	6-cm-wide breccia bounded by parallel high-angle cooling joint faces	Breccia < 0.5 cm Broken bedrock > 0.5 cm	1113 ± 58 2671 ± 158
E161-1 E161-3	35+58	Tptpmn	Fractured rock and underlying breccia zone separated by low-angle cooling joint	Breccia < 0.5 cm Fractured rock, breccia > 0.5 cm	1951 ± 103 2169 ± 80
E175-1 E175-3	35+93	Tptpmn	Sundance fault multigenerational breccia, with 6- to 7-cm domains of cemented breccia	Breccia > 0.5 cm Breccia < 0.5 cm	2840 ± 231 1674 ± 141
E176-1 E176-3	36+55	Tptpmn	Fault gouge from zone of near-vertical fractures	Fractured rock, breccia > 0.5 cm Breccia < 0.5 cm	887 ± 27 604 ± 24
E179-1 E179-3	37+68	Tptpmn	1- to 2-cm-thick breccia bounded by cooling-joint surfaces in fractured tuff	Breccia Fractured wall rock	363 ± 22 397 ± 13
E222-1 E222-2	42+55	Tptpmn	Intersection of thin breccia and shear zones with underlying low-angle cooling joint	Gouge Wall rock	605 ± 18 531 ± 16
E226-1 E226-2	49+56	Tptpmn	0- to 10-cm-wide breccia zone bounded by near-vertical cooling-joint faces	Breccia Wall rock	451 ± 20 456 ± 20
E231-1 E231-2	51+07	Tptpmn	Breccia and broken rock bounded by high-angle cooling joint faces	Wall rock Breccia	709 ± 30 530 ± 26
E242-1 E242-2	56+93	Tptpmn	Breccia/shear zone 10- to 20-cm wide, developed along high-angle cooling joints	Wall rock, breccia > 2 mm Breccia < 2 mm	664 ± 30 1117 ± 49
E256-1 E256-3	59+00	Tptpll	Sheared fracture in broken lithophysal rock	Sheared rock from fracture Wall rock ~1 m from fracture	347 ± 41 361 ± 22
E286-1 E286-4	67+87	Tpcpv	Dune Wash fault, with clayey infillings between hanging wall and rubble zone	Clayey infilling Wall rock	475 ± 22 645 ± 29

Table 3. Continued. ^{36}CL/CL RATIOS OF PAIRED BRECCIA/WALL ROCK SAMPLES

Sample Pair	Station	Lithostratigraphic Unit	Structural Feature	Sample Constituents	Corrected ^{36}Cl/Cl $\times 10^{-15}$ $\pm 1\sigma$
E323-1 E323-3	75+54	Tpcpln	10-cm-wide fine- to medium-grained breccia bounded by cooling joint faces	Breccia Wall rock	465 ± 17 413 ± 19
E324-1 E324-3	75+78	Tpcplnc	10- to 15-cm-wide fault/breccia zone bounded by high-angle cooling joint faces	Breccia Wall rock	418 ± 31 322 ± 13
E325-1 E326-1	76+30	Tpcpln	2-m-wide fault/breccia zone (dm-scale blocks in finer-grained matrix) bounded by broken wall rock	Breccia Wall rock	380 ± 20 423 ± 9
E328-1 E328-3	76+76	Tpcpln	20-cm-wide fault/breccia zone with domains of coarser and finer clasts, bounded by cooling joint faces	Breccia < 1 cm Breccia > 1 cm	334 ± 15 445 ± 20

fault below Diabolus Ridge. The spatial distribution of calcite with respect to ^{36}Cl/Cl values is shown in Figure 2.

The data set, taken as a whole, indicates that bomb-pulse sites may be slightly more likely to contain calcite than sites without unambiguous bomb-pulse signals. This relationship essentially owes its existence to the zone of abundant calcite and bomb-pulse signals between Stations 17+00 and 36+00. A calcite deposit is clearly an indicator that percolating water reached a site where the mineral occurs, but its presence need not be directly related to the hydrologic and structural factors responsible for fast-path behavior. Where these favorable factors exist, the presence of calcite may be taken as an additional but minor predictor of fast flow.

We have observed that calcite deposits in many of the bomb-pulse samples have thicknesses less than about a millimeter. In some cases, this reflects the size of the aperture in which the calcite grew. Some examples exist of thin calcite coatings in samples with sufficient pore space for thicker coatings to have grown; sample E008, a Bow Ridge fault breccia, is one example. These calcite deposits are much thinner than the deposits that have been dated by ^{230}Th/U methods [Paces et al., 1998]. The thin coatings have not been dated by ^{230}Th/U methods because contamination from the bedrock substrate is nearly unavoidable. The thin calcite coatings may record shorter segments of depositional history, or the depositional layers may be thinner due to slower deposition rates than those calculated for other ESF sites with thicker calcite coatings. Slower deposition does not necessarily reflect lower percolation rates, for it may also result from lower evaporation rates or fluid movement too fast to promote nucleation and deposition of calcite. Both of these conditions could be attributes of fast fluid pathways.

Opal

There are three main modes of occurrence for opal-A in the ESF. Opal-A spherules, where present, generally occur only with calcite [Whelan et al., 1998] and, like the calcite, are derived from percolating ground water. Opal-A with a drip-like texture is localized in and below the Tpcplnc/Tpcpv2 transition in the southern ESF and may be a product of alteration during the cooling of the host tuff [Levy et al., 1999]. This type of occurrence is commonly associated with fractures or breccia zones that cut across the Tpcplnc/Tpcpv2 stratigraphic boundary corresponding to the TCw/PTn hydrogeologic boundary (Table 1). The potential role of such transmissive features as fast paths could not be tested because the features lie below areas of net infiltration too low to sustain fast flow. The third mode of occurrence is one in which the identity of the silica as opal-A is uncertain and is queried in Table 2.

Thin layers of silica-rich material with a glassy, opaline appearance are present on broken-rock surfaces in the vicinities of breccia zones in the southern ESF. This material differs from the other opal occurrences by an absence of fluorescence in short-wavelength ultraviolet light, presumably due to very low uranyl content.

Amorphous opal (opal-A) is much less common than calcite in the samples and is associated with only one bomb-pulse sample from the Bow Ridge fault zone (E011, Table 2). Because opal-A generally occurs with calcite [Whelan et al., 1998], the paucity of opal occurrences in our sample set, especially in the bomb-pulse samples, could be linked to the same processes responsible for the lesser thicknesses of calcite deposits in fast pathways.

Clays

Clays, predominantly smectites, are nearly ubiquitous in the major rock units of the ESF. Bulk samples of the devitrified Tiva Canyon and Topopah Spring Tuffs have smectite contents of about 1 to 10 wt %, although values less than 3 wt % are most common [Bish and Chipera, 1989; Chipera et al., 1995]. The clay represented by the bulk analyses of these rocks is disseminated throughout the matrix. As the result of in situ alteration, the matrix clay content of the PTn hydrogeologic unit can be as high as ~95 wt % [Levy et al., 1996].

Except for the samples of various Tpbt and Tpp bedrock, the clays reported in Table 2 are all fracture or breccia-clast coatings or fault fillings, rather than matrix components. The presence of these deposits is assumed to result from aqueous transport of fine clay particles within the fracture network because clays would not be readily released from intact, densely welded rock matrix. The actual sources of the clays and the distances of transport have not been determined. Multiple clay deposits are distinguishable by color and, in some cases, by differences in distribution on the rock surfaces. Some of these distinctive deposits may be derived from different source materials and may have been transported and deposited at different times. The fibrous zeolite mordenite has been identified by X-ray diffraction as a component of some clay deposits in the ESF, but usually is not distinguishable from clay by optical examination. For this reason, clay-size material is reported in Table 2 as "clay/mordenite."

The mineralogic data set, taken as a whole, does not indicate an association between fast paths (identified by the presence of bomb-pulse ^{36}Cl) and the presence of clay. As shown in Figure 3, clay fracture coatings are notably less abundant in an interval between about Stations 18+00 and 25+00. This interval corresponds approximately to the upper lithophysal zone of the Topopah Spring Tuff. Fractures are shorter and less numerous in this zone than in

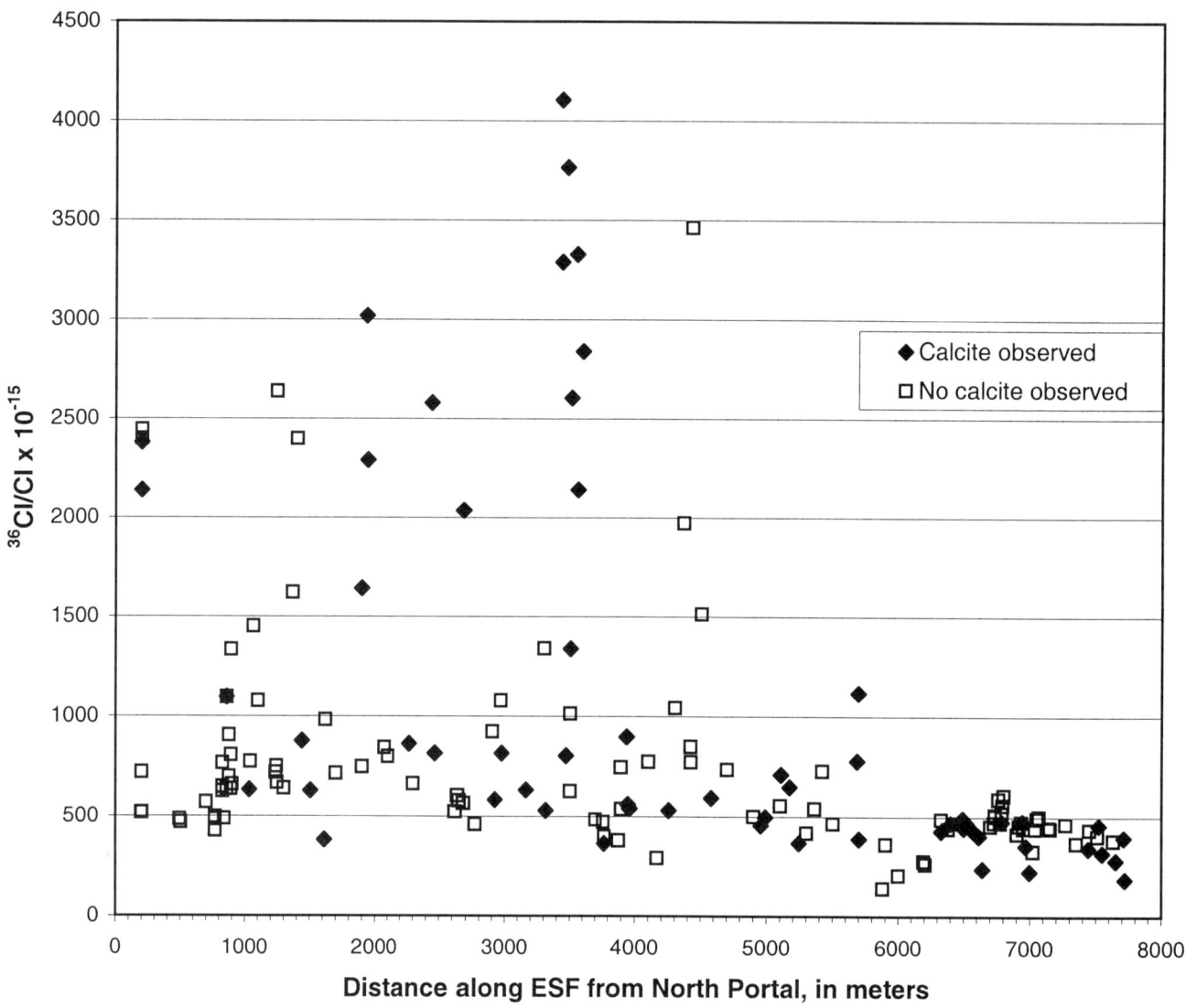

Figure 2. Distribution of calcite and ^{36}Cl/Cl values in the ESF.

the underlying middle nonlithophysal zone [*Sweetkind et al.*, 1997]. Fracture surfaces are rough and irregular in the lithophysal zone, whereas the surfaces are mostly smooth in the nonlithophysal rock [*Buesch et al.*, 1996]. Rough, irregular fracture surfaces tend to inhibit the fluid-film transport processes by which fine clay particles accumulate into macroscopic deposits.

In the interval from Stations 34+28 to 35+93, associated with the footwall zone of the Sundance fault, there is a high incidence of both bomb-pulse ^{36}Cl/Cl values and clay/mordenite. Eight of 11 samples (73%) have the bomb-pulse signature, nine of 11 samples contain clay/mordenite, and seven of the eight bomb-pulse samples (88%) contain clay/mordenite. The clay coatings also tend

to be thicker in this interval than elsewhere. The significance of thick clay coatings as indicators of fracture dilation is described below in relation to the paired breccia/wall-rock samples.

Feldspar, Crystalline Silicas, and Fe-Ti Oxides

This category of syngenetic minerals includes alkali feldspar, tridymite, cristobalite, quartz, opal-CT (opal with short-range cristobalite and tridymite ordering), chalcedony (fibrous microcrystalline quartz), hematite, and other Fe-Ti oxides. Each of the devitrified tuff units exposed in the ESF contains some or all of these phases that formed as it cooled. In particular, feldspar, tridymite,

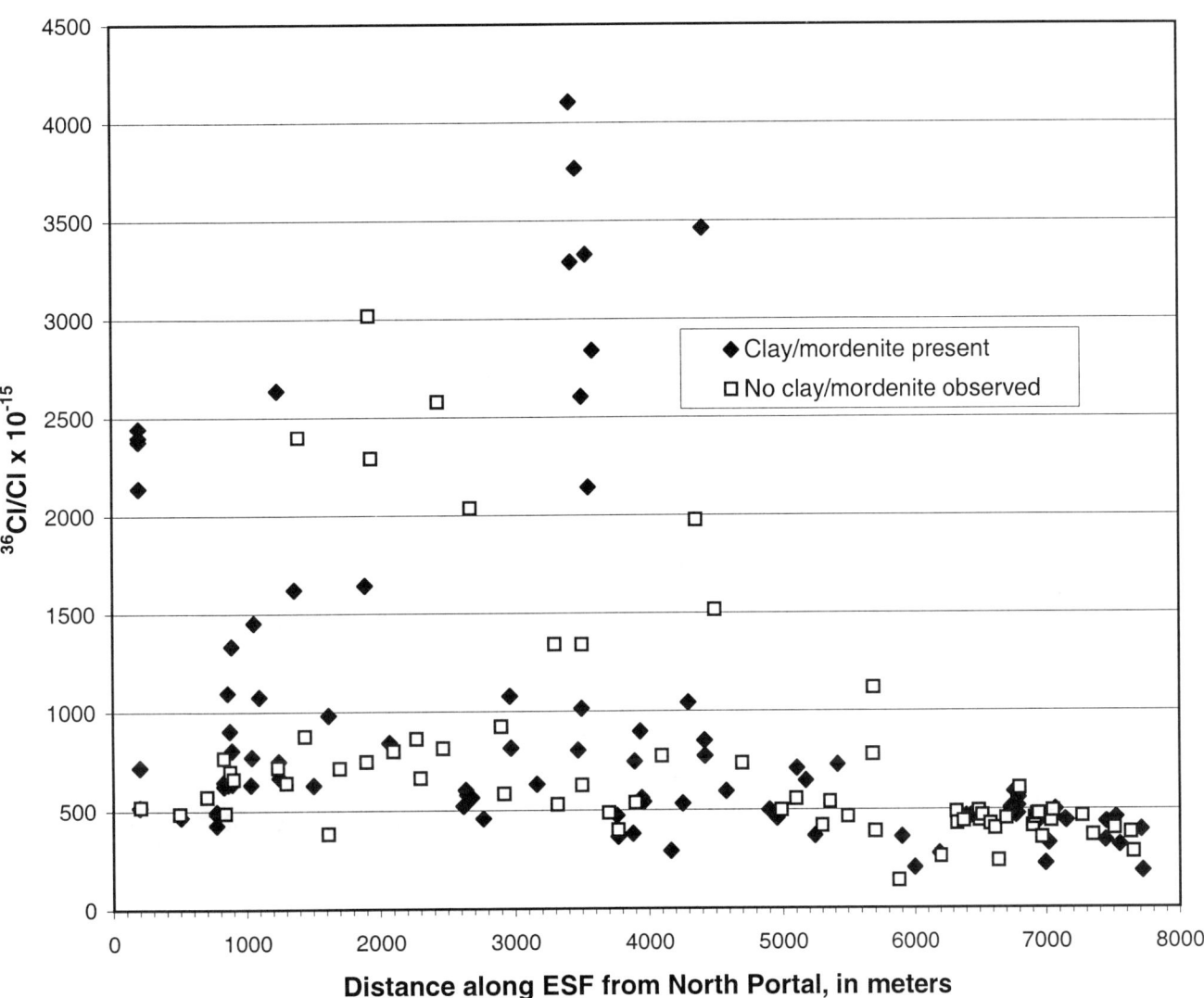

Figure 3. Distribution of clay/mordenite and ^{36}Cl/Cl values in the ESF.

cristobalite, and the Fe-Ti-oxides are known as vapor-phase minerals deposited at high temperatures very early in the cooling history of a tuff [e.g., *Carlos*, 1985]. With respect to fluid pathways, the main significance of syngenetic minerals is that their presence in growth position on fracture walls, breccia fragments, or other secondary pore surfaces establishes the early origins of these transmissive features. Rock fragments from vapor-altered breccias at Station 35+93 (sample E175) have microscopic surface textures with post-breakage growth faces developed on groundmass feldspars. These textures document the syngenetic, intraformational origin of a zone of deformation that is now part of the Sundance fault.

Translocated Particulates

For particulates larger than clay particles, it is possible in many cases to determine whether they are different from the local bedrock and have, therefore, been transported to their present location from elsewhere in the geologic section. It is relatively easy to recognize samples in which the particulates are enriched in vapor-phase or hydrothermal minerals relative to their abundance in the local bedrock, and this enrichment is a good criterion for documenting evidence of particulate transport. The presence of translocated particulates within a flowpath attests

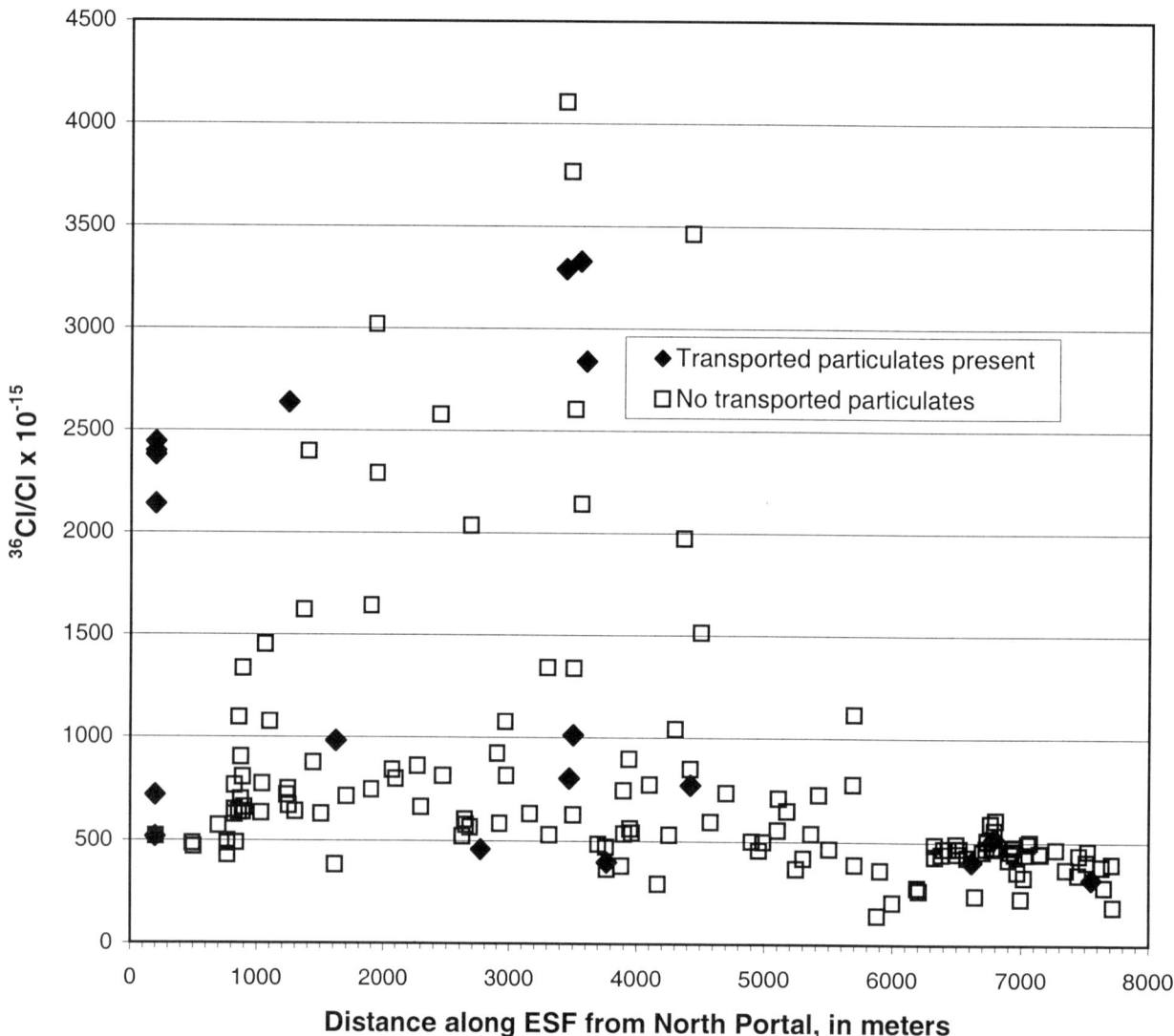

Figure 4. Distribution of transported particulates and ^{36}Cl/Cl values in the ESF.

to the existence of connected pore spaces with apertures large enough to permit the passage of the particulates.

Two zones containing multiple samples with transported particulates exist in the ESF (Table 2 and Figure 4). Centimeter-size fragments of local rock in breccias from the Bow Ridge fault (Station 1+99), a fast path identified by the presence of bomb-pulse ^{36}Cl, are thickly coated by fine vapor-phase particulates probably derived from over-lying rocks of the same ash flow. A second interval with common translocated vapor-phase particulates in fractures and breccias exists between Stations 34+32 and 37+68, encompassing the footwall and hanging-wall zones of the Sundance fault. This interval largely overlaps the locations of numerous bomb-pulse sites restricted to the foot-wall zone of the fault.

The likely intraformational origins of both vapor-phase particulate occurrences suggests that particulate transport in these fractures and breccias operated as local processes only. Deformation that produced the particulate-conducting fractures enhanced local permeability, allowing fluid flow to migrate laterally away from the main fault zone (in the case of the Sundance fault). The deformation, because of its local nature, may not have contributed directly to the formation of fast-flow fault pathways extending from the surface to the ESF.

Manganese Minerals

A transition from uncommon to nearly ubiquitous manganese minerals, corresponding approximately to the

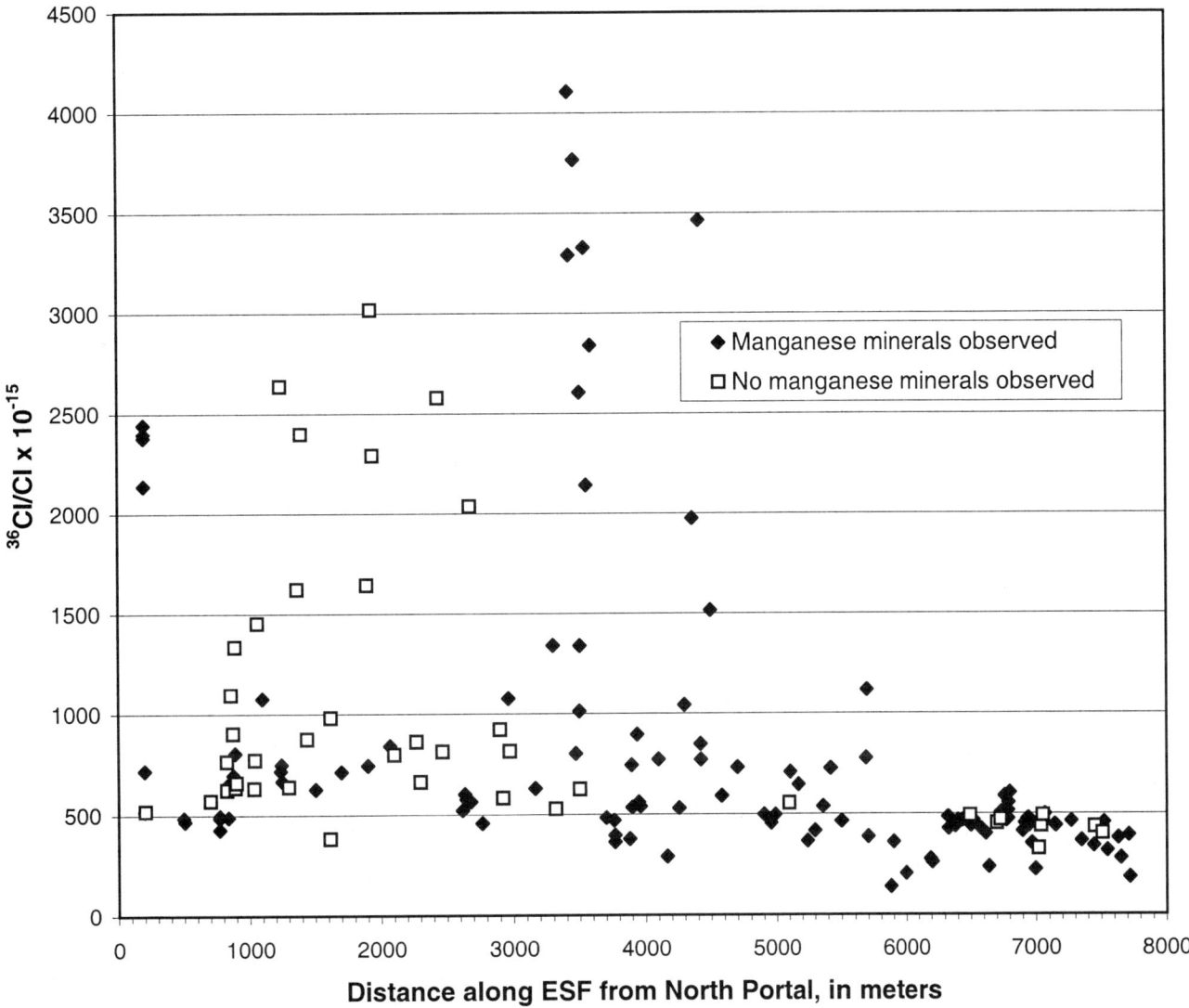

Figure 5. Distribution of manganese minerals and ^{36}Cl/Cl values in the ESF.

boundary between the upper lithophysal and middle nonlithophysal zones of the Topopah Spring Tuff, was observed in data collected from the ESF North Ramp (Table 2 and Figure 5). Data from the South Ramp barely mirror this trend because manganese minerals are common on fractures in both the middle nonlithophysal and upper lithophysal zones. The transition in the North Ramp may be associated with systematic changes in fracture-surface roughness because manganese minerals are especially abundant on smooth-surfaced cooling joints that are best developed in the middle nonlithophysal zone. Smooth surfaces may have promoted the transport of dissolved or colloidal manganese within fluid films that could exist more stably on such surfaces than on rougher surfaces.

The near-uniformity of manganese-mineral occurrences within both subunits in the South Ramp may reflect local development of smoother fracture surfaces in the upper lithophysal zone.

MINERAL DEPOSITION AND FAULT TRANSMISSIVITY

In the course of ^{36}Cl investigations of flow and transport in the ESF, all major faults were sampled for isotopic characterization. Bomb-pulse levels of ^{36}Cl, indicating that a component of percolating water traveled from the surface to the ESF level in less than 50 years, were associated with the traces of the Bow Ridge, Drill Hole

Wash, Sundance, and Ghost Dance (southern exposure on-ly, at Station 57+30) faults. The secondary mineralogy of fault samples, as well as of fractures and breccia zones, is summarized in Table 2 and provides a basis for eval-uating the effectiveness of cementation in reducing transmis-sivity.

In the unsaturated zone as deep as the middle non-lithophysal zone of the Topopah Spring Tuff, common secondary minerals most likely to affect fault trans-missivity include calcite, opal, and clays. The ^{36}Cl data set was used as a basis to compare the mineralogy of fast-path faults, fractures, and breccia zones and non-fast-path transmissive features. The results suggest that the amount of secondary-mineral deposition in faults was insufficient to affect the transmissivity of the faults, at least in the por-tions of the fault flow paths within densely welded tuff. Secondary minerals are not especially abundant within the faults, nor are they more abundant in the faults than in cooling joints or thin, minor-offset breccias with or with-out fast-path isotopic signatures.

There are several possible reasons why the trans-missivity of faults in welded tuffs is relatively unaffected by secondary-mineral deposition. The most important reason probably is that faults at Yucca Mountain do not exist as simple, single-trace features, either at the surface or in the subsurface [Potter et al., 1999, in press]. The existence of multiple fault branches and meters-wide breccia zones means that numerous individual alternative flow paths are associated with a fault. The rate of calcite and opal deposition during the past eight to ten million years has been too low to plug all of the flow paths within fault zones.

An additional reason for the insignificant effect of secondary minerals on fault transmissivity is pertinent to the observation that calcite and opal abundances are lower in faults than in subsidiary features. Conceptual models of calcite deposition in the subsurface stress the slow kinetics of the precipitation process [Paces et al., 1998]. Most or all faults that cut the Paintbrush nonwelded hydrogeologic unit probably are fast paths, even if present net infiltration along their surface traces is too low to support fast path flow during the last 50 years. Fluid flow along the faults in the subsurface may be fast enough to discourage mineral precipitation.

The effectiveness of secondary minerals in modifying the flow characteristics of faults within nonwelded tuffs is less well addressed by the ^{36}Cl data set because fewer exposures of nonwelded tuffs exist in the ESF and the ^{36}Cl sample density is lower in these exposures than in the welded tuffs. The conceptual model of fast-path features requires the presence of faults that cut the nonwelded tuffs between the devitrified portions of the Tiva Canyon and Topopah Spring Tuffs (equivalent to the PTn hydro-geologic unit). None of the ^{36}Cl samples analyzed to date from this interval represent faults with fast-path isotopic signatures, so a basis does not exist at present for com-paring the mineralogy and flow behavior of faults in these nonwelded tuffs.

DATA EVALUATION FOR BRECCIA/WALL-ROCK PAIRED SAMPLES

As noted above, the textural and structural characteristics of breccia zones are difficult to translate into consistent criteria for paired-sample collection. The choice of 0.5 or 0.2 cm as the criterion for separating breccia-zone samples into coarse and fine fractions does not result in the production of separates with true genetic significance because it does not consistently separate "wall rock" from "fault gouge," or "broken but intact rock" from "mineralogically altered gouge," or "bedrock fragments" from "cement." In addition, fine material may be retained with the coarser fraction during sieving because it clumps together or coats larger rock fragments. Based on visual estimates, the coarse fractions of selected samples con-sisted at least 95% of the desired rock-fragment sizes. Small fragments of wall rock, broken off during sample collection, may also have been incorporated into the finer breccia fraction where they are visually indistinguishable. The effects of incomplete size separation will be discussed in the context of the measured isotopic ratios.

Before comparing the isotopic ratios of paired samples, it is also important to note that multiple samples of fractured rock from sites with bomb-pulse ^{36}Cl commonly yield ratios that differ by as much as a factor of two. This is true even for single samples that were processed and analyzed in duplicate and for bomb-pulse sites that were resampled with the intent of matching the original sample as closely as possible. These results probably reflect a combination of genuine isotopic variations along the flow paths and variable release of ^{36}Cl-poor rock chloride during the crushing step of sample preparation.

For us to detect meaningful differences in the isotopic ratios of breccia/wall-rock pairs, the ratios for multiple pairs ideally should define sets with common, consistent relationships to rock texture. The paired analyses of coarse and fine fractions or of broken wall rock and adja-cent breccia from Stations 13+67 through 35+93 show a consistent pattern of higher ^{36}Cl/Cl values in the coarse fractions (Table 3 and Figure 6). All of the coarse-fraction values and about half the fine-fraction values in this inter-val have ^{36}Cl/Cl ratios above 1250×10^{-15}, which is the threshold value indicating the unambiguous presence of

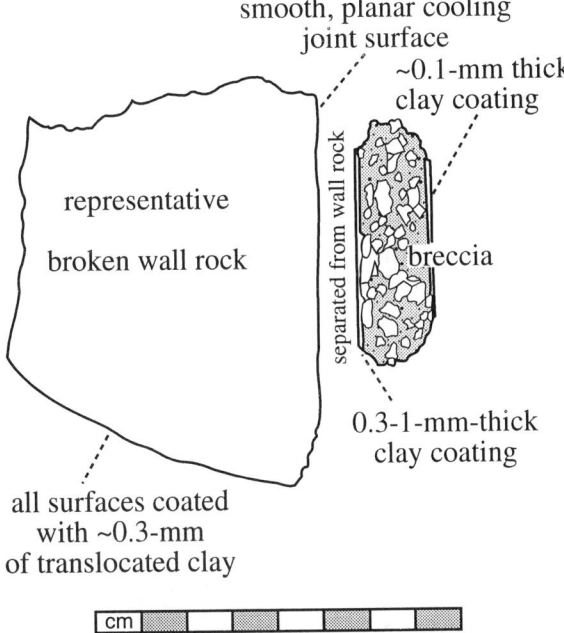

a   E154, Station 34+71

$^{36}Cl/Cl \times 10^{-15}$

Breccia: 803

Wall rock: 3794

smooth, planar cooling
joint surface

~0.1-mm thick
clay coating

separated from wall rock

representative

broken wall rock

breccia

0.3-1-mm-thick
clay coating

all surfaces coated
with ~0.3-mm
of translocated clay

cm

b   E242, Station 56+93

$^{36}Cl/Cl \times 10^{-15}$

<2-mm fraction: 1117 (breccia only)

>2-mm fraction: 664 (wall rock and
breccia)

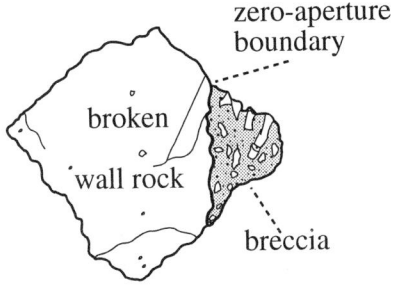

zero-aperture
boundary

broken

wall rock

breccia

Figure 6. Analytical results for paired size separates.

bomb-pulse ^{36}Cl. The consistent results suggest that incomplete or variable segregation of wall rock and breccia during sample collection and preparation did not seriously affect the results for samples from this particular ESF interval. The results, by their very consistency, are somewhat at odds with the variability of duplicate analyses described above. Isotopic ratios for wall-rock fractions should have been more susceptible to modification by rock chloride because these fractions required more crushing than the finer breccia fractions. Given the variability of values for the isotopic ratios of wall rock or coarse-fragment fractions, it is possible that some of these values have been lowered by incorporation of ^{36}Cl-poor chloride during sample preparation.

The sample pair from Station 36+55 shows the same association of higher isotopic ratio with the wall rock as the more northerly samples from Stations 13+67 through 35+93, even though the ratio is below the threshold value for unambiguous bomb-pulse ^{36}Cl (Table 3 and Figure 6). Samples from beyond Station 36+55 have lower values and generally smaller differences between the ratios measured for the different textural separates.

The baseline interpretation supported by the results from Stations 13+67 through 35+93 is that the fluids occupying accessible pore spaces of the fast fluid pathways were not in equilibrium with respect to the ^{36}Cl isotopic signal. Because these samples all represent active fast pathways that have apparently received infiltration from the surface within the past 50 years, and because the isotopic composition of the infiltration they receive has changed significantly over this short period due to input of bomb-pulse ^{36}Cl, isotopic disequilibrium within the fluid pathway is readily detectable. In areas of the ESF that have received no detectable bomb-pulse ^{36}Cl input and where ground-water travel times may have been significantly longer, as indicated by lower $^{36}Cl/Cl$ ratios, the changes of isotopic composition in the infiltrating water moving through these rocks probably occurred so gradually over this period that a lack of isotopic equilibrium between matrix and fracture pore water would not be detectable by a comparison of their ^{36}Cl signals.

TEXTURAL STUDIES OF BRECCIA/WALL-ROCK PAIRED SAMPLES

A substantial amount of the fast flow transmitted by faults into the Topopah Spring Tuff is distributed into cooling joints, many of which serve as the boundaries of minor breccia zones and faults (Table 2). This distribution of fast flow into small fracture-bounded breccia zones is particularly evident within an extensive foot wall zone of the Sundance fault (approximately Stations 33+00 through

35+93). The consistent isotopic results for the paired textural samples from this interval (Table 3 and Figure 6) suggest that the wall rock and breccia portions of these breccia zones have characteristic effects on fluid flow. These second-order transmissive features present an opportunity to investigate local fluid-rock interaction in an environment where the original bedrock hydrologic properties have been highly modified.

Unambiguous bomb-pulse values have not been detected in the southern part of the ESF beyond Sta. 45, although one sample from Sta. 56+93 has a value slightly below the bomb-pulse ^{36}Cl/Cl threshold value of 1250×10^{-15}. Our in-progress statistical analysis of the complete ^{36}Cl/Cl data set for the ESF indicates a high probability that this sample contains bomb-pulse ^{36}Cl. A textural study of the breccia/wall-rock sample pair from Sta. 56+93 is included for comparison with the data from the northern part of the ESF. This sample pair differs from the northerly bomb-pulse pairs in that the ^{36}Cl/Cl ratio is lower in the wall rock than in the breccia.

The main reason for this difference is the prevalence of lower net-infiltration values in the south [*Flint and Flint*, 1994; *Hevesi et al.*, 1994, 1996]. The absence of significant isotopic differences of such a distribution on this study is to restrict our ability to detect changing patterns of ^{36}Cl abundance in wall rock/breccia pairs along the full extent of the ESF. The preliminary results pre-sented here are based on a comparison of a pair from the northern ESF and the pair from the southern part that contains the highest ^{36}Cl/Cl value.

Northern ESF

The wall-rock and breccia subsamples E154-3 and E154-1 (Station 34+71), with ^{36}Cl/Cl values of 3794×10^{-15} and 803×10^{-15}, respectively (Figure 7a), differ substantially with regard to rock fragment sizes. Pervasive multi-directional fracturing of the bedrock has broken it into intact rock domains of ~10 cm dimensions. The breccia deposit, bounded by two cooling joint faces oriented 031/86W, consists of ~1 cm to <0.1 mm fragments and contains additional translocated silt- and clay-size material (<~0.1 mm) near the outer edges of the deposit (Figure 7a). With a relatively fine grain size and a porous clastic texture, the breccia should have hydrologic properties approaching those of a nonwelded tuff in which slower matrix flow predominates over fracture flow. Fluids entering the breccia deposit may travel through tortuous intergranular pathways.

The nature of the interfaces between the rock fracture surfaces and the breccia deposit provides indications of the long-term importance of these features as fluid pathways. The 1-mm-thick clay coating on the outer surfaces of the breccia deposit is several times thicker than the similar coatings on breccia clasts and wall-rock blocks. The clay coatings did not originate *in situ*, but rather had to be transported to their present locations by water. The greater thickness of the clay coating on the breccia deposit could reflect a larger aperture between the wall rock and the breccia deposit than between either wall-rock blocks or breccia clasts, allowing the interface to accommodate a thicker clay coat. In addition, the interface may have received more fluid input than the breccia deposit or the unmodified fractures within the bedrock. Clearly, these two potential factors are not mutually independent because a fluid pathway with larger aperture and greater continuity would likely receive more influx.

The existence of finer-grained translocated material in the outermost few mm of the breccia deposit indicates that the fracture or fractures hosting the deposit experienced mm-scale dilation after the main portion of the deposit was formed. Through the processes of compaction and clay deposition, the breccia gradually developed sufficient cohesion to prevent the component clasts from collapsing into newly opened space as the fracture dilated. Later dilation provided space for transport and deposition of the 1-mm clay coating on the outer surfaces of the breccia deposit.

In summary, repeated dilation occurred as this fracture/breccia system accommodated tectonic extension. By means of this process, a fracture pathway could be maintained along the interface between the bedrock and the breccia that allows at least some percolating water to bypass the breccia. The possible role of the clay coatings in restricting access of fracture fluids to either the breccia or the wall rock is uncertain, but the slightly better adhesion of clay to the breccia than to the wall rock suggests that it may be more effective at inhibiting imbibition into the breccia.

Southern ESF

Paired breccia and wall-rock samples E242-2 and E242-1, from Station 56+93 in the southern part of the ESF, represent the same middle nonlithophysal zone of the Topopah Spring Tuff and the E154 samples. The ^{36}Cl/Cl value of the E242 breccia is the highest obtained from southern ESF samples and may include a very small component of bomb-pulse ^{36}Cl (Table 3). The substantially different matrix and wall-rock values for this pair make it unique among paired samples from south of Station 36+55. The matrix>wall-rock ^{36}Cl/Cl values for this pair are in contrast with the wall-rock>matrix values commonly found at Station 35+93 and northward. Textural and mineralogic characteristics of the sample pair are typical for at least some of the southern ESF, as well.

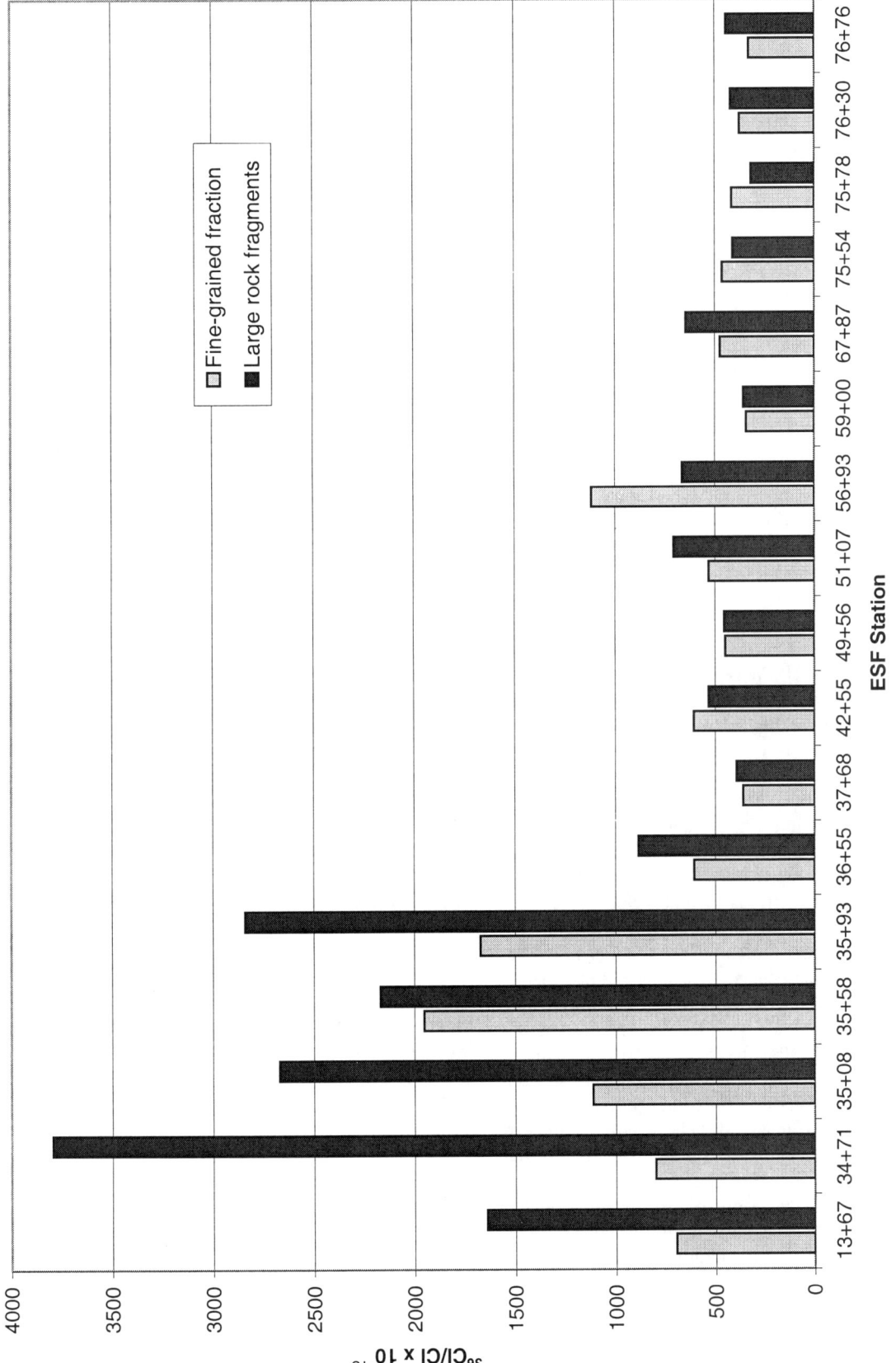

Figure 7. a) Textural relations and secondary mineralogy of wall rock and breccia, sample E154. b) Textural relations of wall rock and breccia, sample E242.

The intact wall rock is densely welded, devitrified tuff. About 10 to 15% of the fractures in the broken wall rock have ≤1-mm-wide bleached margins, with or without vapor-phase mineral deposits <0.5 mm thick. The fracture surfaces are planar to irregular and smooth to rough. Many fracture surfaces have been modified by shearing and breakage, including the development of slickensides. Some polish is present on smooth flat fracture surfaces without vapor-phase coatings.

The breccia is composed of <3-mm fragments, with 95% of the fragments <2 mm. Clay content is estimated to be <5%, and breccia porosity was visually estimated to be <10%. There is no secondary-mineral cementation of the breccia. Some fragments have manganese-mineral coatings ≤0.05 mm thick, with variable coverage in the range of 5 to 100%. This is slightly more abundant than manganese fracture coatings in the wall rock. The boundaries between intact rock and breccia are mostly tight, with no visible aperture or evidence of particulate or mineral deposition along a separated contact (Figure 7b). In some places, original planar cooling-joint surfaces with bleached margins and vapor-phase deposits were broken off the wall rock and incorporated into the breccia, resulting in new, rougher-surfaced wall-rock/breccia boundaries.

SUMMARY

The mineralogic and textural data collected so far have not defined generally applicable, distinctive characteristics of either fast-path or long water-residence sites in the ESF. These special infiltration cases apparently represent relatively minor variations in the hydrologic and geochemical environment of the unsaturated zone, especially when the effects of infiltration, as represented by the mineralogy, are averaged over very long periods of time. Specific examples of fast-path systems, however, may be distinguished by characteristic differences in mineral abundance. The Sundance fault and associated subsidiary flow paths constitute the best example of distinctive mineralogy. The two mineralogic characteristics are the presence of calcite and thick clay/mordenite fracture coatings. Both of these attest to the particularly high fracture connectivity in the Sta. 34+28 to 35+93 interval north of the Sundance fault. The thicknesses and textures of the clay coatings suggest that the apertures of cooling joints were increased by minor tectonic movement shortly after the Topopah Spring Tuff was deposited. The widespread occurrence of calcite precipitates and translocated particulates in this interval also indicates excellent inter-fracture and fracture-fault connectivity. Overall connectivity in the vicinity of the Sundance fault may have been enhanced by multiple episodes of fracture dilation and extension, as suggested by

the mineralogic evidence and by the presence of an additional local fracture set.

Small-scale fluid-flow characteristics of fracture/breccia systems in densely welded tuff may reflect an influence of the local tectonic stress system responsible for the formation of these transmissive features. In tectonic domains where extension predominated and shearing was minor, the cooling joint boundary between wall rock and breccia remained planar and was maintained as an open fracture during multiple episodes of dilation. Deposition of translocated clay in the breccia pore spaces and along the cooling joint surface/breccia boundary helped to isolate the breccia deposit from the periodically re-opened bounding fracture because the clay coatings tended to adhere to the breccia. As a result, downward-percolating water tends to stay in the bounding fracture and bypass the breccia matrix. This process is detectable where bomb-pulse ^{36}Cl is present as higher values of ^{36}Cl/Cl in the wall rock containing the bounding fracture surface than in the breccia. This deformational regime may have been more typical of the rocks penetrated by the northern part of the ESF.

In tectonic domains where shearing predominated over extension, the cooling joint boundary between wall rock and breccia was transformed by breakage into a rough, irregular boundary where the wall rock is in tight contact with the breccia. The breccia itself may be less porous than breccias formed under extensional regimes, but the absence of a fracture gap between the breccia and the wall rock may help maintain fluid flow within the breccia instead of along the wall rock/breccia boundary. This process is detectable where bomb-pulse ^{36}Cl is present as higher values of ^{36}Cl/Cl in the breccia than in the wall rock containing the bounding fracture surface. This deformational regime may have been more typical of the rocks penetrated by the southern part of the ESF.

Acknowledgments. This work was supported and managed by the U. S. Department of Energy, Yucca Mountain Site Characterization Office, as part of the Civilian Radioactive Waste Management Program. Anthony Garcia prepared the illustrations. Reviews by William Murphy, Fred Phillips, David Vaniman, and Steve Beason contributed substantially to the improvement of the manuscript.

REFERENCES

Anna, L., and P. Wallman, Characterizing the fracture network in the unsaturated zone at Yucca Mountain, Nevada, Part 2. Numerical simulation of flow based on a three-dimensional discrete fracture network in fractured reservoirs: characterization and modeling, in *Rocky Mountain Geologists 1997 Guidebook*, edited by T. E. Hoak, A. L. Klawitter, and P. K. Blomquist, 199-208, 1997.

Bentley, H. W., F. M. Phillips, and S. N. Davis, Chlorine-36 in the terrestrial environment, in *Handbook of Environmental Isotope Geochemistry*, vol. IIB, edited by J.C. Fontes and P. Fritz, Elsevier, Amsterdam, 427-480, 1986.

Bish, D. L., and S. J. Chipera, Revised mineralogic summary of Yucca Mountain, Nevada, *Los Alamos National Laboratory Report LA-11497-MS*, 68 p., 1989.

Buesch, D. C., R. W. Spengler, T. C. Moyer, and J. K. Geslin, Proposed stratigraphic nomenclature and macroscopic identification of lithostratigraphic units of the Paintbrush Group exposed at Yucca Mountain, Nevada, *U.S. Geological Survey Open-File Report OFR 94-469*, 47 p., 1996.

Carlos, B.A., Minerals in fractures of the unsaturated zone from drill core USW G-4, Yucca Mountain, Nye County, Nevada, *Los Alamos National Laboratory Report LA-10415-MS*, 55 p., 1985.

Carlos, B. A., S. J. Chipera, and D. L. Bish, Distribution and chemistry of fracture-lining minerals at Yucca Mountain, Nevada, *Los Alamos National Laboratory Report LA-12977-MS*, 92 p., 1995.

Chipera, S. J., D. T. Vaniman, B. A. Carlos, and D. L. Bish, Mineralogic variation in drill core UE-25 UZ#16, Yucca Mountain, Nevada, *Los Alamos National Laboratory Report LA-12810-MS*, 39 p., 1995.

Chung, F. H., Quantitative interpretation of X-ray diffraction patterns of mixtures. I. matrix-flushing method for quantitative multicomponent analysis, *J. Appl. Crystallogr., 7*, 519-525, 1974a.

Chung, F. H., Quantitative interpretation of X-ray diffraction patterns of mixtures. II. adiabatic principle of X-ray diffraction analysis of mixtures, *J. Appl. Crystallogr., 7*, 526-531, 1974b.

Cowan, D. L., V. Priest, and S. S. Levy, ESR dating of quartz From Exile Hill, Nevada, *Appl. Rad. Isot., 44*, 1035-1039, 1993.

Day, W. C., C. J. Potter, D. S. Sweetkind, R. P. Dickerson, and C. J. San Juan, Bedrock geologic map of the central block area, Yucca Mountain, Nye County, Nevada, *U.S. Geol. Surv. Misc. Invest. Map, I-2601*, 1998.

Fabryka-Martin, J. T., A. V. Wolfsberg, S. S. Levy, J. L. Roach, S. T. Winters, L. E. Wolfsberg, D. Elmore, and P. Sharma, Distribution of fast hydrologic paths in the unsaturated zone at Yucca Mountain, in *Proceedings of the Eighth International Conference on on High-Level Radioactive Waste Management*, Am. Nucl. Soc., La Grange Park, Illinois, 93-96, 1998.

Flint, A. L., and L. E. Flint, Spatial distribution of potential near-surface moisture flux at Yucca Mountain, Nevada, in *Proceedings of the Fourth International Conference on High Level Radioactive Waste Management*, Am. Nucl. Soc., La Grange Park, Illinois, 2352-2358, 1994.

Flint, L. E., and A. L. Flint, Shallow infiltration processes at Yucca Mountain, Nevada – neutron logging data 1984-93, *U.S. Geol. Surv. Water Res. Invest. Report 95-4035*, 46 p., 1995.

Glasstone, S., editor, *The Effects of Nuclear Weapons*, Revised Edition, U.S. Atomic Energy Commission, Washington D.C., 730 p., 1962.

Henning, R. J., and L. D. Rickertsen, Unsaturated zone percolation flux at Yucca Mountain, in *Proceedings of the Eighth International Conference on High-Level Radioactive Waste Management*, Am. Nucl. Soc., La Grange Park, Illinois, 133-134, 1998.

Hevesi, J. A., A. L. Flint, and L. E. Flint, Verification of a 1-dimensional model for predicting shallow infiltration at Yucca Mountain, in *Proceedings of the Fifth International Conference on High Level Radioactive Waste Management*, Am. Nucl. Soc., La Grange Park, Illinois, 2323-2332, 1994.

Hevesi, J. A., A. L. Flint, and L. E. Flint, The effect of bedrock properties on the spatial variability of infiltration at Yucca Mountain, Nevada (abstract), *Geol. Soc. Am. Abstr. Prog., 28*, A-522, 1996.

Levy, S. S., Mineralogic alteration history and paleohydrology at Yucca Mountain, Nevada, in *Proceedings of the Second Annual International Conference on High-Level Radioactive Waste Management*, Am. Nucl. Soc. and Am. Soc. Civil. Engin., La Grange, Illinois and New York, 477-485, 1991.

Levy, S. S., Surface-discharging hydrothermal systems at Yucca Mountain – examining the evidence, in *Scientific Basis for Nuclear Waste Management XVI: Mater. Res. Soc. Symp. Proc., 294*, edited by C. G. Interrante and R. T. Pabalan, 543-548, 1993.

Levy, S. S., and J. R. O'Neil, Moderate-temperature zeolitic alteration in a cooling pyroclastic deposit, *Chem. Geol., 76*, 321-326, 1989.

Levy, S. S., D. L. Norman, and S. J. Chipera, Alteration history studies in the Exploratory Studies Facility, Yucca Mountain, Nevada, USA, in *Scientific Basis for Nuclear Waste Management XIX: Mater. Res. Soc. Symp. Proc., 412*, edited by W. M. Murphy and D. A. Knecht, 783-790, 1996.

Levy, S. S., D. L. Bish, and S. J. Chipera, Overview of syngenetic mineralization at Yucca Mountain, Nevada, USA, *Eos Trans. AGU, 80*(17), Spring Meet. Suppl., S9, 1999.

Montazer, P., and W. E. Wilson, Conceptual hydrologic model of flow in the unsaturated zone, Yucca Mountain, Nevada, *U.S. Geol. Surv. Water-Resource Inv. Report 84-4345*, 55 p., 1984.

Neymark, L. A., Y. V. Amelin, J. B. Paces, and Z. E. Peterman, U-Pb age evidence for long-term stability of the unsaturated zone at Yucca Mountain, in *Proceedings of the Eighth International Conference on High-Level Radioactive Waste Management*, Am. Nucl. Soc., La Grange Park, Illinois, 85-87, 1998.

Ortiz, T. S., R. L. Williams, F. B. Nimick, and D. L. South, A three-dimensional model of reference thermal/mechanical and hydrological stratigraphy at Yucca Mountain, southern Nevada, *Sandia National Laboratories Report SAND84-1076*, Albuquerque, New Mexico, 1985.

Paces, J. B., L. A. Neymark, B. D. Marshall, J. F. Whelan, and Z. E. Peterman, Inferences for Yucca Mountain unsaturated-zone hydrology from secondary minerals, in *Proceedings of the Eighth International Conference on High-Level Radioactive Waste Management*, Am. Nucl. Soc., La Grange Park, Illinois, 36-39, 1998.

Plummer, M. A., F. M. Phillips, J. T. Fabryka-Martin, H. J. Turin, P. E. Wigand, and P. Sharma, Chlorine-36 in fossil rat urine: an archive of cosmogenic nuclide deposition during the past 40,000 years, *Science, 277*, 538-541, 1997.

Potter, C. J., R. P. Dickerson, and W. C. Day, Nature and continuity of the Sundance Fault, Yucca Mountain, Nevada, *U.S. Geol. Surv. Open-File Report 98-266*, 1999 (in press).

Sawyer, D. A., R. J. Fleck, M. A. Lanphere, R. G. Warren, D. E. Broxton, and M. R. Hudson, Episodic caldera volcanism in the

Miocene southwestern Nevada volcanic field: revised stratigraphic framework, [40]Ar/[39]Ar geochronology, and implications for magmatism and extension, *Geol. Soc. Am. Bull., 106*, 1304-1318, 1994.

Sweetkind, D. S., S. Williams-Stroud, and J. Coe, Characterizing the fracture network at Yucca Mountain, Nevada, Part 1. Integration of field data for numerical simulations, in *Fractured Reservoirs: Characterization and Modeling,* edited by T. E. Hoak, A. L. Klawitter, and P. K. Blomquist, *Rocky Mountain Association of Geologists 1997 Guidebook*, 185-196, 1997.

Vaniman, D. T., and S. J. Chipera, Paleotransport of lanthanides and strontium recorded in calcite compositions from tuffs at Yucca Mountain, Nevada, USA, *Geochim. Cosmochim. Acta, 60*, 4417–4433, 1996.

Whelan, J. F., and J. S. Stuckless, Paleohydrologic implications of the stable isotopic composition of secondary calcite within the Tertiary volcanic rocks of Yucca Mountain, Nevada, in *Proceedings of the Third Annual International Conference on High Level Radioactive Waste Management,* Am. Nucl. Soc., La Grange, Illinois and New York, 1572-1581, 1992.

Whelan, J. F., R. J. Moscati, S. B. M. Allerton, and B. D. Marshall, Applications of isotope geochemistry to the reconstruction of Yucca Mountain, Nevada, paleohydrology – Status of Investigations: June 1996, *U.S. Geol. Surv. Open-File Report 98-83*, 41 p., 1998.

S. Chipera, S. Levy, and G. WoldeGabriel, Earth and Environmental Sciences Division, MS D462, Los Alamos National Laboratory, Los Alamos, NM, 87545.

J. Fabryka-Martin and J. Roach, Chemical Science and Technology Division, MS J514, Los Alamos National Laboratory, Los Alamos, NM, 87545.

D. Sweetkind, U.S. Geological Survey, MS 421, Box 25046, Federal Center, Denver CO, 80225-0046.

How do Fracture-Vein Systems Form in a Geothermal Reservoir? Examples from Northern Honshu, Japan

Shiro Tamanyu

Department of Geothermal Research, Geological Survey of Japan, Tsukuba, Japan

Productive fractures are classified into a dominant high-angle group, and a less common low-angle group in the Uenotai and Sumikawa geothermal fields and possibly a shallow reservoir of the Kakkonda field in Northern Honshu, Japan. Dominant low-angle fractures with rarer high-angle fractures occur in the deep reservoir at Kakkonda field. The formation mechanisms for these different fractures have been interpreted using the concept of fault-valve behavior. The dominant high-angle fractures in the shallow reservoirs have been interpreted as strike-slip fault by previous workers. However, the author would like to propose another possibility: to interpret them as misoriented tensile fractures formed under a thrust regime, because representative productive fractures have evidence of hydrofracturing, even though others have evidence of strike-slip fault movement. The dominant high-angle fractures are believed to have formed by reactivation of pre-existing high-angle fractures as a result of fault-valve action on unfavorably oriented faults when the stress changed suddenly from lithostatic to hydrostatic. This change may have been triggered by pressurized magmatic water injection. In this paper, these high-angle fractures are regarded as strike-slip fault and/or reactivation of pre-existing fractures. In the deep reservoir at Kakkonda, the dominant low-angle fractures seem to have formed by thrusting and/or tensile fractures as a result of fault-valve action on favorably oriented faults by magmatic fluid injection.

1. INTRODUCTION

Productive fractures in the Uenotai geothermal field, northern Honshu, Japan belong to either a high-angle group, or a less common low-angle group. They appear to have formed as tensile and/or tensile shear fractures under hydrostatic-lithostatic transitional conditions and subhorizontal σ_1 stresses [*Tamanyu et al.*, 1998]. In this paper, fracture distributions and fracture origins at Uenotai are compared to those at Sumikawa and Kakkonda, where extensive exploration data is available.

The morphological features of geothermal drillhole feed zones were investigated using descriptions of drill core

Faults and Subsurface Fluid Flow in the Shallow Crust
Geophysical Monograph 113

samples from the Uenotai and Sumikawa fields and FMI (Formation Micro Imagery) logging data in the Kakkonda field, northern Honshu, Japan. The core samples show that fractures, commonly filled with hydrothermal vein minerals, act as fluid feed zones. Those parts of a fracture-vein system that can be correlated with fluid feed zones are referred to as productive fractures. This paper aims to clarify the relationship between the fluid flow in the reservoir and fracture characteristics. This kind of fracture analysis on core samples or logging data has been tested in The Geysers, USA. *Thompson and Gunderson* [1992] showed that the orientation of steam-bearing fractures within greywacke is generally random, but includes intensive low-angle fracture zones that they believed to be representative of re-opened Franciscan-age structures. In contrast, high productivity in intrusive rocks occurs in narrow, steep-dipping zones that are related to recent strike-slip tectonics. *Hulen et al.*[1995] reported two types of veinlets in drill cores obtained by The Geysers Coring Project. One type consists of randomly-oriented, Franciscan metamorphic quartz -calcite veinlets, and the other consists of high-angle, late Cenozoic hydrothermal veinlets mainly made up of euhedral quartz , bladed calcite, pyrite and epidote. The references of gold deposits in hydrothermal veins also provide the useful information for this work because the environment of gold deposit in veins are pretty similar to geothermal environment.

The locations of the three fields studied are shown in Fig. 1. The Uenotai geothermal field which is located in the southern part of Akita Prefecture, produces 27.5 MW. The Sumikawa geothermal field is located in the northern part of Akita Prefecture, and produces 50 MW. The Kakkonda geothermal field is located in the western part of Iwate Prefecture. The total production from this field is 80 MW.

Magma chambers of Quaternary volcanoes appear to provide the heat that drives fluid circulation [e.g. *Takahashi et al.,* 1996, *Tamanyu et al.,* 1996]. The exploited geothermal reservoirs in these areas are mainly fractured lower Tertiary rocks and the top of the pre-Tertiary basement. However, the fracture zone at the margin of Quaternary granite is exploited as deep steam supply zone in Kakkonda [e.g. *Doi, et. al,* 1995].

2. DEFINITION OF FRACTURE-VEIN SYSTEM

The fractures identified as geothermal fluid feed zones, the so-called productive fractures, are generally accompanied by hydrothermal veins and open-space mineralization.

Meunier [1995] classified hydrothermal veins into three different types (injection vein type, infiltration vein type and drainage vein type) coming from different source regions. He regarded hot springs and geysers in geothermal fields as typical of an injection regime with the hot fluid coming from deep, higher pressure-temperature regimes. The fluid rises to the surface and permeates the surrounding rocks. The fracture-vein systems here identified as fluid feed zones can be regarded as belonging to an injection regime, even though they occur in the deep subsurface and are not obviously related to surface manifestations.

3. MORPHOLOGICAL FEATURES OF FLUID FEED ZONES IN EXPLOITED GEOTHERMAL FIELDS

3.1 Uenotai Geothermal Field

The Uenotai field is about 15 km west from a N-S trending Quaternary volcanic front and situated on the Matsushima-Honjo Belt [*Oide and Onuma,* 1960], which is characterized by a NW trending structural high zone of Tertiary and Quaternary age. The drilled part of the Uenotai geothermal field can be divided into a structural high in the center and a structural low on its periphery with a transitional zone separating the two. The structural high is correlated with high-temperatures and vapor-dominated

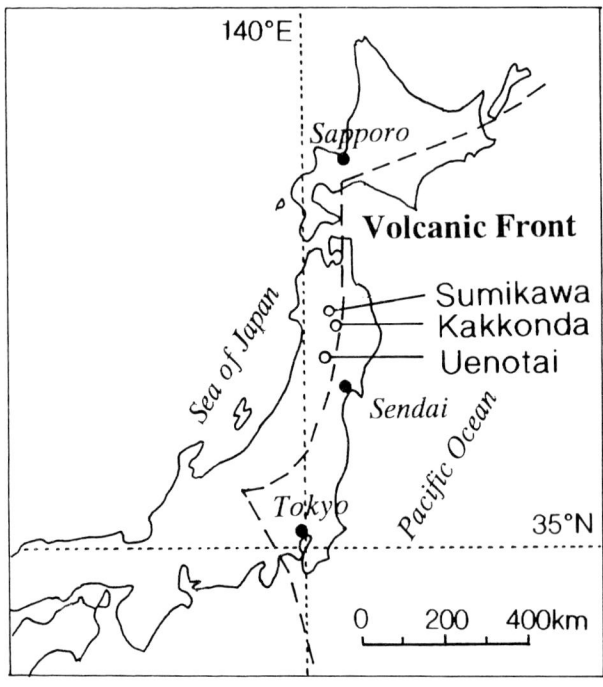

Figure 1. Location map of studied areas.

fluids, whereas the structural low correlates with a medium-temperature and liquid-dominated zone [*Naka et al.,* 1987, *Robertson-Tait et al.,* 1990, *Naka and Okada,* 1992]. Here, The terms "vapor-dominated" and "liquid-dominated" are used to describe systems that discharge mainly vapor or mainly water. The fractures developed in the structural high and its surroundings have essentially the same characteristics, although calcite precipitation is more common in the liquid-dominated system, where it forms in response to boiling. The productive fractures tend to develop around geologic formation boundaries such as Tertiary and pre-Tertiary formation boundary and outer rim of intrusive body, because these parts provide the contrasting physical properties which are easily fractured [*Tamanyu et al.,* 1998].

The characteristics of feed zones in this area, based on studies of core (Plates 1 and 2, Table 1), were described by *Tamanyu et al.*[1998], who emphasized the tensile fracturing associated with increases in pore pressure. A typical example is sketched on the left side of Fig. 2, and shows that the open space along the quartz vein provides the main feed point of geothermal fluids. This feed point can be regarded as a tensile fracture because it is accompanied by a hydrothermal quartz vein and brecciated silicified tuff, which are suggestive of hydrofracturing. Some of other productive fractures indicate strike-slip displacement by virtue of en echelon structure and slickensides.

Table 1. List of lost circulation records and corresponding fractures in the Uenotai field.

Well no.	Lost circulation				Fracture characteristics			
	Depth (m)	Mud lost (l/m)	Formation	Lithology	Fracture type	Width (mm)	Dip (degree)	Filling minerals*
[vapor dominated reservoir]								
T5	911.35	5.0	Basement	Diorite	echelon (partly dissolved)	6.0	85	Ep
	911.68	5.0	Basement	Diorite	single	8.0	60	Ep
	911.93	5.0	Basement	Diorite	brecciated	12.0	35	Q, Cal, Ep (euhedral Q)
T21	1161.35	90.0	Doroyu	Andesite	single	3.0	65	Q,
	1162.38	90.0	Doroyu	Andesite	single	8.0	15	Q
T29	1020.45	100.0	Basement	Granite	single	>10.0	75	Ep, Q
	1021.50	100.0	Basement	Granite	parallel	5.0	70	Q, Ep
KT2	922.00	<1.0	Basement	Breccia	single	<1.0	25	Cal
	1000.05	<1.0	Doroyu	Dacite	single	?	30	Cal
	1025.00	<1.0	Basement	Serpentinite	single	0.0	75	Cal
	1094.06	2.5	Basement	Serpentinite	single (slickenside)	2.0	40	unknown
[liquid dominated reservoir]								
T27	1110.78	unknown	Doroyu	Andesite	single	4.0	65	Ep,Cal,Cly
	1160.00	unknown	Doroyu	Andesite	vein	>12.0	50	Cal
T38	859.35	unknown	Doroyu	Basalt	parallel	25.0	35	Cal
KU1	378.00	150.0	Minase	Dacitic tuff	druse	5.0	30	Q, W
SMT1	1141.90	60.0	dike	Quartz porphyry	dike	?	10	Cal
	1147.65	90.0	dike	Quartz porphyry	network	?	?	Cal, La
YO7	689.10	90.0	Minasegawa	Dacitic tuff	single	1.0	75	non
	1419.00	30.0	Minasegawa	Dacitic tuff	single (slickenside)	1.0	30	non
MS6	1489.00	100.0	Doroyu	Andesite	parallel, druse (partly dissolved)	5.0	70	non

* Ep: epidote, Q: quartz, Cal: calcite, Cly: clay mineral, W: wairakite, La: laumontite

The productive fractures in this field are classified into a dominant high-angle (average: 77°) group and a less common low-angle (average: 29°) group. The bimodal distribution remains after the data have been corrected for sampling bias in a vertical borehole (Table 2, Fig. 3). Non-productive fractures are mostly high-angle features (<45°). The average width of the productive fractures is 4.9 mm for the high-angle group and 4.8 mm for the low-angle group (Table 3).

The fracture-vein system around the Uenotai field has been surveyed mainly on the surface by previous workers, even though it is not clear whether the system is associated with the feed zone. *Tamanyu and Mizugaki* [1993] reported that the fracture-vein system on the Okumaemori-Oyasudake upheaval belt is characterized by high-angle-fractures in NW-SE trend, mid-angle-fractures in N-S trend, and hydrothermal veins in E-W trend. The fracture-vein

system on the Kijiyama subsidence belt is characterized by high-angle-fractures in ENE -WSW and ESE-WNW trends, low-angle-fractures, and hydrothermal veins which are oblique to dominant fracture trends. *Koshiya, et al.*[1994] and *Takemoto, et al.*[1995] reported that the fracture-vein system around the Uenotai field is characterized by two systems: one is NE-SW to E-W trending strike-slip and normal fault, and the other is N-S to NW-SE trending strike-slip and normal faults.

3.2 Sumikawa Geothermal Field

The Sumikawa field is about 20 km west from the N-S trending Quaternary volcanic front of northeast Japan Island, and situated on the north-northeastern slope of the active Mt. Yakeyama volcano. A fracture pattern similar to that at Uenotai occurs at Sumikawa. The reservoir

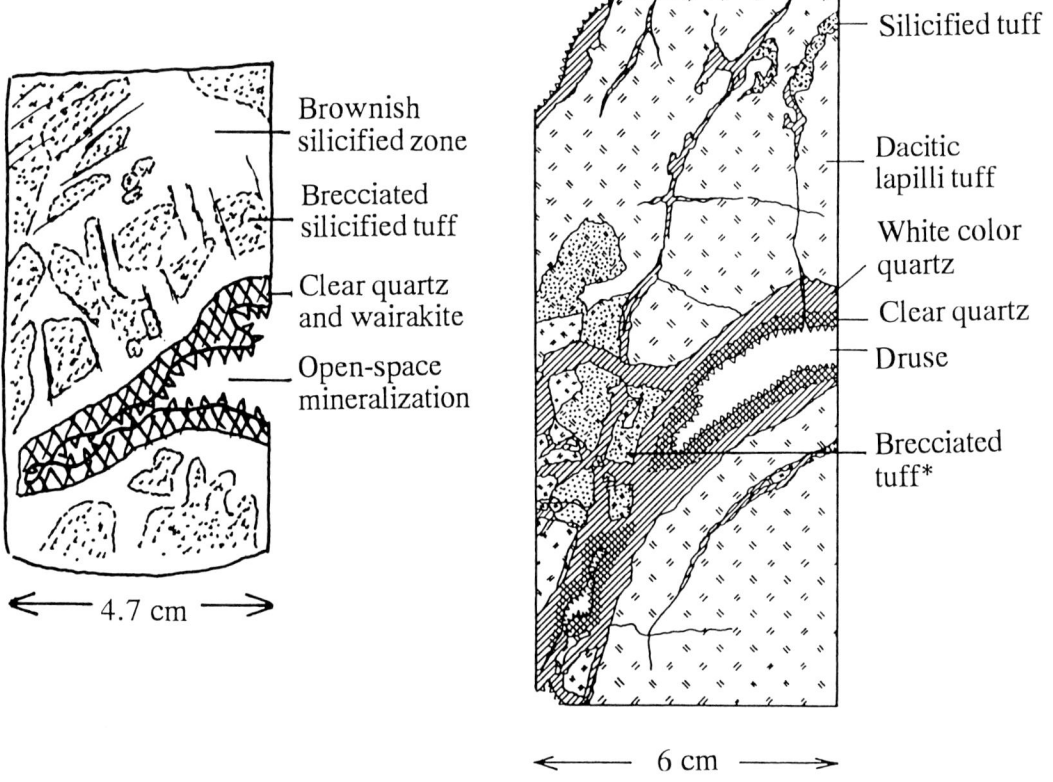

Figure 2. Typical examples of the open fractures from the Uenotai and the Sumikawa fields. Left: sketch of open fracture in a core sample from KU-1 in the Uenotai field. Right: sketch of open fracture in a core sample from S-3 well in the Sumikawa field [*Kubota*, 1985]. Brecciated tuff is regarded as hydrofracturing breccia by the author.

characteristics were described by *Kubota* [1985]. During drilling, circulation losses occurred frequently in the high temperature zone, suggesting a high degree of fracturing. However, fractures are sealed with vein minerals in the low-temperature zone. A typical example of productive fractures is sketched on the right side of Fig. 2, and shows that the open space along the quartz vein provides the main feed point of geothermal fluids [*Kubota*, 1985]. This feed point appears to be tensile fracture because it is accompanied by a hydrothermal quartz vein and brecciated silicified tuff indicating hydrofracturing, even though *Kubota* [1985] does not describe these features.

The angle and width of open fractures in lost circulation zones were measured in drill cores from well S-3 (Fig. 4), and are described in Table 2 and Fig. 5, respectively. The histogram shows that both high- and low-angle fractures are present, with the high-angle fractures dominating. As at Uenotai, a bimodal distribution remains after the data have

been corrected for sampling bias in a vertical borehole. The average width of the productive fractures is 2.8 mm for the high-angle group and 4.6 mm for the low-angle group (Table 3).

The transmissivity of productive fractures was determined by pressure transient tests using water injection, and calculated to be around $3 \times 10^{-12} m^3$. These values are higher than those in the production wells at Ohnuma, 2 km to be east [*Sakai et al.*, 1986]. Circulation losses during drilling occurred on average every 300 m in the > 200°C zone, suggesting that this zone is fairly heavily fractured. These feed points were correlated each other in three dimensions, yielding 5 fracture planes with strikes close to E-W and dips of 54°-67°[*Bamba et al.*, 1995]. The result indicates that productive fractures at Sumikawa have preferential strikes rather than a random distribution as was previously believed. *Bamba and Kubota* [1997] concluded that these fractures were formed by magma injection.

(a) T5

(b)T21

(c)T29

Plate 1. Examples of productive fractures at the vapor-dominated zone in the Uenotai geothermal field, Northern Honshu, Japan [*Tamanyu, et al.*, 1998]. The feed zones identified on these core samples are referred in Table 1.

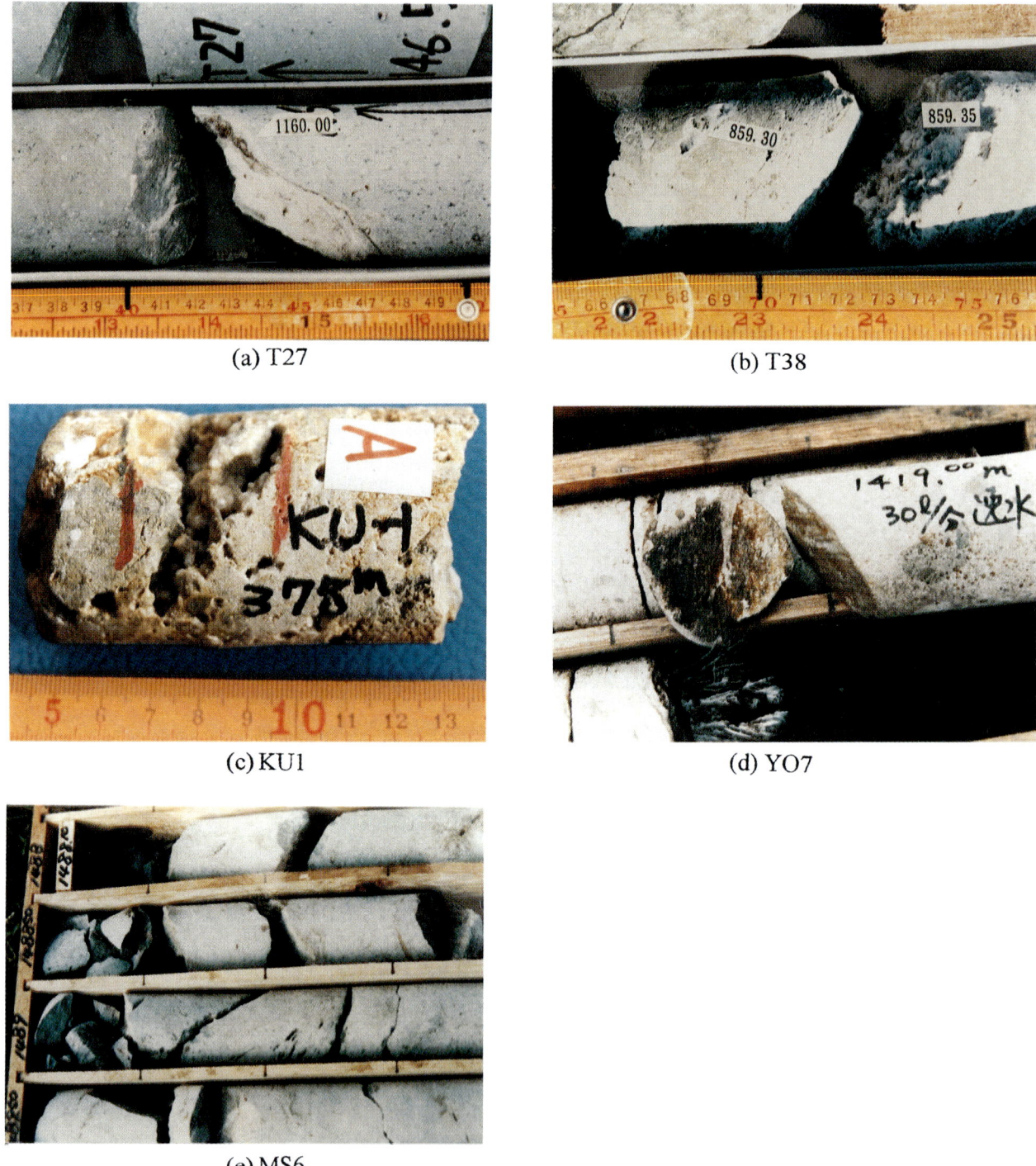

(a) T27

(b) T38

(c) KU1

(d) YO7

(e) MS6

Plate 2. Examples of productive fractures at the liquid-dominated zone in the Uenotai geothermal field, Northern Honshu, Japan [*Tamanyu, et al.,* 1998].

Table 2. Average dip angles of high- and low-angle fractures in the Uenotai and Sumikawa fields.

[Uenotai Field]

Original dip	No	1/cosθ	No * (1/cosθ)	Average dip
85	1	11,474	11,474	
75	3	3,864	11,592	
70	1	2,924	2,924	
65	1	2,366	2,366	
60	1	2,000	2,000	
Subtotal	7		30,356	
(Average dip for High-angle Fractures)			(θ=77 degree)	

40	1	1,305	1,305	
35	1	1,221	1,221	
30	3	1,155	3,465	
25	1	1,103	1,103	
15	1	1,035	1,035	
10	1	1,015	1,015	
Subtotal	8		9,144	
Average dip for Low-angle Fractures			(θ=29 degree)	

[Sumikawa Field]

Original dip	No	1/cosθ	No * (1/cosθ)	Average dip
87	1	19,107	19,107	
85	1	11,474	11,474	
80	3	5,759	17,277	
75	1	3,864	3,864	
70	9	2,924	26,316	
60	4	2,000	8,000	
55	1	1,743	1,743	
50	1	1,556	1,556	
45	2	1,414	2,828	
Subtotal	7		92,165	
Average dip for High-angle Fractures			(θ=76 degree)	

Table 2. Continued.

40	4	1,305	5,220
35	1	1,221	1,221
30	4	1,155	4,620
25	1	1,103	1,103
20	2	1,064	2,128
0	1	1,000	1,000
Subtotal	13		15,292
Average dip for Low-angle Fractures		(θ=32 degree)	

3.3 Kakkonda Geothermal Field

The Kakkonda field is about 10 km west from the N-S trending Quaternary volcanic front of northeast Japan Island. This is the first field in Japan for which it was proven that the young Quaternary granite plays the role of the heat source for the present geothermal system. The New Energy and Industrial Technology Development Organization (NEDO) has carried out the project "Deep-Seated Geothermal Resources Survey" as a part of New Sunshine project in Kakkonda since 1992. This project aims to understand the overall geothermal environment, including shallow systems, and evaluate the possibility of utilizing deep hydrothermal fluids deeper than 3,000 m. Exploration well WD-1a was drilled to 3,729 m depth, encountering Quaternary granite from 2,860 m to the bottom [Yagi et al., 1995]. The fluids discharged from the Kakkonda reservoir have low salinities (< 1 wt.%) and are meteoritic in origin, while Na-K-Ca-Fe type brine with a high salinity over 19 wt.% is trapped inside the granitic pluton of the thermal conduction zone and is considered to be magmatic in origin [Kasai et al., 1996]. The fluid inclusions from hydrothermal veins also suggest mixing of saline brine (magmatic) with low salinity fluid (meteoric) [Sasaki et al., 1998]. The fracture-vein system in this field was surveyed in detail on the surface and three stress fields (F1, F2 and F3) were reconstructed [Koshiya, et al., 1993, Koshiya, et al., 1994]. They concluded that the NW-moderately dipping and NW-steeply dipping systems were formed in the first stage and constrained the outline of the geothermal area trending NW-SE, and that the NE to E-W system has played an important role on the paths of geothermal water after the formation of the two NW systems. The productive fractures in this field were analyzed by Doi et al.[1995] who concluded that they formed by neotectonic stress after the emplacement of the Quaternary Kakkonda Granite.

Fracture analysis on WD-1a, was carried out using spot core samples (13 depths) and FMI logging [Kato et al.,

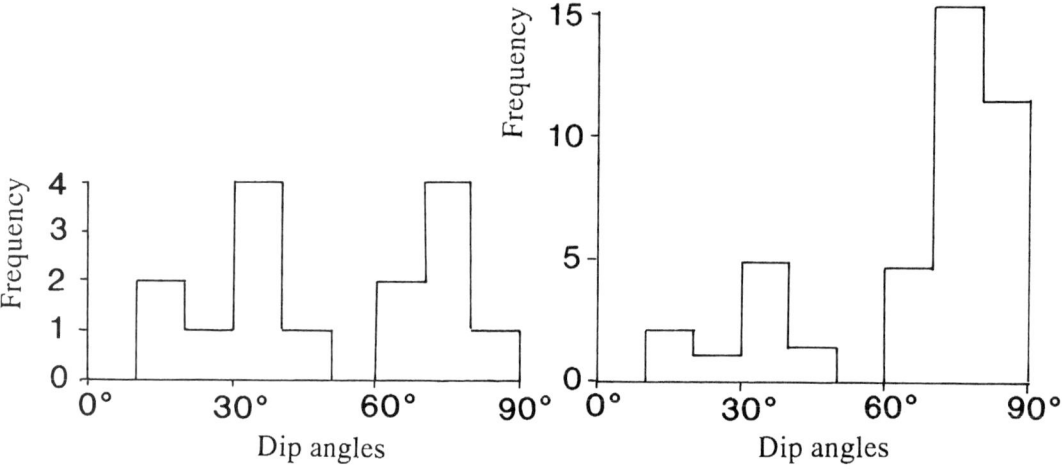

Figure 3. Left: Histogram of dip frequency for the productive fractures in the Uenotai field. Right: Histogram of dip frequency corrected using *Barton and Zoback* [1992] procedure. Vertical axis: frequency. Horizontal axis: dip angles.

1995, *Kato et al.*, 1996]. The depths of the spot core samples do not correspond to lost circulation zones, so it is impossible to directly investigate the morphology of the productive fractures. However, the fractures identified by FMI logging (abbreviated as FMI fractures) can be used to estimate the direction of dip and strike of the productive fractures in the lost circulation horizons (Fig. 6). FMI fractures are classified in terms of degree of resistivity, then called "fracture" for low resistivity and "fracture (closed)" for high resistivity. The π-pole diagram of identified fractures projected on a Schmidt's net indicates that the predominant dips and strike angles change with depth. Above a depth of 1,505 m, roughly corresponding to the bottom of the shallow reservoir [*Doi et al.*, 1995], the fractures seem to be high- and low-angle with dispersed strikes. However, a dominant dip and strike direction can be estimated around the lost circulation depths. At 60-300 m depth, including the lost circulation of 273.30 to 280.80 m depth, NE-trending, high-angle fractures dip SE, and correlated to the youngest system [F3 of *Koshiya et al.*, 1993] of veins and faults identified at the surface [*Kato, et al.*, 1995). Around the lost circulation at 980.7 m depth (±10 m), NNW trending, 20° to 40°SW dipping fractures are identified, and correlated to the oldest system [F1 of *Koshiya et al.*, 1993] of veins and faults identified at the surface [*Kato, et al.*, 1995]. The FMI fractures at 1,320 to 1,350 m, including the lost circulation zones at 1,327 m and 1,345 m, are predominantly NW-SE trending fractures with low dips toward the NE [Fig. 7; *Nagai et al.*, 1997]. The FMI fractures in the shallow part (< 1,500 m depth) are generally characterized by high- and low-angles, whereas in the deep part between 1,505 and 2,653 m, the FMI fractures are characterized by a dominance of low-angle fractures which may correspond to the low-angle fractures in cores from Well-18 drilled close to WD-1 [*Kato et al.*, 1996].

The low-angle fractures might be subdivided into two groups according to the predominant dip direction on Schmidt's nets in Fig. 6. FMI fractures at 1,505-2,155 m depth (including the lost circulation horizons at 1,589 m, 1,900 m, 1,985 m and 2,085 m depth) are characterized by a predominance of NW-SE and E-W trending faults. FMI fractures at 2,155-2,570 m are not associated with lost circulation and are characterized by a predominance of NW-SE, N-S and NE-SW trending faults with low dips toward the east. In summary, the productive fractures in WD-1a are characterized by high and low angle fractures at shallow part (< 1,500 m depth), and low-angle fractures with various trends at deep part (>1,500 m depth) . If the ratio of high to low angle fractures at the lost circulation is almost same as the ratio for all of the FMI fractures on Schmidt's nets in Fig. 6, the shallow part of Kakkonda (60-540 m and 540-1505 m on Schmidt's nets in Fig. 6) seems to have the same characteristics as the Uenotai and Sumikawa fields where high-angle fractures are dominant after correction of sampling bias in a vertical borehole. But the deep part of Kakkonda differs from them in terms of a dominance of low-angle fractures.

Sugihara, et al.[1998] analyzed a microearthquake swarm in Kakkonda, and clarified the hypocenter distribution that is characterized by a NNW-SSE trending vertical zone consisting of a series of small reverse and strike-slip faults. *Kato et al.* [1996] reported the result of the maximum horizontal stress analysis on the basis of orientations of drilling-induced tensile fractures (DIF) and borehole breakouts (BO) from FMI logs at 60-2,653m depth (Fig. 6). These data show the maximum horizontal stress (σ_{Hmax}) does not change from ground surface to 2,653m depth in WD-1a. The DIFs are mainly formed on the borehole wall in a N60-90°E direction that is σ_{Hmax}. This direction is almost the same as the latest maximum

Table 3. Average widths of high- and low-angle fractures in the Uenotai and Sumikawa fields.

[Uenotai Field]

Dip	Width (mm)	1/cosθ	Width * (1/cosθ)	Average width
85	6	11.474	68.844	
75	10	3.864	38.640	
75	1	3.864	3.864	
75	0	3.864	0	
70	5	2.924	14.620	
65	3	2.366	7.098	
60	8	2.000	16.000	
Subtotal		30.356	149.066	149.066/30.356
Average width for high-angle fractures				4.9 mm
40	2	1.305	2.610	
35	12	1.221	14.652	
30	5	1.155	5.775	
30	1	1.155	1.155	
25	1	1.103	1.103	
15	8	1.015	8.120	
Subtotal		6.954	33.415	33.415/6.954
Average width for low-angle fractures				4.8 mm

[Sumikawa Field]

Dip	Width (mm)	1/cosθ	Width * (1/cosθ)	Average width
87	2	19.107	38.214	
85	1	11.474	11.474	
80	5	5.759	28.795	
80	3	5.759	17.277	
80	2	5.759	11.518	
75	1	3.864	19.320	
70	10	2.924	29.240	
70	5	2.924	14.620	
70	3	2.924	8.772	
70	3	2.924	8.772	
70	2	2.924	5.848	
70	2	2.924	5.848	
70	2	2.924	5.848	
70	1	2.924	2.924	
70	1	2.924	2.924	
60	5	2.000	10.000	
60	3	2.000	6.000	
60	3	2.000	6.000	
60	2	2.000	4.000	
55	4	1.743	6.972	
50	2	1.556	3.112	
45	4	1.414	5.656	
45	3	1.414	4.242	
Subtotal		92.165	257.376	257.376/92.165
Average width for high-angle fractures				2.8 mm
40	8	1.305	10.440	
40	5	1.305	6.525	
40	2	1.305	2.610	
40	1	1.305	1.305	
35	1	1.221	1.221	
30	10	1.155	11.550	
30	10	1.155	11.550	
30	6	1.155	6.930	
30	6	1.155	6.930	
25	2	1.103	2.206	
20	2	1.064	2.128	
20	1	1.064	1.064	
0	6	1.000	6.000	
Subtotal		15.292	70.459	70.459/15.292
Average width for low-angle fractures				4.6 mm

Table 3. Continued.

stress inferred from the fracture study by *Koshiya et al.*[1993].

4. CHARACTERIZATION OF PRODUCTIVE FRACTURES

Representative examples of productive fractures are shown for the Uenotai and Sumikawa fields in Fig. 3. These examples show that the open fractures are accompanied by open space mineralization and brecciated

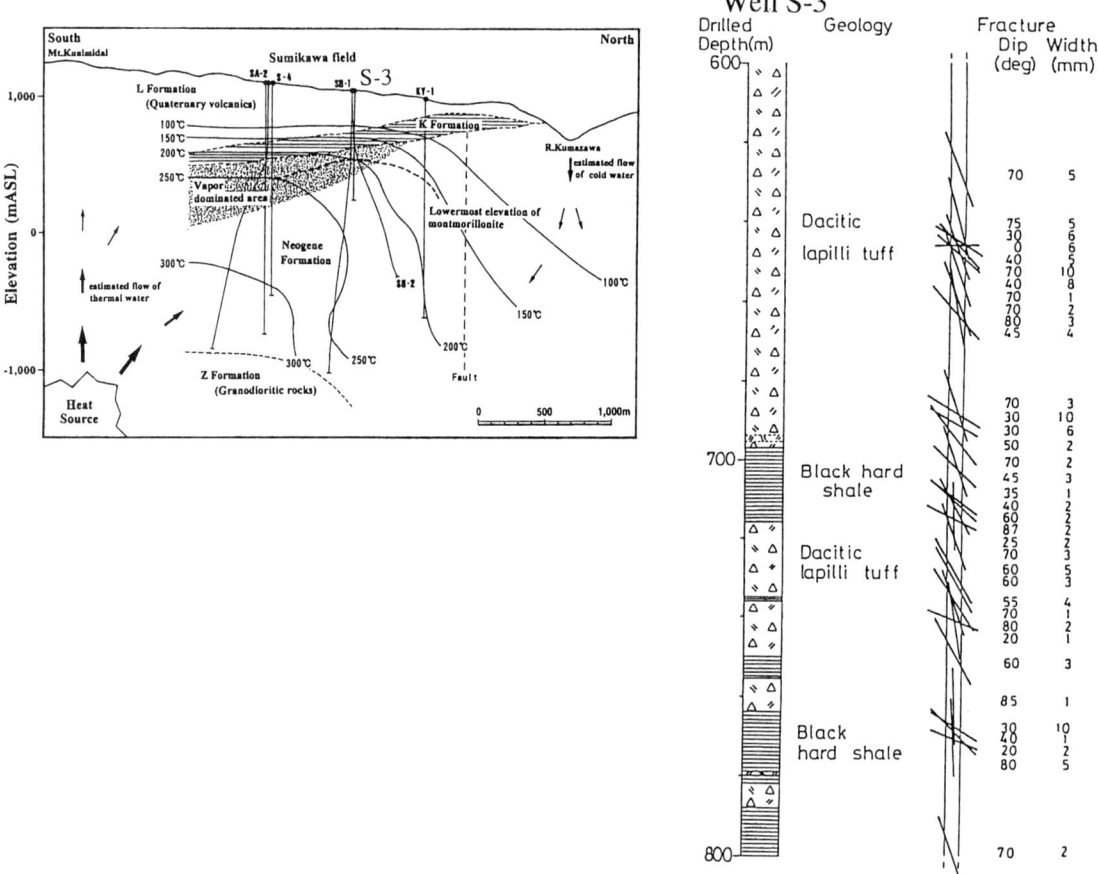

Figure 4. Left: Geothermal cross section of the Sumikawa field [*Bamba and Kubota,* 1997]. Right: Examples of productive fractures on well S-3 in the Sumikawa field [*Kubota,* 1985].

silicified tuff, indicating tensile fractures by hydrofracturing. Some of others show en echelon structure and slickensides, indicating strike-slip faults.

The characteristics of productive fractures are summarized in terms of degree of dip angle in Table 4. The productive fractures in the Uenotai and Sumikawa fields are characterized by dominant high-angle and less common low-angle fractures, a bimodal distribution which remains after the data have been corrected for sampling bias in a vertical borehole (Figs. 3 and 5). The ratio of high- to low-angle fractures is 3.3 : 1.0 in the Uenotai field and 6.0 : 1.0 in the Sumikawa field. The average dip angle is 77° for high-angle and 29° for low-angle fractures at Uenotai, and 76° and 32° respectively at Sumikawa. The average width is 4.9 mm for high-angle, and 4.8 mm for low-angle fractures at Uenotai, and 2.8 mm and 4.6 mm respectively at Sumikawa. The comparison between the width and dip angle for the productive fractures in the Uenotai and Sumikawa fields, indicates that fracture widths are almost the same (4.6 ~ 4.9 mm) except for high-angle fractures at Sumikawa (Fig. 8).

The productive fractures in the shallow (<1,500 m) and deep (>1,500 m) reservoirs at Kakkonda have quite different characteristics (Table 4). The productive fractures in the shallow reservoir dip at high- and low-angles, while those in deep reservoir are mostly low-angle features. The productive fractures in the shallow reservoir at Kakkonda are similar to those of both the Uenotai and Sumikawa fields, and all productive fractures occur at shallower than 1,500 m depth. However, the productive fractures of the deep reservoir at Kakkonda are clearly different from the others.

The productive fractures are mostly accompanied by hydrothermal veins, mineral deposition and dissolution features at Uenotai and Sumikawa, and probably also at Kakkonda (being inferred from closed FMI fractures)

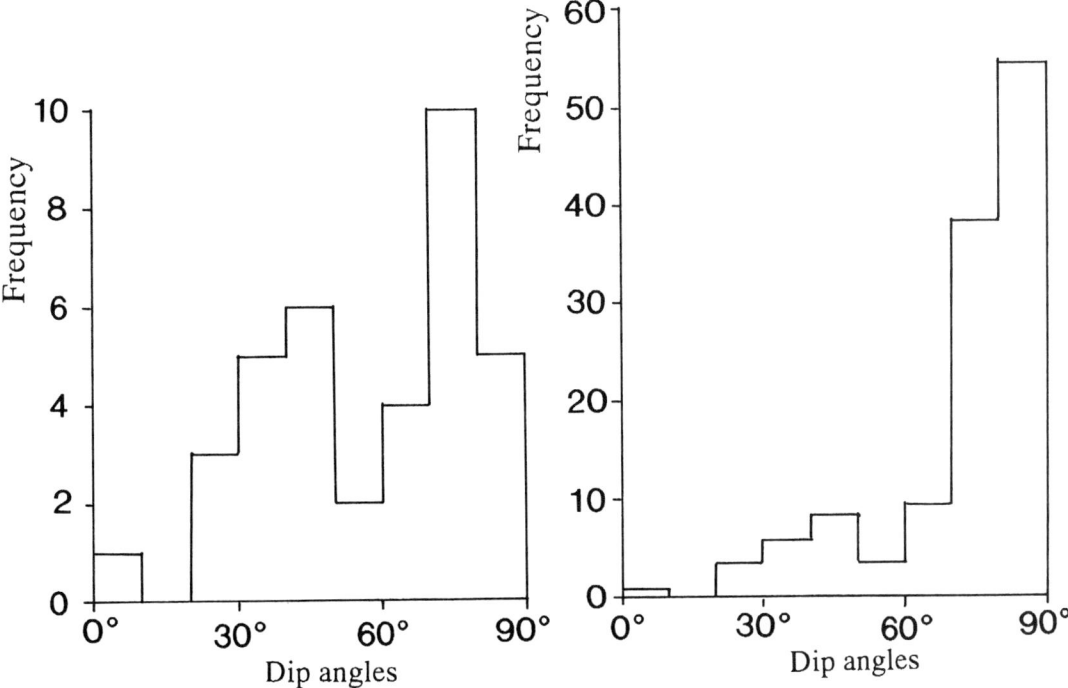

Figure 5. Left: Histogram of dip frequency for the productive fractures in the Sumikawa field. Right: Histogram of dip frequency corrected using *Barton and Zoback* [1992] procedure. Vertical axis: frequency, Horizontal axis: dip angles.

although this could not be directly confirmed. This suggests that productive fractures have been severely influenced by water rock interaction.

5. PRINCIPLE OF BRITTLE FAILURE AND FLUID INJECTION

To investigate the formation mechanism for the productive fractures, the dynamic interplay between brittle failure and fluid flow is reviewed with emphasis on the mechanism of brittle failure, the mechanism of vein formation, and fault-valve behavior.

5.1 Mechanisms of Brittle Failure

Mechanisms of brittle failure are summarized by *Sibson* [1990b]. Three different modes of brittle failure for homogeneous, isotropic and intact rock are shown in relation to loading stress fields and the general failure envelope on the Mohr diagrams in Fig. 9. These are: extension fracturing, extensional shear fracturing, and shear fracturing. To distinguish between stress and strain, the terms extension fracturing, extensional shear fracturing, and shear fracturing are replaced in this paper by tensile fracturing, tensile shear fracturing and compressive shear

fracturing, respectively. The type of failure that occurs depends on the size of the differential stress (the diameter of the stress circle = $(\sigma_1 - \sigma_3)$) in relation to the tensile strength of the rock, T_0 [*Jaeger and Cook*, 1979, *Etheridge*, 1983]. Here, σ_1 is the maximum stress, and σ_3 is the minimum stress.

Tensile fracturing - Tensile fractures form in planes perpendicular to σ_3 when the condition

$$\sigma_3' = - T_0 \tag{1}$$

is met, and the differential stress is sufficiently low to inhibit shear failure such that $(\sigma_1 - \sigma_3) < 4\ T_0)$. σ_3' is the effective normal stress of σ_3. Under zero fluid pressure, σ_3 itself has to be tensile, a condition only likely to prevail close to the earth's surface. More commonly, tensile fractures form by hydraulic fracturing when:

$$P_f = \sigma_3 + T_0 \tag{2}$$

where, again, $(\sigma_1 - \sigma_3) < 4\ T_0$. P_f is the fluid pressure within the rock mass.

Tensile shear fracturing - The second mode of failure is possible when the Mohr circle representing the stress state touches the general failure envelope in the tensile shear

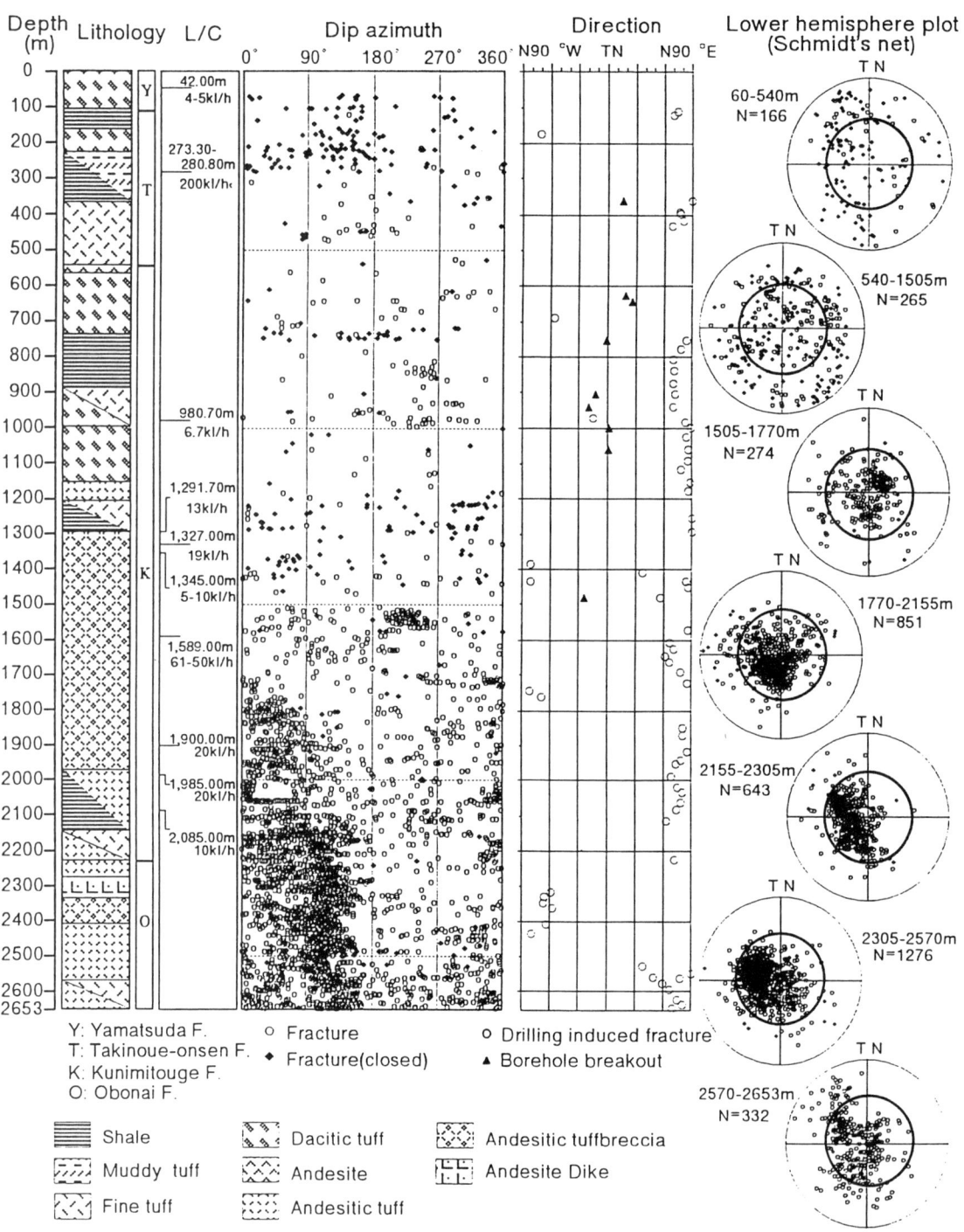

Figure 6. Examples of productive and other fractures detected by FMI in WD-1a drillhole in the Kakkonda geothermal field, Northern Honshu, Japan [*Kato, et al.*, 1996]. L/C is the abbreviation of lost circulation. π pole projection on lower hemisphere of Schmit's net. The inner circle on the Schmit's net is added by the author as a 45°dip angle to distinguish between high- and low-angle.

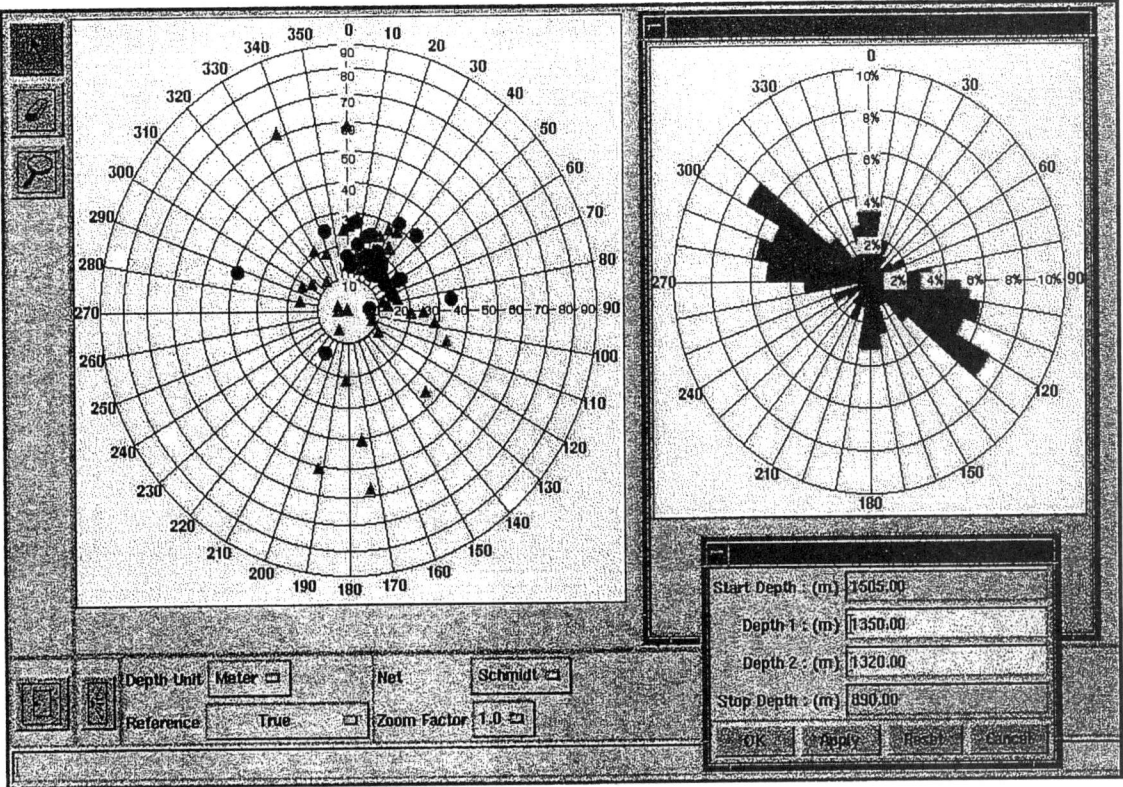

Figure 7. The FMI fractures comparable to the lost circulation at 1,320 - 1,350 m [*Nagai et al.*, 1997]. π pole projection on upper hemisphere of Schmit's net.

regime $(4T_0 < (\sigma_1 - \sigma_3) < {\sim}6T_0)$, so that both compressive shear and tensile occur across the resulting fracture.

Compressive shear fracturing - This is the most common mode of faulting. When $(\sigma_1 - \sigma_3) > {\sim}6\ T_0$, faults form by compressive shear failure along planes containing the σ_2 axis and lying at $\pm(22\text{-}32°)$ to the σ_1 direction, in accordance with the Coulomb criterion

$$\tau = C_0 + \mu_i\sigma_n' = C_0 + \mu_i\ (\sigma_n - P_f) \qquad (3)$$

where the cohesive strength of the rock, $C_0 \approx 2T_0$, the initial friction has values $0.5 < \mu_i < 1.0$ [*Jaeger and Cook*, 1979], and τ and σ_n' are the stress components on the potential failure plane. Compressive shear fracturing may lead to normal faulting when $\sigma_v = \sigma_1$, wrench or strike-slip faulting when $\sigma_v = \sigma_2$, and thrust faulting when $\sigma_v = \sigma_3$.

5.2 Mechanism of Vein Formation

Vein formation by hydraulic tensile fracturing is also described by *Sibson* [1990b]. A large proportion of hydrothermal veins occupy pure tensile fractures and appear to have formed under conditions of low differential stress by successive episodes of hydraulic fracturing and mineral precipitation along planes perpendicular to σ_3 [*Secor*, 1965; *Fyfe et al.*, 1978]. The textures of incremental vein growth characterizing this process, termed crack-seal by *Ramsay* [1980] are being increasingly recognized all over the world [e.g. *Cox and Etheridge*, 1983].

5.3 Fault-Valve Behavior

As described above, the productive fractures are associated with hydrothermal veins along the fractures. Therefore, the fluid injection regime must be carefully considered. The most convincing model for fluid injection is the fault-valve behavior defined by *Sibson* [1990a] (Fig. 10). Faults act as impermeable seals except immediately after failure, when they become highly permeable channelways for fluid discharge, and may behave as fluid-pressure-activated valves wherever they transect a suprahydrostatic fluid pressure gradient. Such fault-valve behavior is caused by abrupt fluctuation in fluid pressure linked to an earthquake cycle. The evidence for fault-valve activity comes from instances of postseismic discharge, the common presence on ancient faults of hydrothermal veins

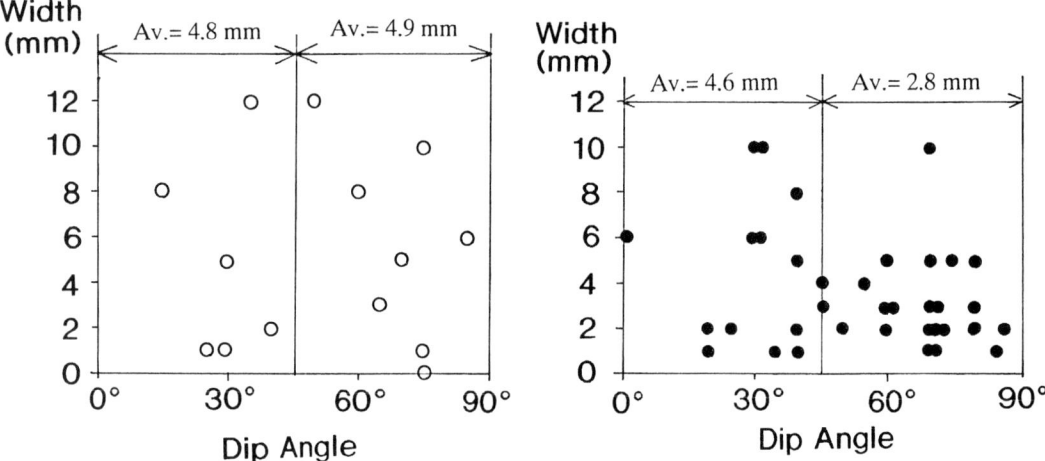

Figure 8. Correlation between width and dip angle of the identified productive fractures in the Uenotai and the Sumikawa fields.

with textural evidence of incremental deposition following episodes of fault slip, and evidence from fluid inclusion studies of fault-hosted hydrothermal minerals that suggests fluid pressure varied in a cyclic manner [*Sibson*, 1992].

In Fig. 10, when translated to a hydrothermal system, the impermeable barrier can be regarded as a cap rock, and "X" as a geothermal reservoir. *Sibson* [1990a] suggested that fault-valve behavior may be responsible for the observation by *Grindley and Browne* [1976] that zones of high geothermal permeability are located near the most recently active faults in New Zealand fields. In these circumstances, hydraulic fracturing produces fault zone breccias which later become choked by hydrothermal mineral deposition. Brecciation and open-space mineralization formed by hydraulic fracturing has been also recognized at Uenotai and Sumikawa.

6. FORMATION MECHANISM OF PRODUCTIVE FRACTURES

The features of productive fractures have been clarified in previous sections. These features help to constrain formation mechanisms and are used to infer the type of brittle failure with reference to the geometry of the regional stress field. In this section the formation mechanism of two fracture sets, one comprising dominant high-angle and less common low-angle fractures and the other comprising dominant low-angle and less common high-angle fractures, is investigated.

6.1 Geometry of Regional Stress Field

The geometry of the regional stress field can be expressed as a map of the trajectories of the maximum horizontal stress (σ_{Hmax}). This kind of map was drawn for the late Quaternary stress field in northern Honshu [Fig. 11;

Nakamura and Uyeda, 1980], using the analysis of focal mechanisms of shallow earthquakes, active faults and recent volcanic dikes given by *Ando* [1979]. The σ_{Hmax} trajectories are generally parallel to the converging direction of the Pacific plate. However, the measured or estimated σ_{Hmax} directions of each studied field differs somewhat from the general trend. Analysis of conjugate fault sets suggests that σ_{Hmax} in the Uenotai field trends NE-SW after 3 Ma [*Tamanyu and Mizugaki*, 1993]. The σ_{Hmax} in the Sumikawa field has not yet been measured, but an E-W trend is presumed from the alignment of Quaternary volcanoes (Hachimantai - Yakeyama - Moriyoshisan) near the field. The σ_{Hmax} in the Kakkonda field was measured as ENE-WSW by shear-wave splitting [*Kaneshima et al.*, 1988] , as NE-SW ~ E-W by conjugate fault sets [*Koshiya et al.*, 1993], and as N60-90°E by drilling induced fractures and bore breakouts [*Kato et al.*, 1996]. These directions are mutually consistent.

6.2 Formation Mechanism of High- and Low-Angle Fractures

As mentioned above, the present or late Quaternary σ_{Hmax} seems to trend from NE-SW to E-W trends in northeast Japan. If most productive fractures originated as tensile fractures under a thrust regime, it is difficult to produce the occurrence of high-angle fractures in intact rock. Therefore, *Tamanyu et al.*[1998] suggested that high-angle fractures formed in response to a temporal change in σ_1 from horizontal to vertical in response to magmatic fluid injection. However, it is difficult to prove this kind of local stress change. These are, however, other mechanisms that can account for high-angle fractures. One possibility is the strike slip fault proposed as F2L and F3L by *Koshiya et al.* [1993]. Another possibility is to reactivate pre-existing high-angle fractures within a thrust regime . The author is

Table 4. The list for counted numbers, corrected ratios, average dip-angles and average width of identified productive fractures in the Uenotai, Sumikawa and Kakkonda fields.

Field Name	No. of fractures		Corrected ratio	Av. angle (degree)		Av. width (mm)	
	High	Low	High : Low	High	Low	High	Low
Uenotai	7	8	3.3 : 1.0	77	29	4.9	4.8
Sumikawa	23	13	6.0 : 1.0	76	32	2.8	4.6
Kakkonda							
Shallow	common*	common*	more : less*	-	-	-	-
Deep	rare*	dominant*	less : more*	-	-	-	-

* Estimated from Fig. 6.

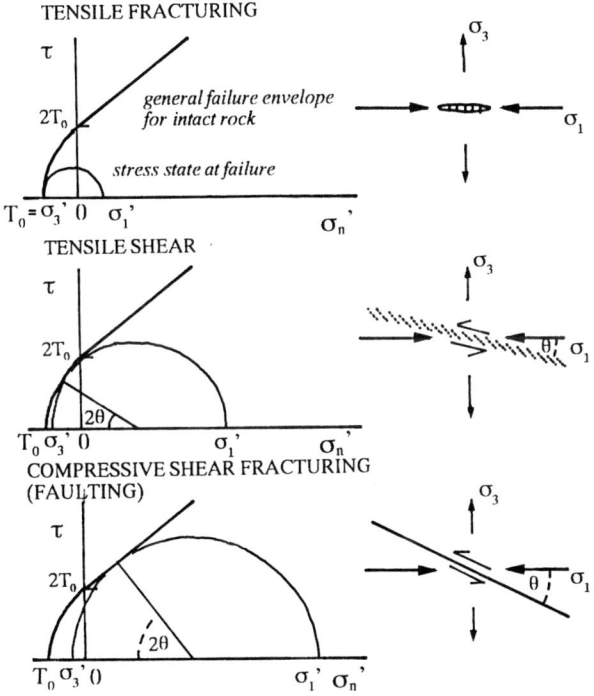

Figure 9. Possible failure models in relation to deviatoric stress field in homogeneous, isotropic rock. Mohr diagrams illustrate the general failure envelope for intact rock, the stress conditions for the three models of brittle failure, and the orientation of the failure surface with regard to the stress field [*Sibson, 1990b*].

interested in the latter idea because most productive fractures have hydrothermal veins and some representative fractures have opening-space mineralization along the fractures with hydrofracturing breccia. This idea was proposed as the concept of unfavorably oriented faults by *Sibson* [1990a](Fig. 12(b)), and described in detail in the following paragraph.

The Uenotai field underwent extension in the early Miocene [Fig. 5 in *Tamanyu and Mizugaki*, 1993]. The main productive horizons here are from the lower Miocene and pre-Tertiary formations. Extensional stress fields in the Miocene and productive horizons are also inferred for many geothermal fields in northeast Japan by previous workers. Unfavorably oriented faults can be expected to occur as high-angle fractures under the Quaternary thrust regime. In fact, the high-angle reverse faults and accompanied hydrothermal mineral veins are observed on the core samples drilled near the Uenotai field. However, it is generally difficult to distinguish reactivated high-angle fractures from high-angle tensile fractures associated with strike-slip faults because displacement mode is commonly unclear. Then, these two possibilities are accepted together in this paper, whereas the accompanied low-angle fractures are simply interpreted as tensile and/or tensile-shear fractures formed under a thrust regime. These are able to be formed at the time of high fluid pressure and low differential stress.

The effect of fault-valve action on unfavorably oriented faults is quoted from *Sibson* [1990a]. He showed that many mesothermal gold-quartz lodes are hosted within granite-greenstone terrains on faults that were undergoing high-angle reverse or reverse-oblique motion at the time of mineralization. A representative example is the gold-bearing quartz-tourmaline vein system at the Sigma Mine in Val d'Or, Quebec, formed during almost pure reverse motion on steeply-dipping (*c.* 70°) shears [*Robert and Brown*, 1986] (Fig. 12(b)). Two main vein-sets occur: lenticular fault-veins lying subparallel to schistosity within the steeply dipping shear zones and flats (subhorizontal tensile veins). These are interpreted as resulting from fault-valve action on severely misoriented high-angle reverse faults, with the flats opening up whenever the prefailure condition of supralithostatic fluid pressures ($P_f > \sigma_3$) was met, whereas the fault-veins developed during episodes of postfailure discharge upwards through the shear zones [*Sibson et al.*, 1988]. Potentially, such valve action could result in a saw-toothed oscillation between slightly supralithostatic and hydrostatic levels of fluid pressure, representing the extreme

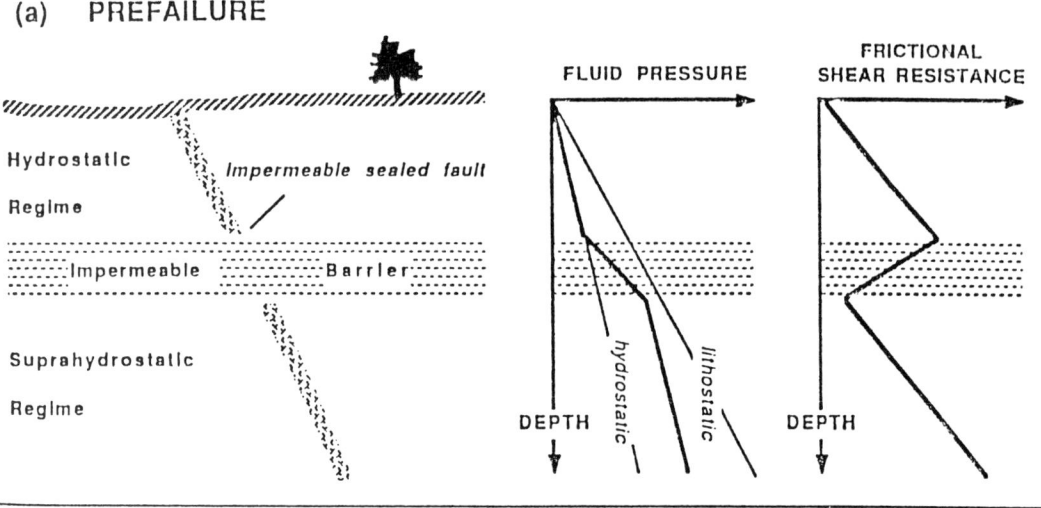

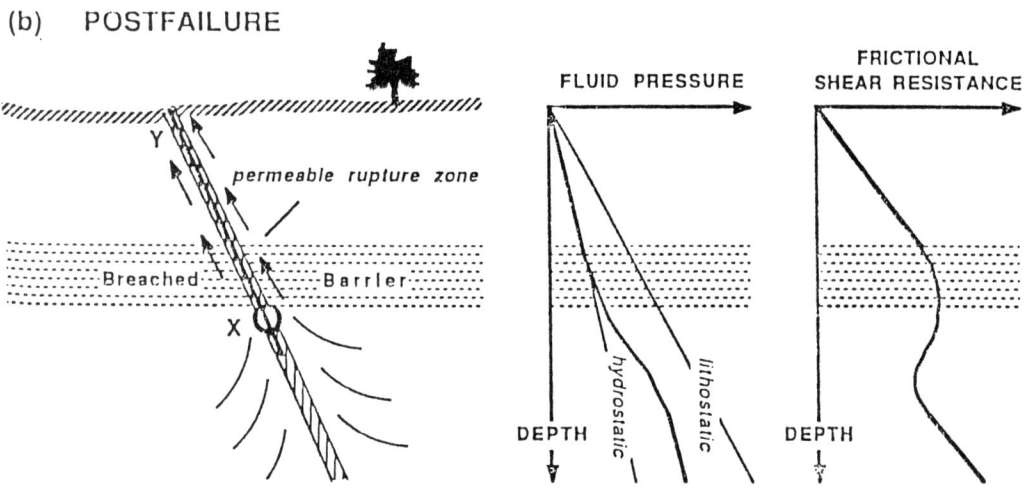

Figure 10.   Potential   for   fault-valve   behavior   (a)   Impermeable   barrier   separating   hydrostatic   and suprahydrostatic fluid pressure regimes. (b) Breaching of barrier by fault rupture X-Y, leading to an upwards discharge of fluids [*Sibson,* 1990a].

form of fault-valve behavior (Fig. 12(b)). Stress conditions for this behavior at large reactivation angles are illustrated in the Mohr diagram. Note that fault reactivation of this kind necessarily occurs under low differential stress, otherwise more favorably oriented faults will form. Steep fault-veins in several of these deposits are, in fact, quite commonly disrupted by late shallow-dipping Andersonian thrusts, implying that reactivation of these unfavorably oriented structures persisted only while fluid pressures could be restored to approximately lithostatic levels before differential stress built up to a level sufficient to induce shear failure in the surrounding intact rock.

6.3 Formation mechanism of dominant low-angle fractures

The productive fractures in the deep reservoir in the Kakkonda field are characterized by dominant low-angle fractures. The reason why high-angle fractures are rare may be explained by the existence of favorably oriented faults and low-angle tensile fractures under thrust regime.

Fault-valve action on favorably oriented faults is described by *Sibson* [1990a]. Regions of elevated fluid pressure are likely to be loci for the formation of new faults because of the marked local reduction in crustal strength. In such circumstances, fault-valve action may follow fault

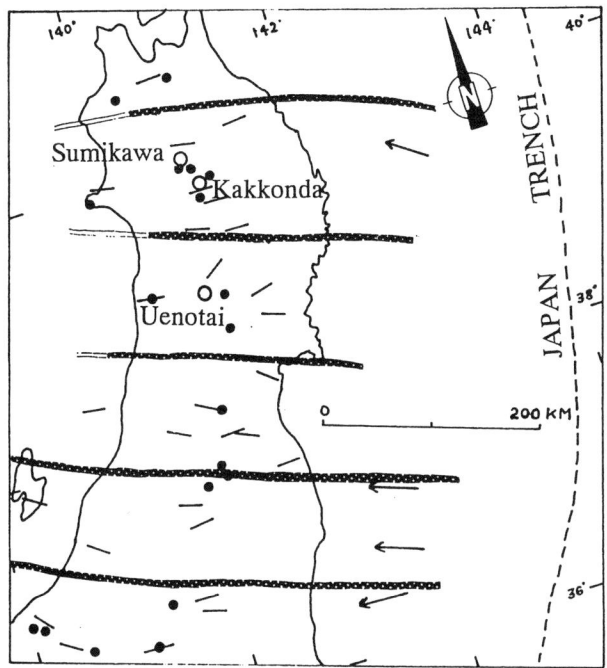

Figure 11. A σ_{Hmax} trajectory map of present-day northern Honshu, Japan [*Nakamura and Uyeda*, 1980], drawn on the basis of data given by *Ando* [1979]. Arrows indicate slip vectors for low-angle thrust events. Dots: active volcanoes from *Kuno* [1960]. Where σ_{Hmax} is judged to be σ_1, trajectories are dark lines; double lines are for the transitional portion between the places where σ_{Hmax} is σ_1 and σ_{Hmax} is σ_2. The location of the studied fields is marked as open circle by the author.

inception. As an example, consider the mesothermal gold-quartz veins of Grass Valley, California, which occupy a conjugate set of thrust faults cutting a greenstone-dominant assemblage intruded by an early Cretaceous granodiorite [*Johnstone*, 1940, *Bohlke and Kistler*, 1986]. The faults intersect each other at 50-60° (Fig. 12(a)) and are zones of intense fracturing with varying proportions of gouge, breccia, and hydrothermal quartz and carbonate. It seems probable that the fault system hosting the veins developed under suprahydrostatic fluid pressures, with the thrusts acting as self-sealing fault-valves through the hydrothermal precipitation accompanying each discharge episode and drop in fluid pressure. Effective stress conditions at failure and immediately postfailure for an Andersonian thrust fault acting as a fault-valve are illustrated in Fig.12(a). From the Mohr diagram, it is clear that shear failure under suprahydrostatic fluid pressures, leading to fault-valve discharge, causes reductions in both the differential stress and the fluid pressure. For failure to recur, either differential stress must build up to a much greater level under

hydrostatic fluid pressure, or suprahydrostatic fluid pressure must reaccumulate beneath sealed portions of the faults.

Considering a possible fault-valve action, the productive fractures in the deep reservoir at Kakkonda may be explained by favorably oriented faults which accompanied low-angle thrust faults. One of the dominant trends of these low-angle faults is NW-SE to N-S which cross σ_{Hmax} at right angles. These features indicate that the low-angle fractures should be thrusts. The occurrence of favorably oriented faults in the Kakkonda deep reservoir may be related to the fact that the deeper part is more purely lithostatic, whereas the shallower part, is generally hydrostatic. Thus, favorably oriented faults should occur more readily in the deep reservoir than in the shallow reservoir. The absence of high-angle fractures may also be attributed to healing of pre-exisiting fractures by contact metamorphism. The metamorphism extending over the 1500 - 2,860 m depth range in WD-1 drillhole was caused by injection of the Quaternary granitic pluton. The low-angle fractures have various trends and FMI closed fractures are filled with highly resistive material. These are interpreted as tensile and/or tensile-shear fractures accompanied with hydrothermal mineral deposit, and they can be formed at the time of high fluid pressure and low differential stress under thrust regime.

7. CONCLUSION

The formation process of the productive fractures is summarized in Fig. 13. The vertical axis is the degree of differential stress, and the type of brittle failure is classified into compressive shear, tensile shear, and tensile regions by point A (σ_1-σ_3=6T$_0$) and point B (σ_1-σ_3=4T$_0$). The vertical axis is positively correlated to the values of σ_1 and inversely to the values of fluid pressure. In the case of the shallow reservoir, the high-angle fractures predominate. These are interpreted to occur by two formation mechanism: one is as shear fractures under a wrench fault regime and the other is unfavorably oriented faults under thrust regime. The latter can occur at or just after a fault-valve rupture, triggered by a suprahydrostatic fluid pressure gradient (such as might be caused by magmatic fluid injection, but has not yet be proved), or as unfavorable reactivation under low differential stress. At the same time or later, low-angles developed as tensile hydraulic fractures and/or tensile-shear fractures under the regional horizontal compressive stress regime as fluid pressures (P$_f$) declined. Water-rock interaction occurred simultaneously with fracture formation, and resulted in mineral precipitation which finally sealed off the fracture aperture. In the case of the deep reservoir, where σ_1 is generally greater and P$_f$ smaller than in shallow reservoir, the low-angle fractures occurred at or just after a fault-valve rupture, as conjugate thrust faults under high differential stress. However, low-angle fractures may also occur by tensile hydraulic fracturing under as fluid pressure (P$_f$) returned to their original values.

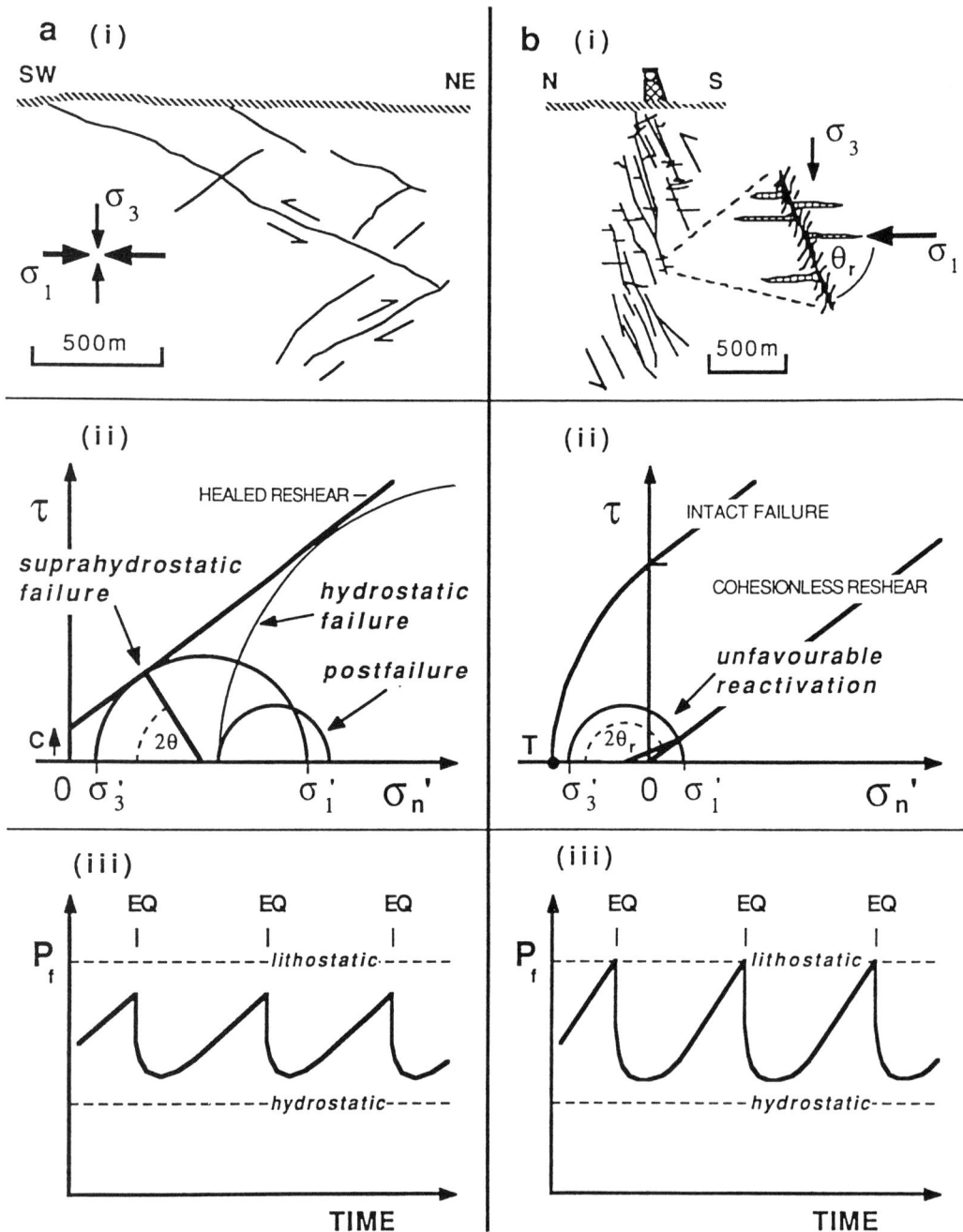

Figure 12. Examples of vein systems attributable to fault-valve behavior [*Sibson*, 1990a].
(a) Favourably oriented faults: (i) section through North Star and subsidiary Au-quartz lodes occupying a set of conjugate thrust faults, Grass Valley, California [after *Johnston*, 1940]; (ii) Mohr diagram illustrating stress conditions at failure and postfailure; (iii) fluid pressure cycles inferred to accompany successive earthquake (EQ) ruptures.
(b) Unfavourably oriented faults: (i) section through Au-quartz lodes hosted by a system of high-angle reverse faults and associated extension fractures ($\theta_r \approx 70°$), Sigma Mine, Val d'Or, Quebec [after *Robert and Brown*, 1986]; (ii) Mohr diagram illustrating stress condition which allows reactivation of a severely misoriented faults without causing failure of surrounding intact rock; (iii) resultant fluid pressure cycling accompanying successive earthquake (EQ) ruptures.

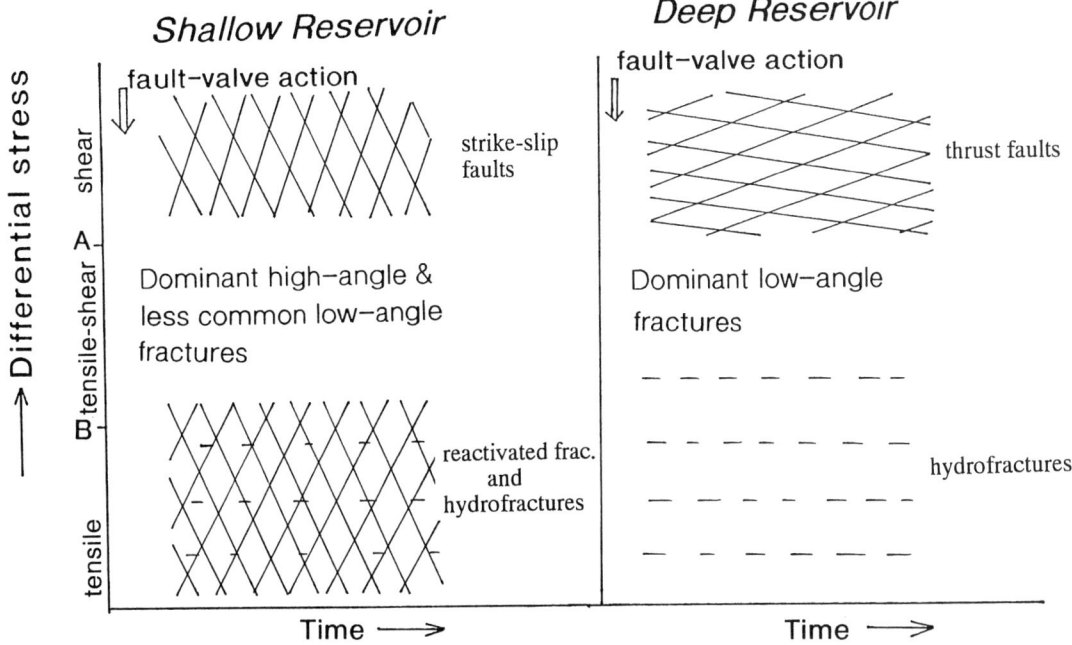

Figure 13. Developing history of productive fractures. A: $\sigma_1-\sigma_3=6T_0$. B: $\sigma_1-\sigma_3=4T_0$. Time unit is arbitrary.

It can be concluded that the geothermal reservoirs have been developed and maintained by these continuous fracturings (repeated fault-valve rupture, hydraulic tensile fracturing accompanied by strike-slip fault, unfavorably oriented fractures, and thrust faults).

This conceptual model is based on a limited data set of non-oriented core samples and FMI logging data. A more rigorous model would require oriented core samples which were not available. The features like what of the productive fracture within extensional stress field should be also investigated.

Acknowledgments. The author expresses his appreciation to C.P. Wood of IGNS, New Zealand for a critical reading of this draft. The author also extends his thanks to two anonymous reviewers for their constructive comments and W.C. Haneberg for his comments as the editor of this volume. T. Nagai of Schlumberger K.K. kindly offerd her original print for Fig. 7. This work was carried out at the Institute of Geological & Nuclear Sciences in New Zealand as a senior research fellow funded by the Japan Science and Technology Corporation. The author thanks these two organizations for their kind support for his study.

REFERENCES

Ando, M., The stress field of the Japan islands in the last 0.5 million years, *Earthquake Monthly Symposium, Tokyo University, 7*, 541-546, 1979.

Bamba, M., Y. Kubota, A. Ueda, Y. Kihara, and Y. Yamagishi, Fracture pattern simulated statistically in the Sumikawa geothermal system, northeast Japan, *Proceedings, World Geothermal Congress, 1995, Florence*, vol. 3, International Geothermal Association, 1709-1713, 1995.

Bamba, M., and Y. Kubota, Geothermal conceptual model from the viewpoint of the thermal history in the north Hachimantai-Yakeyama area, *Chinetsu, 34*, 1-13, 1997 [in Japanese with English abstract].

Barton, C.A., and M.D. Zoback, Self-similar distribution and properties of macroscopic fractures at depth in crystalline rock in the Cajon Pass Scientific Drill Hole, *J. Geophys. Res., 97*, 5181-5200, 1992.

Bohlke, J.K., and R.W. Kistler, Rb-Sr, K-Ar, and stable isotope evidence for the ages and sources of fluid components of gold-bearing quartz veins in the northern Sierra Nevada foothills metamorphic belt, California, *Economic Geology, 81*, 296-322, 1986.

Cox, S.F., and M.A. Etheridge, Crack-seal fibre growth mechanisms and their significance in the development of oriented layer silicate microstructures, *Tectonophysics, 92*, 147-170, 1983.

fracturing after emplacement of Quaternary granitic pluton in the Kakkonda geothermal field, Japan, *Transaction, Geothermal Resources Council, 19*, 297-303, 1995.

Etheridge, M.A., Differential stress magnitudes during regional deformation and metamorphism: upper bound imposed by tensile fracturing, *Geology, 1*, 231-234, 1983.

Fyfe, W.S., N.J. Prince, and A.B. Thompson, *Fluids in the Earth's Crust*, Elsevier, Amsterdam, 383 pp., 1978.

Grindley, G.W., and P.R.L. Browne, Structural and hydrological factors controlling the permeabilities of some hot-water geothermal fields, *Proceedings, Second United Nations Symposium on the Development and Use of Geothermal Resources*, San Francisco, May, 1975, vol. 1, Washington, D. C., U.S. Government Printing Office (Lawrence Berkeley Laboratory, University of California), 377-386, 1976.

Hulen, J.B., B.A. Koenig, and D.L. Nielson, The Geysers coring project, Sonoma County, California, USA -Summary and initial results, *Proceedings, World Geothermal Congress, 1995*, Florence, vol. 2, International Geothermal Association, 1415-1420, 1995.

Jaeger, J.G., and N.G.W. Cook, *Fundamentals of Rock Mechanics*, 3rd edition., Chapman & Hall, London 593 pp, 1979.

Johnston, W.D., The gold quartz veins of Grass Valley, California, *US Geological Survey Professional Paper, 194*, 101 pp., 1940.

Kaneshima, S., H. Ito, and M. Sugihara, Shear-wave splitting observed above small earthquakes in a geothermal area of Japan, *Geophysical Journal, 94*, 399-411, 1988.

Kasai, K., Y. Sakagawa, S. Miyazaki, M. Sasaki, and T. Uchida, Supersaline brine obtained from Quaternary Kakkonda granite by the NEDO's deep geothermal well WD-1a in the Kakkonda geothermal field, Japan, *Transaction, Geothermal Resources Council, 20*, 623-629, 1996.

Kato, O., N. Doi, T. Akazawsa, Y. Sakagawa, M. Yagi, and H. Muraoka, Characteristics of fracture systems based on FMI logs and cores in well WD-1 in the Kakkonda geothermal fields, Japan, *Transaction, Geothermal Resources Council, 19*, 317-322, 1995.

Kato, O., N. Doi, K. Ikeuchi, T. Kondo, H. Kamenosono, M. Yagi, and T. Uchida, Characteristics of temperature curves and fracture systems in Quaternary granite and Tertiary pyroclastic rocks of NEDO WD-1a in the Kakkonda geothermal fields, Japan, *Proceedings, 8th International Symposium on the Observation of the Continental Crust Through Drilling*, Tsukuba, Japan, 241-246, 1996.

Koshiya, S., K. Okami, Y. Kikuchi, T. Hirayama, Y. Hayasaka, M. Uzawa, K. Honma, and N. Doi, Fracture system developed in the Takinoue geothermal area, *Journal of Geothermal Research Society Japan, 15*, 109-139, 1993 [in Japanese with English abstract].

Koshiya, S., K. Okami, Y. Hayasaka, M. Uzawa, Y. Kikuchi, T. Hirayama, and N. Doi, On the hydrothermal mineral veins developed in the Takinoue geothermal area, Northeast Honshu, Japan, *Journal of Geothermal Research Society Japan, 16*, 1-24, 1994 [in Japanese with English abstract].

Kubota, Y., Conceptual model of the North Hachimantai-Yakeyama geothermal area, *Journal of Geothermal Research Society Japan, 7*, 231-245, 1985 [in Japanese with English abstract].

Kuno, H., *Catalogue of the active volcanoes of the world including solfatara fields, part II (Japan, Taiwan and Marianas)*, International Association of Volcanology, Rome, 332 pp., 1960.

Meunier, A., Hydrothermal alteration by veins, in *Origin and mineralogy of clays*, edited by B. Velde, pp. 247-267, Springer, 1995.

Nagai, T., H. Yamamoto, O. Kato, Y. Sakagawa, and T. Uchida, Lithology and fracture analysis with wireline logging tools for geothermal exploration -Examples at WD-1, Kakkonda geothermal field with FMI and other tools, *Proceedings, 19th NZ Geothermal Workshop*, University of Auckland, 129-134, 1997.

Naka, T., and H. Okada, Exploration and development of Uenotai geothermal field, Akita prefecture, northeastern Japan, *Mining Geology, 42*, 223-240, 1992 [in Japanese with English abstract].

Naka, T., R. Takeuchi, S. Iwata, and A. Fukunaga, Exploration and exploitation of Uenotai geothermal field, Akita, Japan, *Chinetsu*, Japan Geothermal Energy Association, *24*, 113-135, 1987 [in Japanese with English abstract].

Nakamura, K., and S. Uyeda, Stress gradient in arc - back arc regions and plate subduction, *J. Geophys. Res., 85*, 6419-6428, 1980.

Oide, K., and K. Onuma, Igneous activity of the "Green Tuff Region" in Northeast Honshu, Japan, *Chikyu Kagaku, 50-51*, 36-55, 1960 [in Japanese with English abstract].

Ramsay, J.G., The crack-seal mechanism of rock deformation, *Nature, 244*, 135-139, 1980.

Robert, F., and A.C. Brown, Archean gold-bearing quartz veins at the Sigma Mine, Abitibi greenstone belt, Quebec: Part I. Geologic relations and formation of the vein system, *Economic Geology, 81*, 578-592, 1986.

Robertson-Tait, A., C.W. Klein, J.R. McNitt, T. Naka, R. Takeuchi, S. Iwata, Y. Saeki, and T. Inoue, Heat source and fluid migration concepts at the Uenotai geothermal field, Akita prefecture, Japan, *Transaction, Geothermal Resources Council, 14(II)*, 1325-1331, 1990.

Sakai, Y., Y. Kubota, and K. Hatakeyama, Geothermal exploration at Sumikawa, North Hachimantai, Akita, *Chinetsu*, Japan Geothermal Energy Association, *23*, 281-302, 1986 [in Japanese with English abstract].

Sasaki, M., K. Fujimoto, T. Sawaki, H. Tsukamoto, H. Muraoka, M. Sasada, T. Ohtani, M. Yagi, M. Kurosawa, N. Doi, O. Kato, K. Kasai, K. Komatsu, and Y. Muramatsu, Characterization of magmatic/meteoric transition zone at the Kakkonda geothermal system, northeast Japan, in *Water Rock Interaction (WRI-9)* edited by G.B. Arehart, and J. Hulston, A. A. Balkema. pp. 483-486, 1998.

Secor, D.T., Role of fluid pressure in jointing, *American Journal of Science, 263*, 633-646, 1965.

Sibson, R.H., Conditions for fault-valve behavior, in *Deformation Mechanisms, Rheology and Tectonics*, edited by R.J. Knipe, and E.H. Rutter, *Geol. Soc. Spec. Publ., 54*, pp. 15-28, 1990a,.

Sibson, R.H., Chapter 4. Faulting and fluid flow, in *MAC Short Course on Fluids in Tectonically Active Regimes of the Continental Crust, Mineralogical Association of Canada, Short Course Handbook*, vol. 18, edited by B.E. Nesbitt, Vancouver, Canada, pp. 93-132, 1990b.

Sibson, R.H., Implications of fault-valve behavior for rupture nucleation and recurrence, in *Earthquake Source Physics and Earthquake Precursors*, edited by T. Mikumo, K. Aki, M. Ohnaka, L.J. Ruff, and P.K.P. Spudich, *Tectonophysics, 211*, pp. 283-293, 1992.

Sibson, R.H., F. Robert, and K.H. Poulsen, High-angle reverse faults, fluid pressure cycling and mesothermal gold-quartz deposits *Geology, 16*, 551-555, 1988.

Sugihara, M., T. Tosha, and Y. Nishi, An empirical green's part of Kakkonda geothermal reservoir, Japan, *Geothermics, 27*, 691-704, 1998.

Takahashi, M., Y. Murata, M. Komazawa, and S. Tamanyu, Geothermal resources map of Akita area, *Geological survey of Japan Miscellaneous maps series (31-2)*, scale 1:500,000, 1 sheet and its text, 162 pp., 1996 [in Japanese with English abstract].

Tamanyu, S., and K. Mizugaki, The fracture system related with geothermal fluid flows -Examples in the Yuzawa-Ogachi geothermal field, Akita, Japan-, *Journal of the Geothermal Research Society of Japan, 15*, 253-274, 1993 [in Japanese with English abstract].

Tamanyu, S., M. Takahashi, Y. Murata, K. Kimbara, M. Kawamura, and H. Yamaguchi, Geothermal resources map of the Tohoku volcanic arc, Northeast Japan, *Transaction, Geothermal Resources Council, 20*, 401-405, 1996.

Tamanyu, S., S. Fujiwara, J. Ishikawa, and H. Jingu, Fracture system related to geothermal reservoir based on core samples of slim holes -Example from the Uenotai geothermal field, Northern Honshu, Japan-, *Geothermics, 27*, 143-166, 1998.

Thompson, R.C., and R.P. Gunderson, The orientation of steam-bearing fractures at The Geysers geothermal field, *Geothermal Resources Council, Monograph on The Geysers Geothermal Field, Special Report, 17*, 65-68, 1991.

Yagi, M., H. Muraoka, N. Doi, and S. Miyazaki, "Deep-Seated Geothermal Resources Survey" overview, *Transaction, Geothermal Resources Council, 19*, 377-382, 1995.

_____

S. Tamanyu, Department of Geothermal Research, Geological Survey of Japan, Tsukuba, 305-8567, Japan

Hydrogeothermal Studies on the Southern Part of Sandia National Laboratories/Kirtland Air Force Base—Data Regarding Ground-Water Flow Across the Boundary of an Intermontane Basin

Marshall Reiter

New Mexico Bureau of Mines and Mineral Resources, Socorro, New Mexico

Subsurface temperature measurements made in drill holes on the Albuquerque bench provide data concerning ground-water flow from the eastern mountains toward the Albuquerque basin. The study area on the southern part of Sandia National Laboratories/Kirtland Air Force Base is traversed by a number of faults parallel to the Rio Grande rift as well as cross-cutting faults. At one or more of the rift-parallel faults very steep gradients are noted in the ground-water table elevation. The temperature data lie along three profiles traversing these faults and the sediments on either side. As with most geothermal studies the present data are taken in drill holes completed for other purposes and are therefore limited both in location and depth. Along the southern profile the temperature data suggest that the Sandia fault is acting as an effective seal at shallow depths. Along the middle profile the data suggest that the Sandia-Tijeras fault complex is transmissive and that the ground-water flow in the sediments occurs along thin depth zones or nearly horizontal channels. These data are used to hypothesize that flow across the fault occurs near or below the water table on the downthrown block. Data along a northern profile are inconclusive as to whether or not the Sandia and Tijeras faults are transmissive; however, a low linear temperature gradient at one site suggests channel flow associated with the Tijeras arroyo. Although the data are limited, they appear to indicate that ground water flows across the faults preferentially at select locations and along thin horizontal zones in the sediments bordering the faults. Models of ground-water flow in such areas should incorporate the probable large vertical differences in hydraulic conductivity, in both the sediments and the faults, in order to provide accurate flow characteristics.

INTRODUCTION

Ground-water flow has the capacity to noticeably perturb subsurface temperatures. For example, downward flow of cool ground water in recharge areas will cool shallow subsurface temperatures below what might be expected for conduction only geothermal gradients; upward flow of ground water in discharge areas can raise shallow temperatures above values expected for conduction conditions. As early as 1939, Bullard recognized that heat-flow data from different depth intervals may vary because of climatic variation and ground-water flow. Regional and local ground-water flow patterns have been studied by

Faults and Subsurface Fluid Flow in the Shallow Crust
Geophysical Monograph 113

calculating heat-flow values over various depth intervals 1983; *Garven and Freeze*, 1984]. *Morgan et al.* [1981], *Ingebritsen et al.* [1989], *Witcher* [1988], *Barroll and Reiter* [1990] and *Reiter and Jordan* [1996] are a few of the studies showing the potential for positive geothermal anomalies to be caused by deep ground water transporting heat toward the surface.

From accurate temperature logs, information regarding the vertical component of specific discharge, horizontal flow, well bore flow, and fracture flow may be gained. *Bredehoeft and Papadopulos* [1965] and *Mansure and Reiter* [1979] present methods of estimating one-dimensional constant vertical flow across a layer. *Reiter et al.* [1989] show that small vertical flows of about 5×10^{-9} m s^{-1} (0.15 m/yr) over a few hundred meters will cause large heat-flow variations. Although vertical ground-water flow is commonly involved with the perturbation of subsurface temperature measurements, horizontal flow, borehole flow, and fracture flow can also be detected from accurate temperature measurements [*e.g. Birch*, 1947; *Lewis and Beck*, 1977; *Drury et al.*, 1984; *Ziagos and Blackwell*, 1986; *Jessop*, 1987]. If horizontal temperature gradients and ground-water head data are available, estimates of horizontal flow and hydraulic conductivity can be made [*McCord et al.*, 1992]. Fluid flow along thin channels or fractures often causes abrupt changes and anomalies in the temperature gradient because the channel or fracture is transporting ground water at a different temperature from the unperturbed conductive temperature regime. Fracture flow may involve a vertical component whereby cooler or warmer water is brought to a site; or alternatively, a preferential flow path along which water at a different temperature moves largely horizontally.

In this study a number of temperature logs have been made in an extensively faulted area bordering the Rio Grande rift on the east. Although the data are limited in depth and location, some insight can be gained into the nature of ground-water flow recharging in the mountains and generally moving westward toward the rift valley across a complex of graben defining faults. Although many more sites for temperature data at greater depths are needed to better understand the ground-water flow characteristics along the borders of the Albuquerque basin, the present data do provide an interesting glimpse of the potential diversity and characteristics of ground-water flow along an intermontane basin.

BRIEF DESCRIPTION OF THE HYDROGEOLOGY OF THE STUDY AREA

The study area is located southeast of Albuquerque on Sandia National Laboratories/Kirtland Air Force Base (Figure 1). The area is part of the Albuquerque bench which is shown in a seismically derived cross section by *Russell and Snelson* [1994]. The bench straddles a region of extensive faulting between pre-Cambrian rocks exposed at the surface in the Manzanita Mountains to the east and the Albuquerque graben block defined by the Rio Grande fault to the west. Considerable hydrogeologic information regarding the study area is published in a report by *Gillentine et al.* [1995]. *Kelley* [1977] presents a tectonic map of the Albuquerque basin showing much of the fault complexity in the area. Fission-track data in the uplifts, along with approximate age control of the basin fanglomerates, suggest rapid uplift of the rift shoulders with synchronous rapid deposition of major fanglomerates, especially in late Miocene time [*May et al.*, 1994].

In addition to the rift bounding faults along the study area, the Tijeras fault traverses the region from northeast to southwest, and appears to continue across the Albuquerque basin as the Tijeras accommodation zone [*Woodward*, 1984]. This zone separates the Albuquerque basin into two subbasins; a northern basin with east-dipping beds and ~17% extension, and a southern basin with west-dipping beds and at least ~28% extension [*Russell and Snelson*, 1994]. *Reiter et al.* [1986] suggest the Tijeras accommodation zone is a deep crustal feature where faults and fractures both parallel and transverse to the Rio Grande rift intersect to provide avenues for rising magmas, and/or very deep ground water, to bring heat upward and produce a modest positive geothermal anomaly with a heat flow of ~105 mW m^{-2} [other areas of the Albuquerque basin have heat flows of ~80 mW m^{-2}; *Reiter et al.*, 1986].

Water table contours show that ground water in the study area generally has the potential to flow from the uplifts on the east toward the Rio Grande on the west [*Bjorklund and Maxwell*, 1961; *Titus*, 1961; *Gillentine et al.*, 1995], although pumping over the past few decades has modified the water table contours somewhat in the western part of the study area. *Hawley et al.* [1995] present a cross section illustrating a hypothetical distribution of lithofacies from the Rio Grande to the eastern uplifts and *Lozinsky* [1994] discusses the depositional history of the Albuquerque basin, indicating that through-flowing drainage appears to have occurred at ~5 Ma. Along this sedimentologically complicated slope of the Albuquerque bench facies changes probably cause complicated variations in hydraulic conductivity; numerous faults, some apparently intersecting, add still more complications in understanding the ground-water flow pattern. *Forster and Smith* [1988] describe the sensitivity of water-table elevations to the factors controlling ground-water flow in mountainous terrain and relate that although water-table elevations are likely to be high in most mountainous areas, exceptions

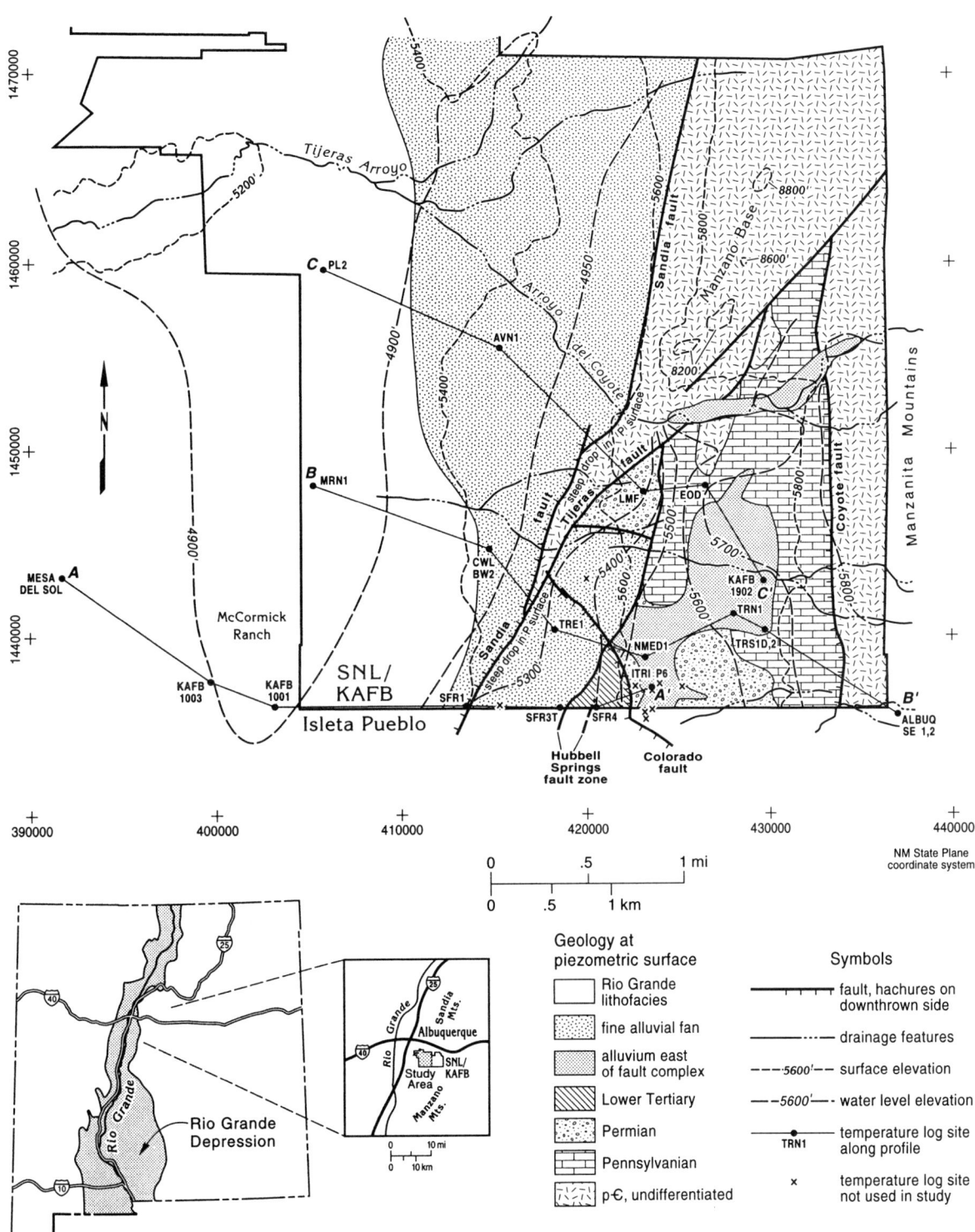

Figure 1. Map of study area on Sandia National Laboratories/Kirtland Air Force Base. Base map Gillentine et al. [1995] and Sandia National Laboratories [1996]. Correlation of the Santa Fe Group across major faults in cross sections (below) is uncertain. The region demonstrates a complicated structural framework with Laramide thrust faults and syn-rift normal faults likely superimposed on Paleozoic and pre-Cambrian fractures [Karlstrom et al., 1997].

may be expected in regions of high permeability and arid climate. Highly transmissive extensive faults and fractures will have important influences on ground-water flow in mountainous areas [*e.g., Levens et al.,* 1994].

The most obvious ground-water trend in the study area is the very steep drop in the ground-water table across the Sandia and/or Tijeras faults (Figure 1), which is as much as ~122 m. Evidence for this drop in the ground-water table can be seen from water levels in the drill holes along the profiles shown in Figure 1. *Haneberg* [1995] analyzes steady state ground-water flow across idealized faults where relative permeabilities between faults and country rock are varied. He applies the analysis to the Hubbell Springs fault, which continues south from the southern part of the study area, suggesting a relatively non-transmissive fault is consistent with the observed steep water table change. A number of studies suggest faults can act as zones of either high or low permeability with respect to adjacent country rock [*e.g., Smith,* 1966; *Knipe,* 1993; *Levens et al.,* 1994; *Antonellini and Aydin,* 1994; *Matthäi and Roberts,* 1996]. *Sibson* [1994] suggests that flow along faults may be considerable at times and also intermittent, where fracture permeability may be associated with episodic fault slip. *Bodner and Sharp* [1988] propose that high temperature bands corresponding to growth faults in portions of the Gulf Coast basin, result from the upward advection of deep warm fluids along the fault zone.

In the present study I present subsurface temperature data which relate to ground-water flow across the study area. I believe the data indicate locales of little water movement across faults in the region as well as locales where ground-water movement across faults seems likely. The data also appear to relate to the nature of flow in the sediments, showing zones of preferential flow. Because of the limited depth of the present data, the study portrays only the relatively shallow ground-water system.

DATA PRESENTATION

Methodology

Subsurface temperature measurements were made below the water table with instrumentation similar to that described by *Reiter et al.* [1980]. In the present study the temperature was recorded going down the well at two meter intervals in water, except Mesa del Sol where the interval was five meters. Measurements in air were typically taken with a faster time-response temperature tool to attempt to compensate for the relatively small heat capacity of air. Comparing temperatures measured at the same depth by the water and air tools provides some idea of the absolute accuracy of the measurements, which is a few tenths of a degree K. Much more important is the relative precision of different measurements with the same

tool, which is likely to be at least an order of magnitude better than the absolute accuracy. In stable sections of wells, reproducibility of temperatures over several weeks is observed to be a few thousandths of a degree K. It is the relative accuracy of the measurement that is important in analyzing ground-water perturbations to the conductive temperatures. The observed effects of ground-water movement on the subsurface temperatures presented in this paper are almost always much greater than the probable measurement uncertainties.

Southern profile, from west to east

The subsurface temperature data are presented along three west to east profiles shown in Figure 1. The first data group to be discussed are along the southern profile, AA′ in Figure 1. There are seven sites along the profile where useful data were obtained (Figure 2); temperature data for the seven appropriate sites are shown in Figures 3a, 3b, and 3c. Completion data are also given in Table 1 for drill holes where water flow is noticed.

There are no observable first-order perturbations in six of the temperature logs (Figures 3a and 3b) although three of these logs have very limited data below the ground-water table. The very low temperature gradient at ITRIP6 (Figures 2 and 3b) compared to the gradients at other wells, may suggest a zone of cool water flow below the temperature data. The cooler temperatures measured at ITRIP6 are consistent with the site being closer to mountain recharge. The topography near site ITRIP6 is relatively smooth and the site is about 1.5 km from the Manzanita Mountain front, so topography itself should not produce the observed cool temperatures. The well at site SFR3T probably passes through a branch of the Hubbell Springs fault zone (Figure 2), but there do not appear to be any perturbations induced by ground-water flow along this fault (Figure 3a). The small undulations in temperature data for sites SFR3T and Mesa del Sol are quite common in sedimentary sections where clay and sand content vary, and probably result largely from thermal conductivity variations.

The most noticeable feature along profile AA′ is the steep drop in water level elevation at the Sandia fault (Figure 2). The site closest to the Sandia fault is SFR1 and if ground water were moving across the fault, one might expect to notice some evidence of downward ground-water flow in the temperature data at SFR1. From the temperature log (Figure 3c) it appears that ground water is, however, moving upward in formation from ~112 m to ~90 m at site SFR1. Curvature in the temperature log may resemble somewhat the effects of upward flow in the well or annulus, but there does not appear to be a characteristic

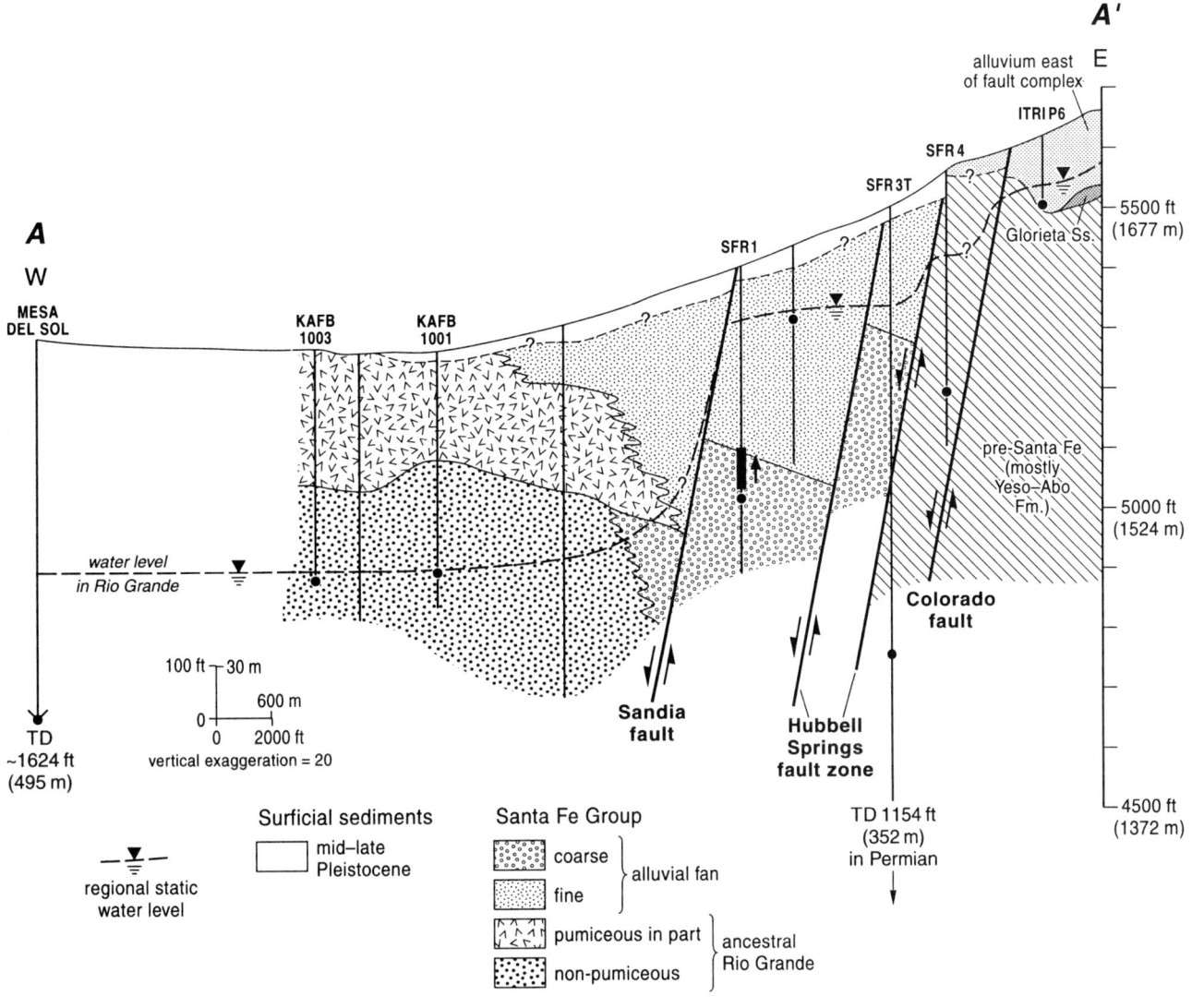

Figure 2. Geologic cross section along profile AA' (Figure 1), after cross sections and data presented in Gillentine et al. [1995]. Wells and drill tests indicated by solid vertical lines; sites of temperature logs labeled; black dot indicates depth of temperature log; water flow at site SFR1 indicated over widened-line section of well (arrow indicates flow direction). Note the vertical exaggeration causes considerable distortion of the figure; faults are kept slanted at high angles to differentiate from wells.

step toward a lower temperature due to water flowing out of the annulus at the sealed depth [*e.g., see Drury et al.,* 1984]. The temperature log was taken inside casing so there is no place for water entering the screen to escape above the seal (Table 1). These considerations support the suggestion that water is flowing upward in formation.

In order to estimate the vertical specific discharge at site SFR1, a plot of temperature gradient versus temperature can be made [after *Mansure and Reiter,* 1979; Figure 4 this paper]. Assuming a constant thermal conductivity across the zone of interest [the thermal conductivity of sediments

in the Albuquerque basin seems to vary little; *Reiter et al.,* 1986], one can calculate the vertical specific discharge from the slope of the least mean squares fit to the data in Figure 4 and the equation

$$\rho c v_z = k \Delta (\Delta T / \Delta z) / \Delta T \qquad (1)$$

[*Mansure and Reiter,* 1979; *Reiter et al.,* 1989], where ρ is fluid density (1000 kg m^{-3}), c is heat capacity per unit mass of fluid (4,184 J kg^{-1} K^{-1}), k is thermal conductivity of sediments [2.35 W m^{-1} K^{-1}, estimate from *Reiter et al.,*

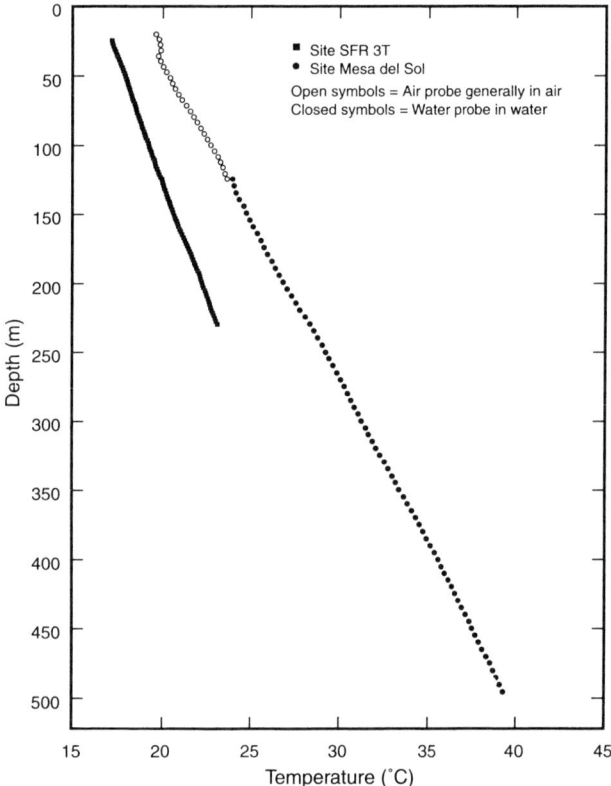

Figure 3a. Temperature data from the two deepest wells along profile AA' (Figures 1 and 2).

1986], and $\Delta(\Delta T/\Delta z)/\Delta T$ is the slope in Figure 4. From equation 1 an upward specific discharge of 6.2×10^{-8} m s^{-1} ($\sim$ 2 m/yr) between 112 m and 90 m is estimated. The temperatures across the zone of constant flow can be defined by the equation

$$T = T_1 + (T_2 - T_1)\left[e^{\beta Z / L} - 1\right] / \left(e^{\beta} - 1\right) \qquad (2)$$

[after *Bredehoeft and Papadopulos, 1965; Mansure and Reiter, 1979; Reiter et al., 1989*] where $\beta = c\rho v_z/L$ and T_1 and T_2 are the temperatures at the bottom and top of the zone of flow. Using T_1 and T_2 as measured at 112 m and 90 m, estimates of temperatures across the zone of flow are compared with measurements (Figure 5); the maximum difference in the temperatures is a few hundredths of a degree K with most data showing better agreement. These small temperature differences could result from changes in rock thermal conductivity and additional types of ground-water flow, perhaps in the annulus between the casing and the rock, that are unrecognized. From the available data it appears that upward flow in the formation is likely from

$\sim$112 m to $\sim$90 m at site SFR1, and this conclusion would seem to contraindicate any downward flow which might be possible along the steep drop in water table across the Sandia fault, at least for the logged depth interval. With present data, which are scarce, it would appear that the Sandia fault at this location acts as a very effective ground-water seal at shallow depths. Deeper flow across the fault may be occurring and temperature data at greater depths would be valuable in resolving this question.

Middle Profile, BB'

Consider next the middle profile which goes from west to east and is defined by BB' in Figure 1. A geologic cross section along BB' (Figure 1) is shown in Figure 6. Ten wells are indicated on the cross section, nine of which were logged for temperatures. Again there appears to be a very steep drop in the water-table elevation associated with the Sandia-Tijeras fault complex. A smaller drop in the piezometric surface also occurs near the Hubbell Springs fault (Figure 6). Temperature data for nine wells along

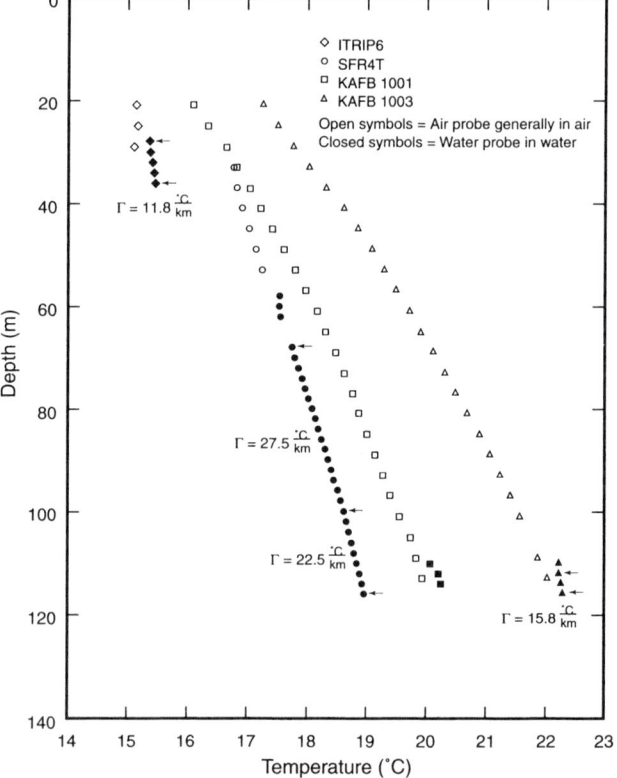

Figure 3b. Temperature data from four intermediate and shallow-depth wells along AA' (Figures 1 and 2) which do not show noticeable ground-water perturbations, Γ is temperature gradient over indicated sections of log.

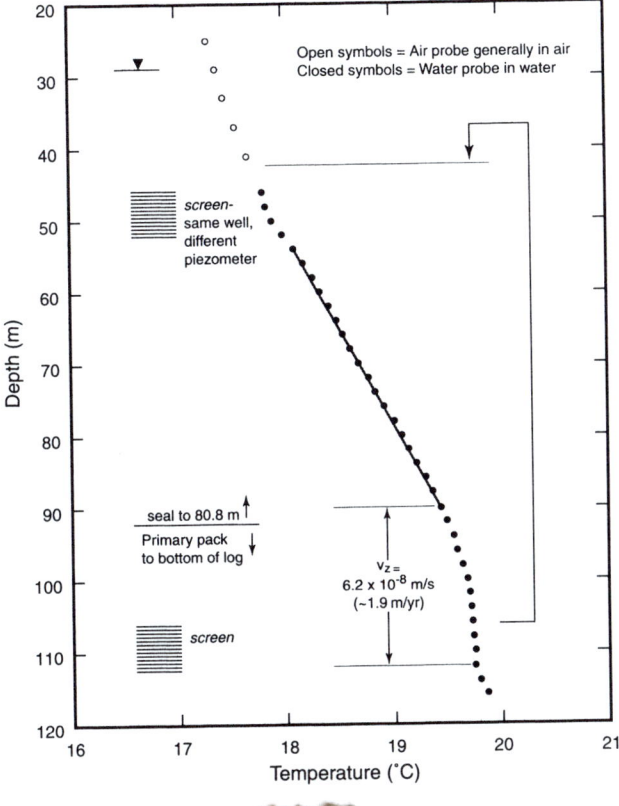

Figure 3c. Temperature data from site SFR1 on profile AA′ (Figures 1 and 2), which show a noticeable perturbation due to upward ground-water flow.

and 1.6°C/km; Figures 7a and 7b). The very small gradient at site CWLBW2 continues from 180 m to 228 m where another abrupt change in gradient occurs; data at site TRE1 are of insufficient depth to detect any deeper abrupt gradient change. The linear gradients with abrupt changes probably indicate steady-state temperature conditions resulting from fracture or thin-channel flow at the depths where gradient changes occur [*e.g., Ziagos and Blackwell,* 1986]. The anomalous temperature gradient zones are also screened zones which always creates some uncertainty as to the cause of observed temperature perturbations (see Table 1).

A noticeable discontinuity in the temperature log occurs at 228 m depth at site CWLBW2. The measured temperature at this depth is ~1.2°C less than would be the case if an undisturbed gradient were present from the top to the bottom of the log (i.e. along 25.5°C/km, see Figure 7a). I suggest that cool water is moving at 228 m depth and causing the notable change in the temperature gradient. An approximation of flow in the postulated cool-water zone can be made by assuming that the difference in conductive heat flow between the layers above and below 228 m results from cool, largely horizontal, water movement. First, however, the log at the bottom of the well needs to be examined. There is a curved section in the temperature log suggesting downward vertical flow from 228 m to 258 m (Figure 7a), which is superimposed on a relatively large temperature gradient (48.4°C/km). Using the technique described by equation 1 and illustrated in Figure 4, I calculate a vertical downward specific discharge of ~2.36 × 10^{-8} m s^{-1} (~0.8 m/yr) from 228 m to 258 m depth. From equation 2 and the formulation in *Ramey* [1962], I calculate temperatures between 228 m and 258 m while varying flow in the well bore itself; the solid curved line (Figure 7a) represents the best visual fit with a wellbore flow of ~5 * 10^{-3} kg s^{-1} (well bore radius ~10 cm). The formation flow produces most of the observed curvature in the data.

profile BB′ are shown in Figures 7a–7d. The two sites nearest the Sandia-Tijeras fault complex are CWLBW2 and TRE1; the temperature logs for these two sites indicate large disturbances probably caused by ground-water flow (Figures 7a and 7b). Logs at both sites show an upper zone with a linear gradient somewhat less than might be expected as regionally representative (~17.5°C/km) and an abrupt change to a very low gradient at greater depths (4.3

Table 1. Completion data for wells where water flow is noted in temperature log.

Profile	Well	Completion TD (m)	DTW (m)	Screened Interval (m)	Annulus Seal (m)	Gr. Sur. Elev. (m)	Location—NM State Plane Coor. Sys. Easting	Location—NM State Plane Coor. Sys. Northing
A.	SFR1	110.4	50.0	95.1-107.3	70.1-84.5	1675.0	418472.52	1436169.84
B.	CWLBW2	298.8	151.3	149.4-298.8		1656.0	414834.06	1444713.92
B.	TRE1	93.0	54.0	77.7-89.9	33.8-40.5	1674.5	418296.45	1440327.53
B.	TRN1	106.7	25.0	97.6-103.7	89.3-93.0	1747.2	428065.30	1441151.27
B.	TRS1D	96.5	36.3	81.1-93.3	65.5-75.3	1760.6	429359.20	1440276.95
	TRS2	64.0	37.8	50.3-62.5	0.5-10.4	1760.8	429718.69	1440316.38
B	ALBQSE1*							
	ALBQSE2*							

*Well temperatures logged in the early 1970s; detailed information as given for other wells not recorded.

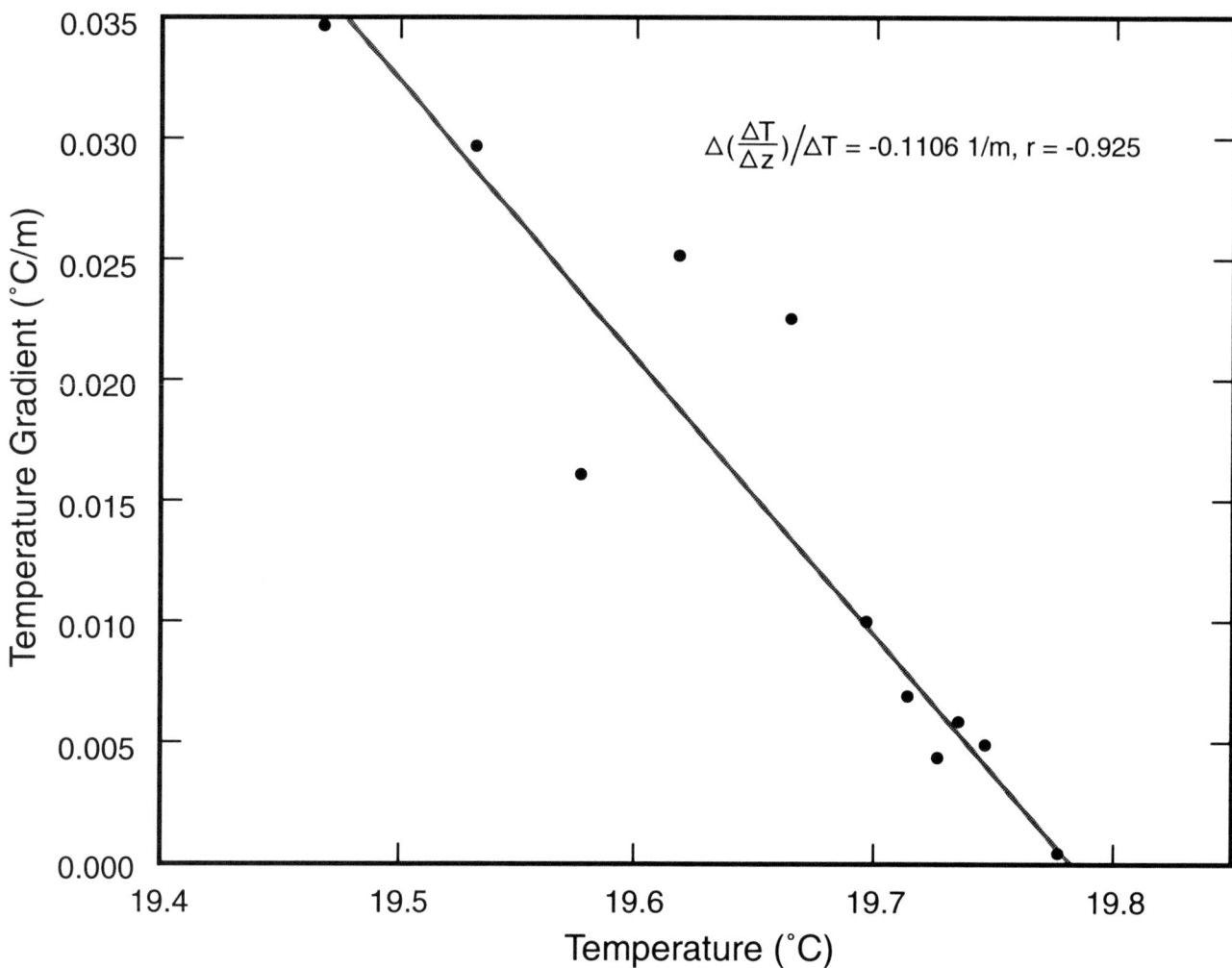

Figure 4. Plot of ($\Delta T/\Delta z$) vs T for site SFR1, depth 112 m to 90 m. The slope of the best fit (least mean squares) straight line is used to calculate the vertical specific discharge in equation 1.

Returning to the largely horizontal flow at 228 m, one needs to estimate the heat flow above and below the flow zone in order to calculate the heat removed by water flow. This is straightforward above 228 m because the gradient is linear, so $Q_1 \cong$ (4.25°C km^{-1}) × (2.35 W m^{-1} K^{-1}). Because water is flowing downward from ~228 m and the gradient is changing, using the gradient between 228 m and 230 m will provide a most appropriate estimate of the upward heat flux at 228 m. Therefore $Q_2 \cong$ 24.5°C km^{-1} × 2.35 W m^{-1} K^{-1}. The difference in heat flow, ΔQ, is ~48 mW m^{-2}. This heat-flow difference may be equated to the heat removed by water flow through a cube; it can be shown that

$$\frac{\Delta Q}{\Delta Z} = V_x \bullet \rho \bullet c \bullet \Gamma_x \qquad (3)$$

[*McCord et al.*, 1992] where V_x is specific discharge along the flow channel, Γ_x is temperature gradient along channel and $\Delta Z = 1$ m for this case. With present data coverage Γ_x is difficult to estimate. I will attempt to estimate Γ_x by suggesting that flow with a downward component is occurring from site TRE1 to site CWLBW2 (Figure 6); and that the flow is occurring along a channel which is indicated at CWLBW2 at the base of the low temperature gradient zone (228 m) and inferred at TRE1 because the data are not deep enough to see the bottom of the low temperature gradient zone (Figures 7a and 7b; the temperature log in the upper half of CWLBW2 is similar in character to the log at TRE1). If such is the case the channel temperature is 21.73°C at site CWLBW2 and ~17.6°C at site TRE1; the resulting gradient along the flow channel is then ~4.8°C km^{-1} (i.e. [21.7°C - 17.6°C]/0.85

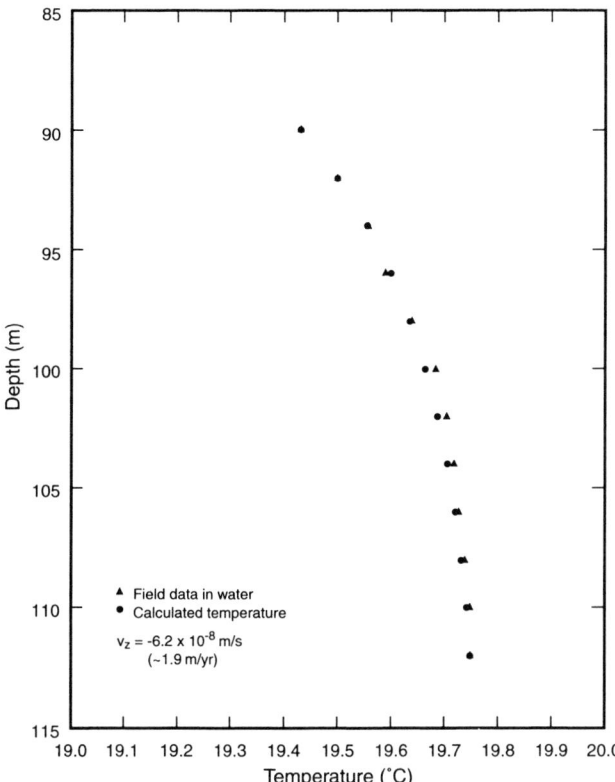

Figure 5. Comparison of calculated and measured temperatures across the flow zone in the SFR-1 well.

km). Using this value along with the parameters given above in equation 3 one calculates $V_x \approx 2.4 \times 10^{-6}$ m^3 m^{-2} s^{-1} (~80 m/yr) or ~2.2 gal m^{-2} hr^{-1}.

The shallow gradient discontinuity for the two sites (~76 m at TRE1 and ~180 m at CWLBW2) may be interpreted as channel flow of relatively warm water. At CWLBW2 a linear (non-disturbed) gradient between the temperature at 228 m, suggested to be held constant by channel flow, and the water table temperature at ~154 m, would require temperatures to be less than observed over the depth range 228 m to 154 m. Channel flow with an upward component might bring relatively warm water across the wells at the shallow depths showing gradient breaks. At CWLBW2 flow along a channel with a 10 m depth change could increase the temperature where the channel crosses the well by ~0.2°C if the background gradient is 25.5°C/km (Figure 7a).

Data at sites MNR1, MNED1, and TRN1 do not show obvious effects of ground-water flow (Figure 7b). The cause of high temperature gradients at site TRN1 is uncertain. Low thermal conductivity of claystones in the Yeso-Abo formations could explain the high temperature gradients; however, the thermal conductivity may or may not be appreciably different from the thermal conductivity of sediments at other sites [e.g. the average thermal conductivity of claystone is ~2.4 W m^{-1} K^{-1} and the range is 1.7 to 3.4 W m^{-1} K^{-1}; *Kappelmeyer and Halnel*, 1974; whereas the average thermal conductivity of sediments in the Albuquerque basin is ~2.4 W m^{-1} K^{-1}; *Reiter et al.*, 1986]. It is possible that low hydraulic conductivity in the Yeso-Abo formations (Figure 6) restricts ground-water flow so that a conductive temperature gradient is present; the resulting heat flow (2.4 W m^{-1} K^{-1} × 47.3°C km^{-1} ≈ 114 mW m^{-2}) is consistent with relatively high heat flows along the Tijeras accommodation zone [*Reiter et al.*, 1986]. Alternatively, the small hump in the water table at site TRN1 would be consistent with upward flow at the contact of the Madera limestone (high hydraulic conductivity) and the Yeso-Abo formations (low hydraulic conductivity) as indicated in *Haneberg* [1995]. If upward flow along the contact were deep enough then heating of shallow depths at site TRN1 may be possible. At site NMED1 the low temperature gradient may suggest cool water flow at deeper depths. As with most sites in the study deeper data would be very valuable.

Temperature data from sites TRS1D and TRS2 do show dramatic deviation from conduction gradients (Figure 7c). The lithology beneath the water table is typically alternating sandstone, siltstone and limestone. The beds dip ~45° down-to-the-west and so lithologies at the two sites, about ~100 m apart, do not correlate directly. Large discontinuities in the temperature gradients are often associated with fractures noted in the caliper logs (Figure 7c). At site TRS1D, notes in the lithologic log state that >50 gpm water production occurs in the limestone between ~46.6 m and ~48.8 m; this depth correlates with a temperature gradient change (Figure 7c). Negative gradients seen in the upper part of the log at site TRS2 appear to be associated with fractures or bedding zones and associated cool water flow. At site TRS1D the negative temperature gradient observed between 58 m and 70 m is most likely caused by cool water flow in channels within the limestone between ~67 m and ~72 m. The curved portion of the log from 86 m to 96 m at site TRE1D occurs over zones of interbedded limestone and shale; using the methods above, the solid curved line in Figure 7c is derived to represent theoretical temperatures between 86 m and 96 m resulting from the prescribed downward specific discharge.

The last two logs along profile BB′ are shown in Figure 7d (sites Albuq. SE 1 and 2). Most of the logged depth intervals at the two sites are in pre-Cambrian rocks (Figure 6). The general shape of both the temperature logs indicates upward ground-water flow in formation from at least ~170 m depth. The steps in the temperature log at site

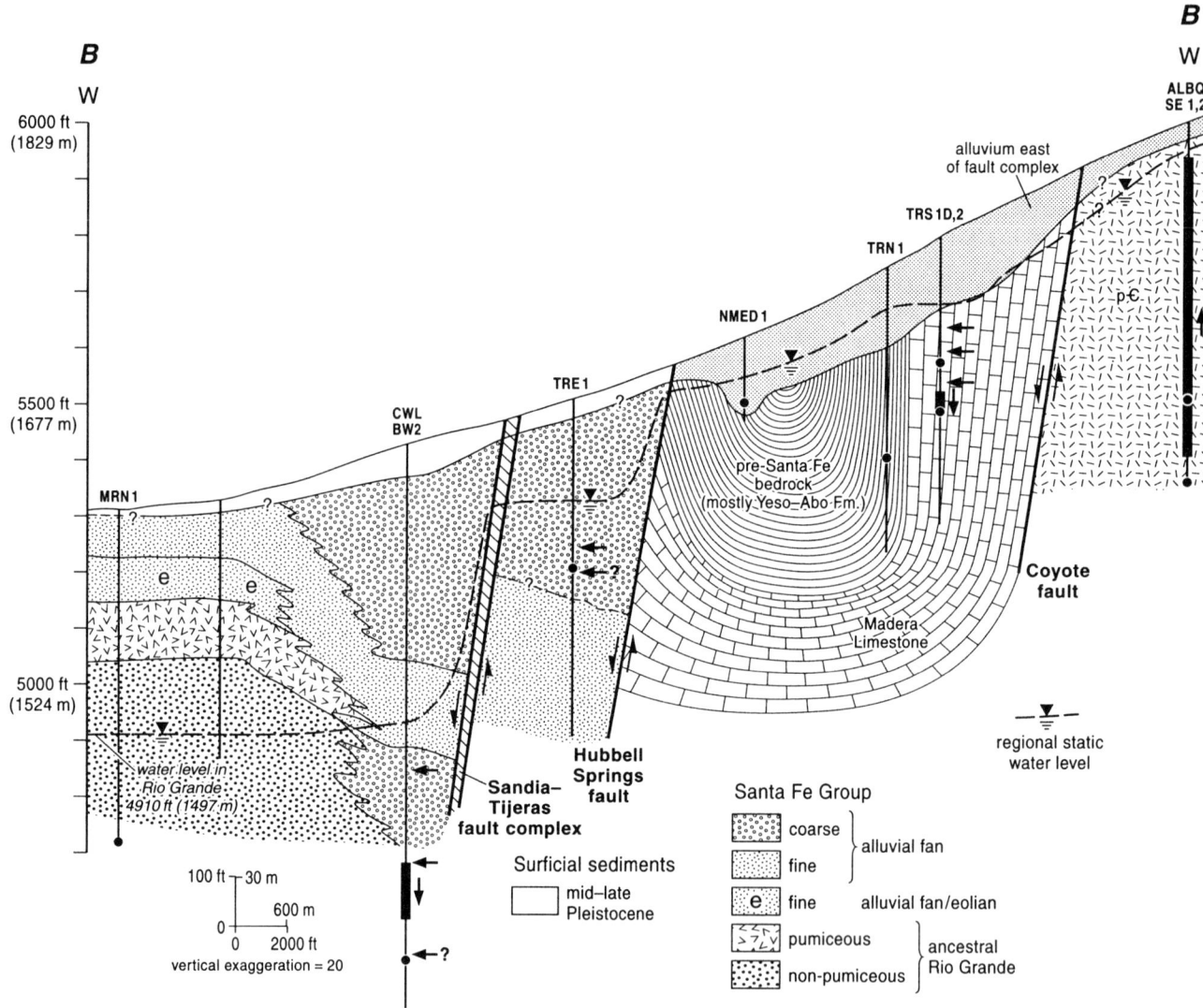

Figure 6. Geologic cross section along profile BB' (Figure 1), after cross sections and data presented in Gillentine [1995] and information in Karlstrom et al. [1997]. Sandia Tijeras fault complex after Hawley et al. [1995]. Wells and drill tests indicated by vertical solid lines; sites of temperature logs labeled, black dot indicates depth of temperature log; horizontal arrows at sites CWLBW2, TRE1, and TRS1D, 2, indicate channels of horizontal flow; wide line at wells CWLBW2, TRS1D, 2 and Albuq. S.E.1, 2, show zones of flow in formation with vertical arrows indicating direction of flow. Note vertical exaggeration causes considerable distortion of figure; faults are kept slanted at high angles to differentiate from wells.

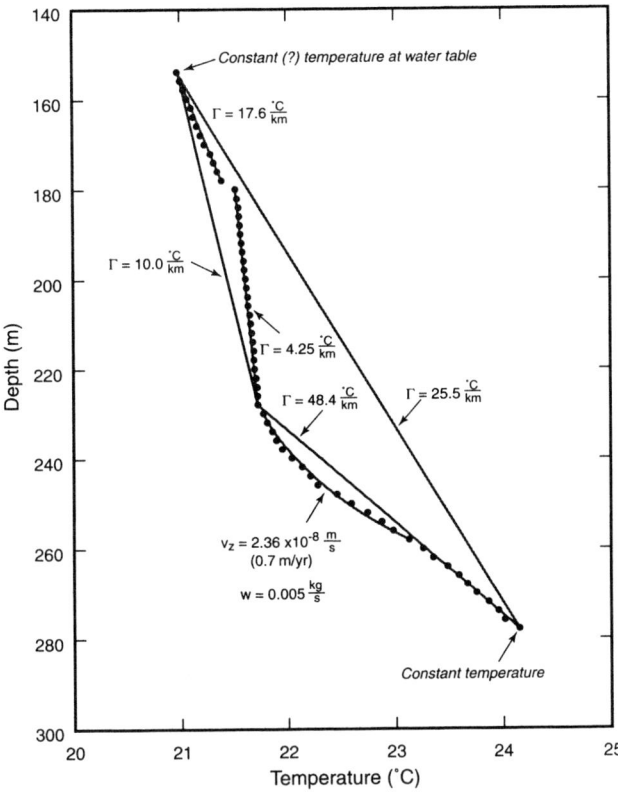

Figure 7a. Temperature data in water at site CWLBW2 (see Figure 6). Estimate of downward specific discharge across formation, V_z, and down well flow, W, with resulting temperatures (curved solid line) shown for 228 m to 258 m (curved data). Temperature gradient discontinuities at ~180 m and ~228 m, and linear gradients, imply steady-state channel or zonal flow; Γ is temperature gradient of indicated interval.

Albuq. SE 2 suggest fracture flow across the bore hole (e.g., at 130 m and 160 m depth). The log at site Albuq. SE 1 is less erratic and allows a more straightforward estimate of the upward specific discharge. Temperatures calculated using this flow estimate are indicated by the solid curved line between 20 m and 150 m depth at site Albuq. SE 1 (Figure 7d).

Profile CC', northern profile

A geologic cross section along the northern profile CC' is given in Figure 8. Along profile CC' there are two steep drops in water table elevation, each ~107 m, over a distance of at most ~3.7 km. The western step in water-table elevation is associated with the Sandia and Tijeras faults; the eastern step probably results from a significant contrast in hydraulic conductivity between the thick

claystones in the Yeso-Abo formations and the limestones in the Madera Formation (Figure 8). Temperature data show no obvious disturbances caused by ground-water flow (Figure 9). Because the data do not extend as deep below the water table as along profiles AA' and BB' it is difficult to determine the hydrogeologic behavior of the Sandia and Tijeras faults. However, the measured gradient at site PL2, several kms west of the steep ground-water step is quite low and linear, suggesting cool water flow in a deeper zone, perhaps associated with flow along the Tijeras and Coyote arroyos (Figure 1). Flow at depth along the Arroyo del Coyote may also lower the gradient observed at ANV1, but deeper data are needed to confirm the possible flow.

CONCLUSIONS

The limited data available in the present study allows one to make some preliminary observations concerning the

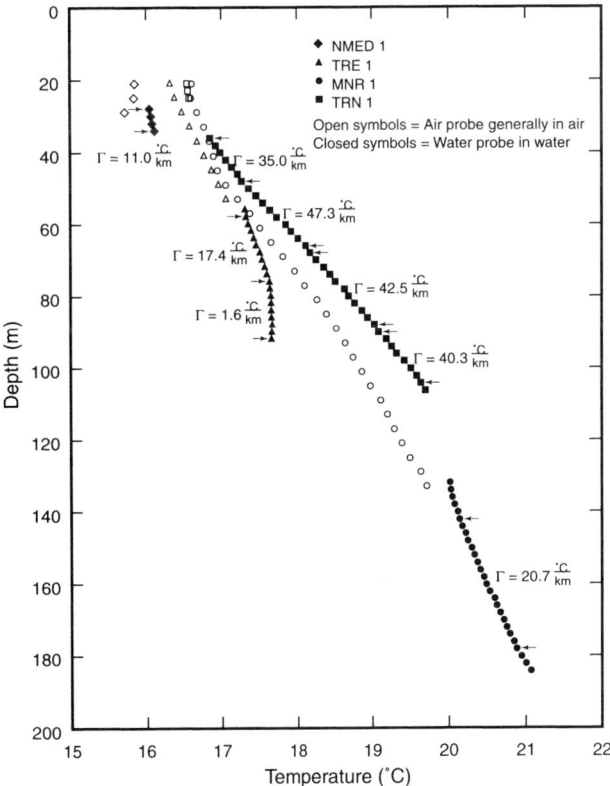

Figure 7b. Temperature data for sites NMED1, TRE1, MNR1, and TRN1 (see Figure 6); Γ is temperature gradient over indicated interval. Note temperature gradient discontinuity at site TRE1, and low gradient at site NMED1.

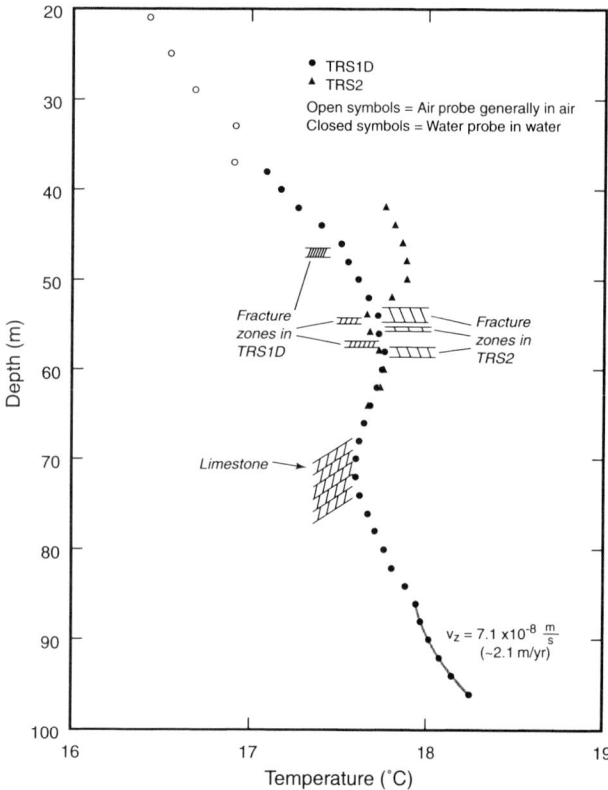

Figure 7c. Temperature data for sites TRS1D and TRS2 (see Figure 6). Fractures derived from caliper logs shown for both sites (hatchured zones), limestone zone shown for site TRS1D.

hydraulic conductivity of the faults on the boundary of the Albuquerque basin. Along a southern profile the data suggest that the Sandia fault is acting as an effective seal over relatively shallow depths (profile AA′, Figures 1 and 2). Along the middle profile, the data imply that water is crossing the Sandia-Tijeras fault complex (profile BB′, Figures 1 and 6). Along the northern profile the data are too limited to permit inferences about cross-fault ground-water flow (profile CC′ Figures 1 and 8). Ground-water movement from the mountains east of the Albuquerque bench toward the Albuquerque basin may occur across the major faults along preferential paths where the hydraulic conductivity of the fault is enhanced. Enhanced hydraulic conductivity may occur at the intersection of faults and/or fractures. Perhaps this is the situation along the middle profile (BB′), where the Sandia and Colorado faults appear to intersect the Tijeras fault just south and north of the profile, respectively (Figure 1).

The thermal data also have implications for hydraulic conditions in the sediments bordering the faults. Along the profile (AA′) at a site near the Sandia fault, the data demonstrate upward ground-water flow in the formation over a twenty meter interval. Along the middle profile (BB′), where the data in the sediments suggest that ground water is crossing the Sandia-Tijeras fault complex, it appears that the ground-water flow is occurring primarily along relatively thin channels in the sediments on both sides of the faults. This implies that the hydraulic properties of the sediments are quite variable with depth.

Figure 10 shows a simplified hydrogeologic cross section between sites CWLBW2 and TRE1 along profile BB′ (Figure 1). If the channeled flow proposed at the two drill sites is connected by a straight path, then flow across the Sandia-Tijeras fault complex occurs just below the ground-water table on the downthrown block. This would imply that the hydraulic properties of the Sandia-Tijeras fault complex also change greatly with depth. Data nearer the faults will be necessary to confirm the possible flow paths as shown in Figure 10. A significant caveat is that the

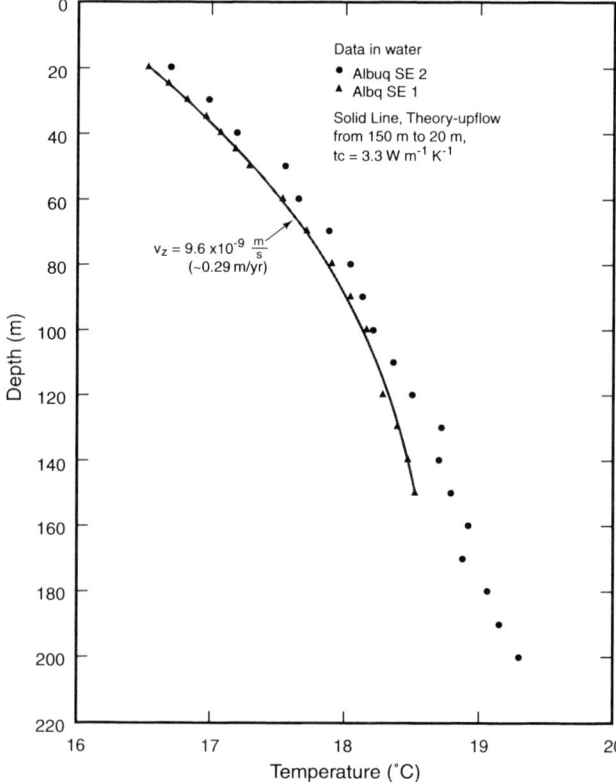

Figure 7d. Temperature data for sites Albuq. SE 1 and 2 (Figure 6). Both data sets convex upward from ~150 m and 170 m. Steps in data for Albuq. SE 2 show fracture flow, upward specific discharge calculated at Albuq. SE 1 shown by solid line. Thermal conductivity (tc) from Reiter et al. [1975].

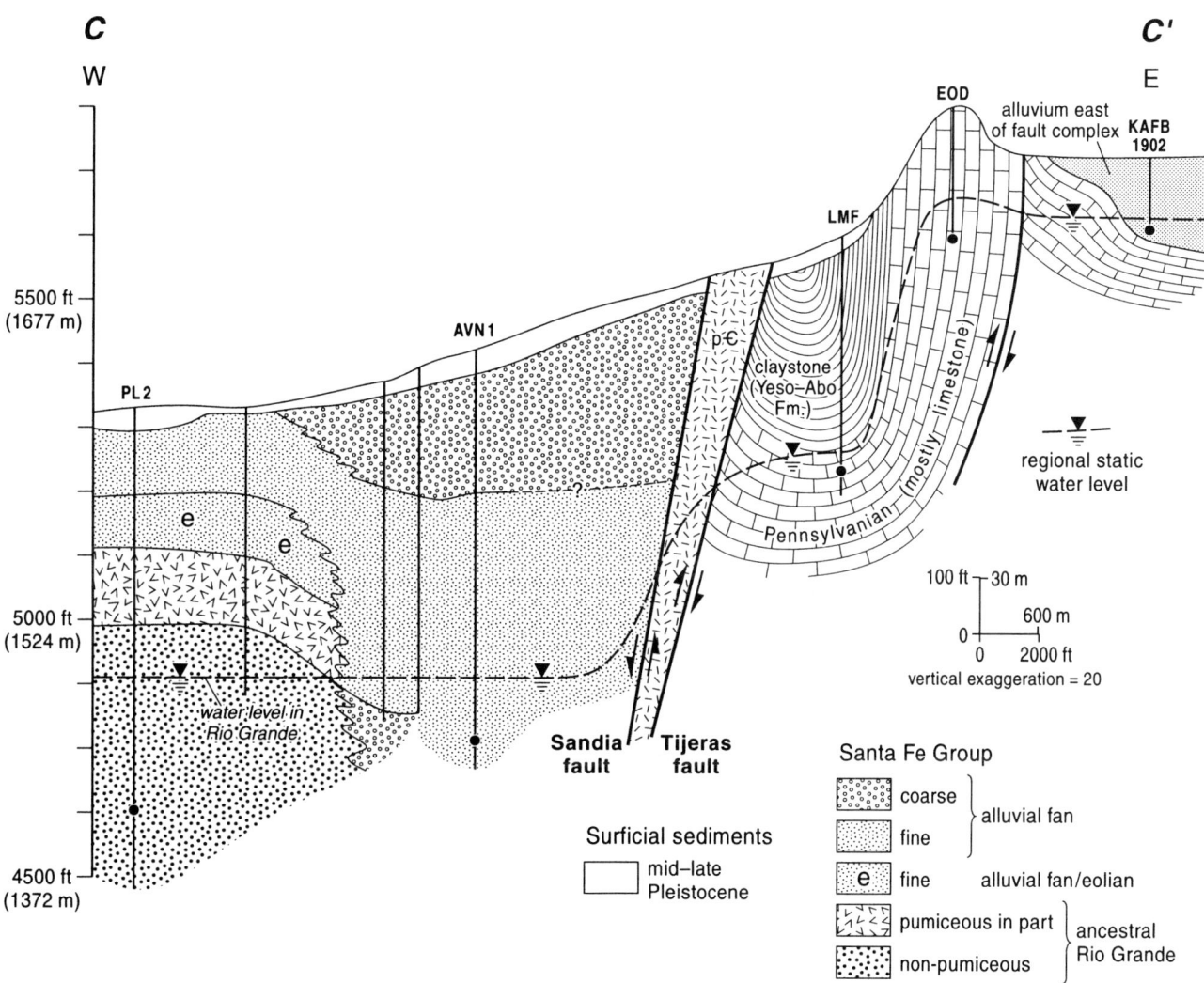

Figure 8. Geologic cross section along CC′ (Figure 1). Cross section after cross sections and data presented in Gillentine [1995] and information in Karlstrom et al. [1997]. Wells and drill tests indicated by solid vertical lines; sites of temperature logs labeled; black dot indicates depth of log. Note vertical exaggeration causes considerable distortion of figure; faults are kept slanted at high angles to differentiate from wells.

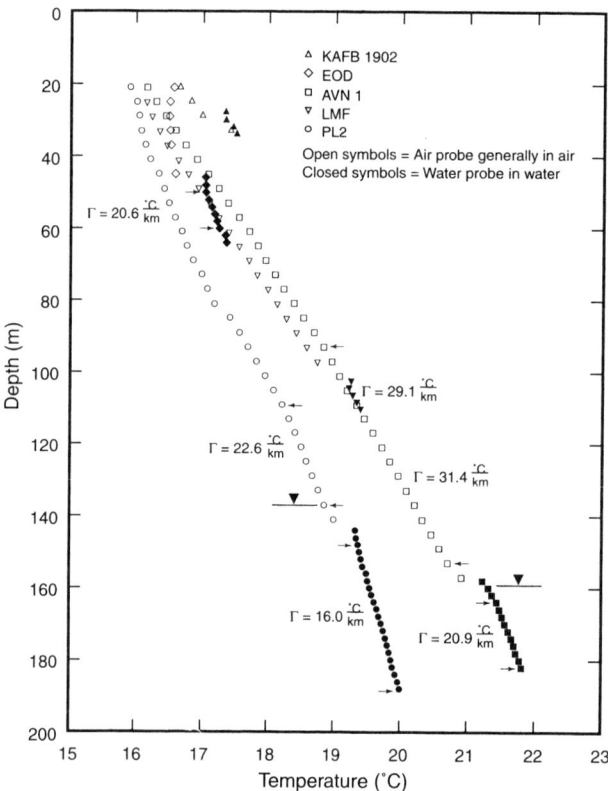

Figure 9. Temperature data for sites along CC′ (see Figures 1, 8), Γ is temperature gradient over indicated section of log.

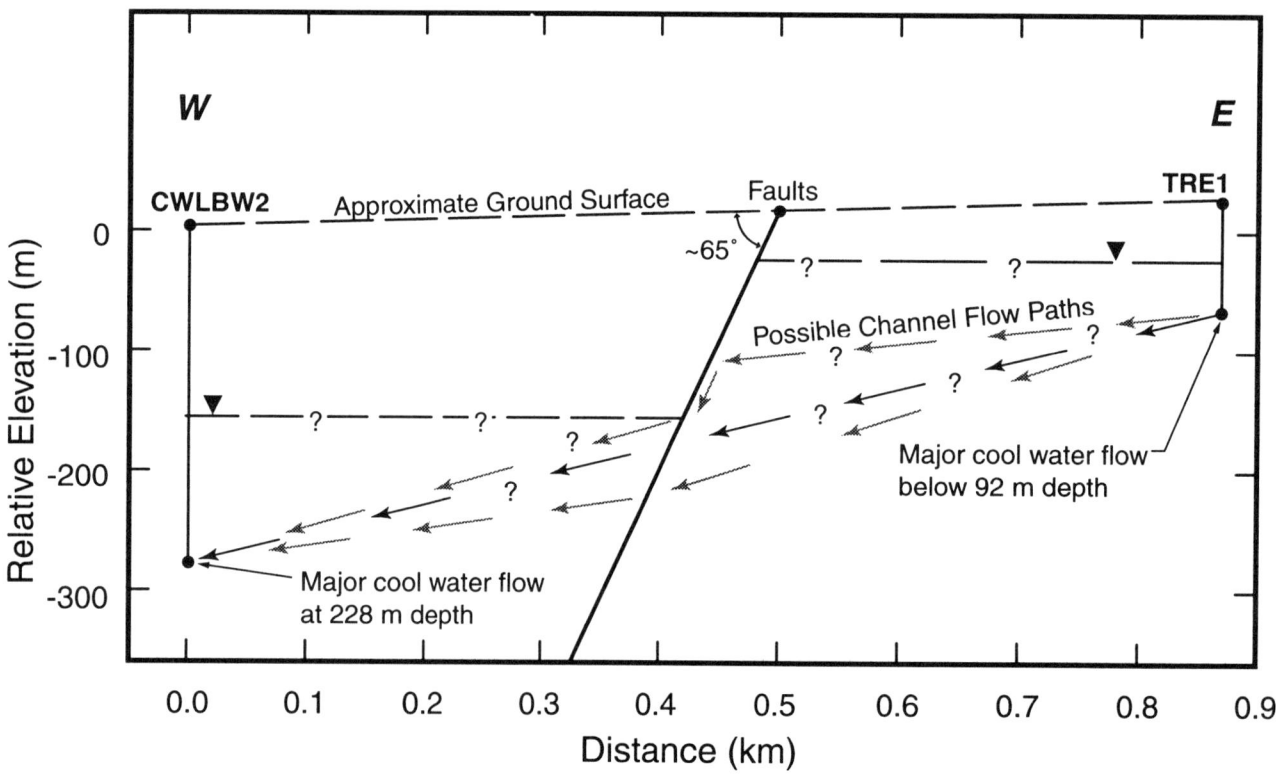

Figure 10. Idealized cross section between sites CWLBW2 and TRE1 showing possible channel flow paths.

zones of flow indicated for the two sites nearest the Sandia-Tijeras fault complex along profile BB' are also associated with screened intervals (Table 1); future heat-flow tests should be grouted over these intervals. The low linear temperature gradients observed at several additional sites in the study area suggest channeled ground-water flow at depth; this is especially true at site PL2 (along profile CC'), where flow associated with the Tijeras arroyo and possibly the Arroyo del Coyote, is anticipated (Figure 1).

I suggest that the ground-water flow across the faults that border the Albuquerque basin may be very focused, both laterally and vertically. Flow models should incorporate the variability of hydraulic conditions to reasonably represent ground-water flow patterns.

Acknowledgments. I thank a number of people involved with the work at SNL/KAFB who helped gain access to drill holes for temperature logging: S. Collins (SNL), M. Holmes (KAFB), D. Johnson (KAFB), E. Storms (Intera), W. Stone and W. McDonald (NMED), J. Lache (Lovelace, ITRI), and L. Logan (N. M. State Engineer Office). S. Collins also aided the study by providing valuable information regarding the drill tests. I thank M. Whitworth (NMBMMR) for computer advice and guidance on data reduction, and also for interesting discussions on the ground-water flow characteristics of the study area. R. Chamberlin guided me through the drawing of the cross sections east of the Sandia-Tijeras faults. W. Haneberg, R. Chamberlin, and B. McPherson had many helpful comments on the manuscript. Lynne Hemenway typed the manuscript, Becky Titus and Kent DeGruyter drafted the figures. I also thank Steve Ingebritsen and an anonymous reviewer for many helpful suggestions to improve the manuscript.

REFERENCES

Antonellini, M., and A.Aydin, Effects of faulting on fluid flow in porous sandstones: Petrophysical properties, *American Association of Petroleum Geologists Bulletin, 78,* 355–377, 1994.

Barroll, M. W., and M. Reiter, Analysis of the Socorro hydrogeothermal system, central New Mexico, *Journal of Geophysical Research, B95,* 21, 949–21,963, 1990.

Birch, F., Temperature and heat flow in a well near Colorado Springs; *American Journal of Science, 245,* 733–753, 1947.

Bjorklund, L. J., and B. W. Maxwell, Availability of ground water in the Albuquerque area, Bernalillo and Sandoval counties, New Mexico, *New Mexico State Engineer Technical Report 21, Santa Fe, 117 p.,* 1961.

Bodner, D. P., and J. M. Sharp Jr., Temperature variations in south Texas subsurface, *American Association of Petroleum Geologists Bulletin, 72,* 21–32, 1988.

Bredehoeft, J. D., and I. S. Papadopulos, Rates of vertical ground-water movement estimated from the earth's thermal profile, *Water Resources Research, 1,* 325–328, 1965.

Bullard, E. C., Heat flow in South Africa, *Proceedings of the Royal Society of London, Series A, 173,* 474–502, 1939.

Drury, M. J., A. M.Jessop, and T. J. Lewis, The detection of ground-water flow by precise temperature measurements in bore holes, *Geothermics, 13,* 163–174, 1984.

Forster, C., and L. Smith, Ground-water flow systems in mountainous terrain. 2. Controlling factors, *Water Resources Research, 24,* 1011–1023, 1988.

Garven, G., and R. A. Freeze, Theoretical analysis of the role of ground-water flow in the genesis of stratabound ore deposits, *American Journal of Science, 284,* 1085–1124, 1984.

Gillentine, J., S. E. C. McKitrick, Thomas, D. Van Hart, C. Hitchcock, K. Kelson, J. Noler, and T. Sawyer, Conceptual geologic model of the Sandia National Laboratories and Kirtland Air Force Base: prepared for Sandia National Laboratories by Gram Inc. and William Lettis and Associates Inc., Albuquerque, 1995.

Haneberg, W. C., Steady-state ground-water flow across idealized faults, *Water Resources Research, 31,* 1815–1820, 1995.

Hawley, J. W., C. S. Haase, and R. P. Lozinsky, An underground view of the Albuquerque basin, *Annual New Mexico Water Conference, 39th Proceedings, New Mexico Water Resources Research Institute, Las Cruces,* 37–55, 1995.

Ingebritsen, S. E., D. R. Sherrod, and R. H. Mariner, Heat flow and hydrothermal circulation in the Cascade Range, north-central Oregon, *Science, 243,* 1458–1462, 1989.

Jessop, A. M., Estimation of lateral water flow in an aquifer by thermal logging, *Geothermics, 16,* 117–126, 1987.

Kappelmeyer, O, and R. Haenel, Geothermics with special reference to application, *Gebruder borntraeger, Berlin,* 238 p, 1974.

Karlstrom, K. E., R. M. Chamberlin, S. D. Connell, C. Brown, M. Nyman, W. J. Cavin, M. A. Parchman, C. Cook, and J. Sterling, Geology of the Mount Washington 7.5 minute quadrangle, Bernalillo and Valencia counties, New Mexico, *New Mexico Bureau of Mines and Mineral Resources Open-file Report DGM 8, 55 p.,* 1997.

Kelley, V. C., Geology of Albuquerque basin, New Mexico, *New Mexico Bureau of Mines and Mineral Resources Memoir 33, 59 p,* 1977.

Kilty, K., and D. S. Chapman, Convective heat transfer in selected geologic situations, *Ground Water, 18,* 386–394, 1980.

Knipe, R. J., The influence of fault zone processes and diagenesis on fluid flow, *American Association of Petroleum Geologists Studies in Geology, 36,* 135–148. 1993.

Levens, R. L., R. E. Williams, and D. R. Ralston, Hydrogeologic role of geologic structures. Part I the paradigm, *Journal of Hydrology, 156,* 227–243, 1994.

Lewis, T. J., and A. D. Beck, Analysis of heat flow—Detailed observations in many holes in a small area, *Tectonophysics, 41,* 41–59, 1977.

Lozinsky, R. P., Cenozoic stratigraphy, sandstone petrology, and depositional history of the Albuquerque basin, central New Mexico, *Geological Society of America Special Paper 291*, 73–81, 1994.

Mansure, A. J., and M. Reiter, A vertical ground-water movement correction for heat flow, *Journal of Geophysical Research, 84*, 3490–3496, 1979.

Matthäi, S. K., and S. G. Roberts, The influence of fault permeability on single-phase fluid flow near fault-sand intersections: Results from steady-state high-resolution models of pressure-driven fluid flow, *American Association of Petroleum Geologists Bulletin*, 1763–1779, 1996.

May, S. J., S. A. Kelley, and L. R. Russell, Footwall unloading and rift shoulder uplifts in the Albuquerque basin: Their relations to syn-rift fanglomerates and apatite fission-track ages, *Geological Society of America Special Paper 291*, 125–134, 1994.

McCord, J., M. Reiter, and F. Phillips, Heat flow data suggest large ground-water fluxes through Fruitland coals of the northern San Juan basin, Colorado-New Mexico, *Geology, 20*, 419–422, 1992.

Morgan, P., V. Harder, C. A. Swanberg, and P. H. Daggett, A ground-water convection model for Rio Grande rift geothermal resources, *Transactions of Geothermic Resources Council, 5*, 193–196, 1981

Ramey, H. J. Jr., Well bore heat transmission, *Journal of Petroleum Technology, 14*, p. 427–435, 1962.

Reiter, M., J. K. Costain, and J. Minier, Heat flow data and vertical ground-water movement, examples from southwestern Virginia, *Journal of Geophysical Research, 94*, 12,423–12,431, 1989.

Reiter, M., C. L. Edwards, H. Hartman, and C. Weidman, Terrestrial heat flow along the Rio Grande rift, New Mexico and southern Colorado, *Geological Society of America Bulletin, 86*, 811–818, 1975.

Reiter, M., R. E. Eggleston, B. R. Broadwell, and J. Minier, Terrestrial heat-flow estimates from deep petroleum tests along the Rio Grande rift in central and southern New Mexico and southern Colorado, *Geological Society of America Bulletin, 86*, 811–818, 1975.

Reiter, M., R. E. Eggleston, B. R. Broadwell, and J. Minier, Terrestrial heat-flow estimates from deep petroleum tests along the Rio Grande rift in central and southern New Mexico, *Journal of Geophysical Research, 91*, 6225–6245, 1986.

Reiter, M., and D. L. Jordan, Hydrogeothermal studies across the Pecos River Valley, southeast New Mexico, *Geological Society of America Bulletin, 108*, 747–756, 1996.

Reiter, M., A. J. Mansure, and B. K. Peterson, Precision continuous temperature logging and correlations with other types of logs, *Geophysics, 5*, 1857–1868, 1980.

Russell, L. R., and S. Snelson, Structure and Tectonics of the Albuquerque basin segment of the Rio Grande rift: Insights from reflection seismic data, *Geological Society of America Special Paper 291*, 83–112, 1994.

Sandia National Laboratories, SNL ER sites, Air Force IRP sites and well locations at Kirtland Air Force Base, plate 2.a m l, 1996.

Sibson, R. H., Crustal stress, faulting and fluid flow, in *Geofluids; Origin, Migration and Evolution of Fluids in Sedimentary Basins*, edited by J. Parnell, Geological Society of America Special Publication 78, 69–84, 1994,

Smith, D. A., Theoretical considerations of sealing and non-sealing faults, *American Association of Petroleum Geologists Bulletin, 50*, 363–374, 1966.

Smith, L., and D. S. Chapman, On the thermal effect of ground-water flow: 1. Regional scale systems, *Journal of Geophysical Research, 88*, 593–608, 1983.

Titus, F. B. Jr., Ground-water geology of the Rio Grande trough in north-central New Mexico, with sections on the Jemez Caldera and the Lucero uplift, *New Mexico Geological Society, Guidebook 12*, 186–192, 1961.

Witcher, J. C., Geothermal resources of southwestern Arizona, *New Mexico Geological Society, Guidebook 39*, 191–197, 1988.

Woodward, L. A., Basement control of Tertiary intrusions and association mineral deposits along the Tijeras-Cañoncito fault system, New Mexico, *Geology,12*, 531–533. 1984.

Ziagos, J. P., and D. D. Blackwell, A model for the transient temperature effects of horizontal fluid flow in geothermal systems, *Journal of Volcanology and Geothermal Research, 27*, 371–397, 1986.

Marshall Reiter, New Mexico Bureau of Mines and Mineral Resources, New Mexico Institute of Mining and Technology, Socorro, NM 87801